TRANSFORMER
Principles and Applications

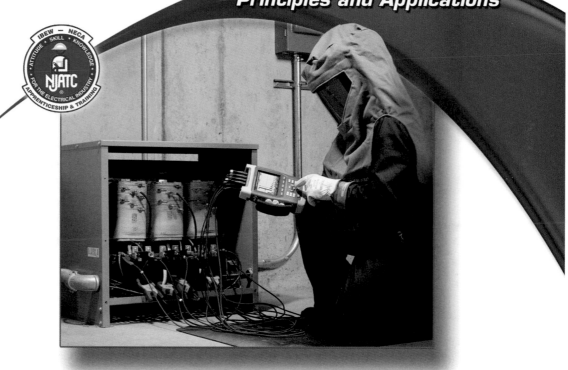

AMERICAN TECHNICAL PUBLISHERS, INC.
HOMEWOOD, ILLINOIS 60430-4600

Transformer Principles and Applications contains procedures commonly practiced in industry and the trade. Specific procedures vary with each task and must be performed by a qualified person. For maximum safety, always refer to specific manufacturer recommendations, insurance regulations, specific job site and plant procedures, applicable federal, state, and local regulations, and any authority having jurisdiction. The material contained is intended to be an educational resource for the user. Neither American Technical Publishers, nor the National Joint Apprenticeship & Training Committee for the Electrical Industry is liable for any claims, losses, or damages, including property damage or personal injury incurred by reliance on this information.

American Technical Publishers, Inc., Editorial Staff

Editor in Chief:
  Jonathan F. Gosse
Production Manager:
  Peter A. Zurlis
Art Manager:
  James M. Clarke
Technical Editor:
  Eric F. Borreson
Copy Editor:
  Richard S. Stein
Cover Design:
  Carl R. Hansen
Illustration/Layout:
  Mark S. Maxwell
  Thomas E. Zabinski
  Peter J. Jurek
  Eric T. Comiza
  William J. Sinclair
CD-ROM Development:
  Carl R. Hansen
  Christopher J. Bell

1 2 3 4 5 6 7 8 9 – 06 – 9

Printed in the United States of America

ISBN 978-0-8269-1604-4

 This book is printed on recycled paper.

Acknowledgments

Technical information and assistance was provided by the following companies, organizations, and individuals:

ABB, Inc.
ABB Power T&D Company, Inc.
Advanced Transformer Co.
Ametek®
ASCO Valve, Inc.
ASI Robicon
Baldor Motors and Drives
Cooper Wiring Devices
FLIR Systems
Fluke Corporation
Furnas Electric Co.
GE Motors and Industrial Systems
Library of Congress
Megger Group Limited
Mine Safety Appliances Co.
MTE Corporation
Oseco, Inc.
Salisbury
Square D Company
Square D-Schneider Electric

NJATC Acknowledgments
Technical Editor
     William R. Ball, NJATC Staff

Technical Reviewers
     Jim Dewig, Evansville Electrical JATC (Evansville, IN)
     Laura Jenkins, Portland Electrical JATC (Portland, OR)
     Brent Tyroff, South Texas Electrical JATC (San Antonio, TX)

Table of Contents

CD-ROM Contents

- *Using this CD-ROM*
- *Quick Quizzes™*
- *Illustrated Glossary*
- *Transformer Resources*
- *Media Clips*
- *Reference Material*

Introduction

Transformer Principles and Applications provides a comprehensive overview of transformer operation, maintenance, installation, and troubleshooting. This textbook is designed to develop basic competencies of electrical apprentices and beginning learners.

Transformer Principles and Applications begins with a thorough discussion of magnets, magnetism, electromagnetism, and how these apply to transformer operation. Subsequent chapters include the latest information on how transformers are used to reduce harmful effects of harmonics, and how reactors and isolation transformers are used to improve the power quality available to electronic equipment. Many different types of specialized transformers are explained. Installation, maintenance, and troubleshooting of transformers are discussed in detail. The text presents correct safety procedures in compliance with the National Electric Code® (NEC®) and National Fire Protection Association (NFPA 70E).

Transformer Principles and Applications contains 13 chapters. At the end of each chapter, a chapter Summary, Definitions, and Review Questions help learners review key concepts and reinforce common operation, maintenance, and troubleshooting aspects of typical transformer installations. Answers to odd-numbered questions are included at the end of the book.

Key terms are italicized and defined in the text for additional clarity. Tech Facts, Safety Tips, and vignettes throughout the text provide information that enhances text content. An extensive Glossary and Appendix provide useful, easy-to-find information. A comprehensive Table of Contents and Index simplify navigation and make finding desired information easy.

The *Transformer Principles and Applications* CD-ROM located in the back of the book is designed as a self-study aid to enhance information presented in the book, and includes Quick Quizzes™, an Illustrated Glossary, Media Clips, Transformer Resources, and Reference Material. The Quick Quizzes™ provide an interactive review of topics in a chapter. The Illustrated Glossary provides a helpful reference to terms commonly used in industry. The Media Clips are a collection of video clips and animated graphics. Transformer Resources provides access to a collection of key supplemental charts and tables. Reference Material provides access to Internet links to manufacturer, association, and American Tech resources. Downloadable digital audio files contain supplemental learning material that provides a review of key topics in each chapter. Clicking on the American Tech web site button (www.go2atp.com) or the American Tech logo accesses information on related electrical training products.

Features

The chapter Table of Contents makes it easy to find relevant information.

Definitions are emphasized throughout to ensure understanding of important concepts.

Technical vignettes provide supplemental facts related to the topic discussed.

Chapter introductions provide an overview of key content found in the chapter.

Industrial application photographs supplement the text and illustrations.

Tech Facts are used to provide supplemental background information of interest to technicians.

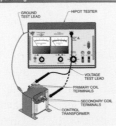

Detailed drawings detail common testing procedures.

Equipment from leading manufacturers is depicted throughout.

About the Authors

Mr. Otto Taylor has over 40 years of experience as an electrician, business executive, and JATC Instructor. After completing his formal education at Central Vocational Tech, he started working in the service side of the electrical industry. He rose to the position of superintendent of a large company before he moved on to a position with a small motor repair facility. Under his guidance, the company rose to prominence serving the large steel mills in the Chicago area. As an Executive Vice President, he attended many educational seminars offered by major manufacturers in the motor and magnet industry. Serving as a JATC Instructor offered a chance to share the knowledge gained over many years in this field. He continues as a Motors Lab Instructor at the South Bend, Indiana, Electrical JATC and the NTI in Knoxville.

Mr. Jim Overmyer has over 40 years of experience as an electrician and JATC Instructor. He, too, began work in the service side of the electrical industry. He completed his apprenticeship at Indiana Vocational Tech and later earned a degree in electronics from the Radio, Electronics, and Television Schools (RETS). Upon completion of his electronics degree, he started teaching for the local IBEW motor winders apprenticeship committee. His employment by a large motor repair facility gave him the opportunity to work directly on the equipment and the opportunity to attend the many seminars offered by leading companies in the field. Serving as a JATC Instructor has offered him an opportunity to give back to the industry that has been so good to him. He continues teaching fourth-year curriculum at the South Bend, Indiana, Electrical JATC and at the NTI in Knoxville.

Mr. Ron Michaelis has over 35 years of experience as an electrician, electrical inspector, and JATC Instructor and Training Director. He completed his apprenticeship at Indiana Vocational Tech and enrolled at Indiana University South Bend to continue his education. He has worked in locations ranging from the Alaska pipeline to the oil fields in Saudi Arabia. He is now the South Bend Electrical JATC Training Director. He also serves as the electrical inspector for a large township in Michigan. His affiliation with Indiana University continues today, as the students at the South Bend JATC are required to attend classes staffed by the University.

Magnetism and Electromagnetism

Magnets are surrounded by a magnetic field. Electromagnets are surrounded by a similar electromagnetic field. The strength of an electromagnetic field depends on the amount of current flowing through a coil, the number of turns in the coil, and the material used in the core. Iron and iron alloys are often used in electromagnets and in transformer cores because they are good conductors of magnetic flux. The units used to measure the properties of a magnetic circuit are analogous to the units used with an electric circuit.

MAGNETISM

Magnetism is a force that acts at a distance and is caused by a magnetic field. The early Greeks discovered magnetism when they noticed that a certain type of mineral attracted bits of iron. This mineral was first found in Asia Minor in the province of Magnesia. The mineral was named magnetite after the place where it was discovered.

Magnets

A *magnet* is a substance that produces a magnetic field and attracts ferromagnetic materials. A *magnetic field* is a force produced by a magnet that interacts with other magnets or other magnetic fields. A *ferromagnetic material* is a material that is easily magnetized. Three common naturally occurring ferromagnetic metals are iron, nickel, and cobalt. Less common ferromagnetic materials include neodymium, samarium, and other rare earth elements. Most magnets are constructed of alloys of various ferromagnetic materials.

A *permanent magnet* is a magnet that can hold its magnetism for a long period of time. Permanent magnets include natural magnets (magnetite) and manufactured magnets. The most common permanent magnets are horseshoe magnets, compasses, and bar magnets. **See Figure 1-1.** A *temporary magnet* is a magnet that retains only trace amounts of magnetism after a magnetizing force has been removed. The most common temporary magnets are coils

DEFINITION

Magnetism is a force that acts at a distance and is caused by a magnetic field.

A magnet is a substance that produces a magnetic field and attracts ferromagnetic materials.

A magnetic field is a force produced by a magnet that interacts with other magnets or other magnetic fields.

A ferromagnetic material is a material that is easily magnetized.

A permanent magnet is a magnet that can hold its magnetism for a long period of time.

of wire used as electromagnets. *Retentivity* is a measure of the ability of a magnet to retain magnetism after the magnetizing force has been removed. Magnets are commonly used in electric motors and generators, speakers and microphones, and many industrial products.

Magnetic Flux. Michael Faraday proposed using magnetic flux to visualize a magnetic field. *Magnetic flux,* or *field flux,* is the imaginary lines of force that make up the total quantity of an electromagnetic field. A denser magnetic flux makes a stronger magnetic force. Magnetic flux is most dense at the ends of a magnet. For this reason, the magnetic force is strongest at the ends of a magnet.

All magnets have a north (N) and a south (S) pole. The magnetic flux leaves the north pole and enters the south pole of a magnet. The basic law of magnetism states that unlike magnetic poles (N and S) attract each other and like magnetic poles (N and N or S and S) repel each other. **See Figure 1-2.** The force of attraction between two magnets increases as the distance between the magnets decreases. Likewise, the force of attraction between two magnets decreases as the distance between the magnets increases.

Polarity. The polarity of a bar magnet can be determined by suspending it from a point overhead and allowing it to turn freely. The end of the bar magnet that points north is the north pole of the magnet. Since unlike magnetic poles attract, the Earth's magnetic field interacts with the bar magnet's magnetic field so that the bar magnet's north pole points toward the Earth's south magnetic pole. From experience, we all know that the north pole of a magnet points toward the Earth's geographic north pole. This shows that the Earth's geographic north pole is actually the Earth's south magnetic pole.

A bar magnet is surrounded by its own magnetic field that exits the north pole and enters the south pole. This can be confirmed by placing a small compass at different positions around the magnet. The needle of the compass marked N aligns with the field to point to the opposite pole. **See Figure 1-3.**

> ## ! DEFINITION
>
> A *temporary magnet* is a magnet that retains only trace amounts of magnetism after a magnetizing force has been removed.
>
> *Retentivity* is a measure of the ability of a magnet to retain magnetism after the magnetizing force has been removed.
>
> *Magnetic flux,* or *field flux,* is the imaginary lines of force that make up the total quantity of an electromagnetic field.

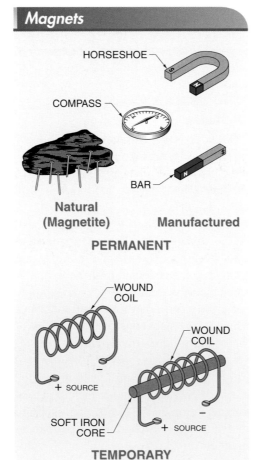

Figure 1-1. *Permanent magnets include natural magnets and manufactured magnets. Temporary magnets include wound coils of wire with a source of electricity.*

ASCO Valve, Inc.

Solenoids use electromagnetic fields to move a plunger to open or close a valve like this gas shutoff valve.

Molecular Theory of Magnetism

The molecular theory of magnetism states that ferromagnetic materials are made up of a very large number of molecular domains acting like molecular magnets that can be arranged in either an organized or disorganized manner. **See Figure 1-4.** A material is magnetic if it has organized molecular magnets so that the individual fields add together. A material is nonmagnetic if it has disorganized molecular magnets so that the individual fields cancel each other. The individual magnetic domains consist of many atoms or molecules that are individually aligned so their electrons all spin in the same direction. The moving charges of the electrons create a magnetic field.

The molecular theory of magnetism explains how certain materials used in control devices react to magnetic fields. For example, it explains why hard steel is used for permanent magnets, while soft iron is used for the temporary magnets found in control devices. Permanent magnets can be manufactured by using another magnetic field to align the magnetic domains during manufacture. Hard steel is difficult to magnetize and demagnetize, making it a good permanent magnet. The dense molecular structure of hard steel does not easily disorganize once a magnetizing force has been removed. Hard steel has high retentivity. However, permanent magnets may be demagnetized by a sharp blow or by heat that causes the molecular arrangement to become disorganized. Vibrations over an extended time may also demagnetize a permanent magnet.

Magnetic domains in temporary magnets are aligned by a magnetic field created by electric current flowing through a coil. Soft iron is ideal for use as a temporary magnet in control devices because it does not retain residual magnetism very easily. By not retaining residual magnetism, a temporary magnet can be turned off while de-energized, such as a coil in a magnetic starter.

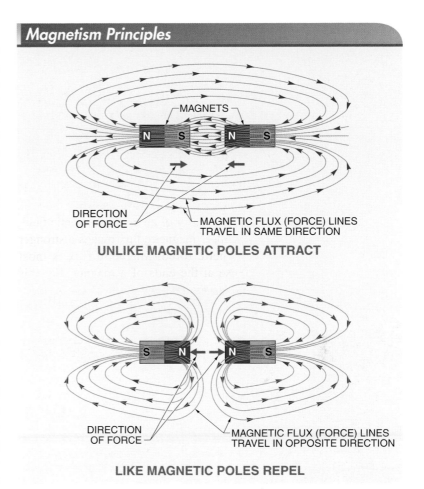

UNLIKE MAGNETIC POLES ATTRACT

LIKE MAGNETIC POLES REPEL

Figure 1-2. With two magnets, unlike magnetic poles attract and like magnetic poles repel.

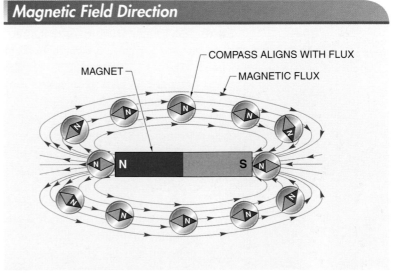

Figure 1-3. A compass aligns itself with a magnetic field.

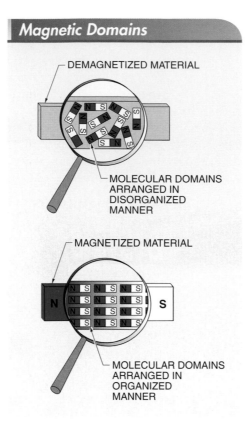

DEFINITION

Electromagnetism is magnetism produced when electricity passes through a conductor.

Figure 1-4. *The molecular theory of magnetism states that ferromagnetic materials are made up of a very large number of molecular domains that can be arranged in either an organized or disorganized manner.*

ELECTROMAGNETISM

Electromagnetism is magnetism produced when electricity passes through a conductor. Electromagnetism is a temporary magnetic force because the magnetic field is present only as long as current flows. The electromagnetic field is reduced to zero when the current flow stops.

The electromagnetic field around a straight conductor is relatively weak and must be concentrated for use in a transformer. An electromagnetic field is concentrated by increasing the amount of current flowing through the conductor, wrapping the conductor into a coil, or wrapping the conductor around an iron core. A strong, concentrated electromagnetic field is developed when a conductor is wrapped into a coil.

The strength of the electromagnetic field is directly proportional to the number of turns in the coil and the amount of current flowing through the conductor. An iron core increases the strength of the electromagnetic field by concentrating the field. **See Figure 1-5.**

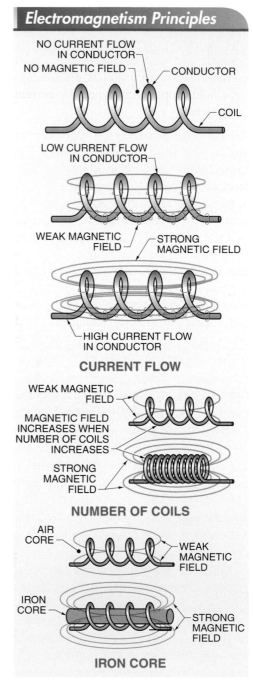

Figure 1-5. *Electromagnetic field strength is increased by increasing the current flow, increasing the number of coils, or adding an iron core.*

Direct Current. Direct current applied to a conductor starts at zero and goes to its maximum value almost instantly. The electromagnetic field around the conductor also starts at zero and goes to its maximum strength almost instantly. The current and strength of the electromagnetic field remain at their maximum values as long as the load reactance does not change. The current and the strength of the electromagnetic field increase if the resistance of the circuit decreases. The current and the electromagnetic field drop to zero when the current is removed.

Alternating Current. Alternating current applied to a conductor causes the current to continuously vary in magnitude and the electromagnetic field to continuously vary in strength. Current flow and electromagnetic field strength are at their maximum value at the positive and negative peaks of the AC current waveform. The current is zero and no electromagnetic field is produced at the zero points of the AC waveform. The direction of current flow and polarity of the electromagnetic field change every time the current passes the zero point of the AC waveform.

Left-Hand Rules

The left-hand rule for conductors can be used to determine the direction of magnetic flux around a conductor. When a conductor is wrapped with the left hand with the thumb in the direction of the current flow, the fingers point in the direction of the magnetic field. The direction of current is from negative to positive and can be determined by the polarity of the source. **See Figure 1-6.**

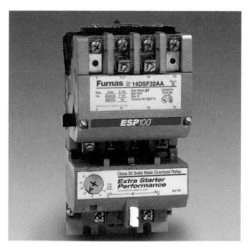

Furnas Electric Co.
A magnetic motor starter uses an electromagnetic field to close the contacts and connect the motor to the source.

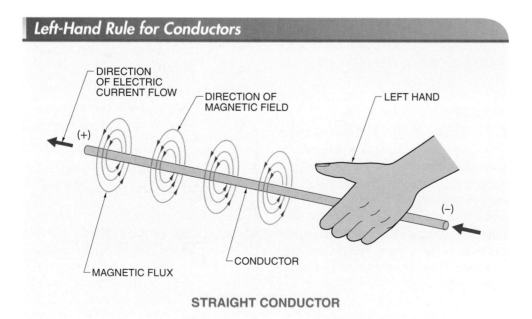

Figure 1-6. *The left-hand rule for conductors states that when the thumb is pointed in the direction of current flow, the fingers point in the direction of the magnetic field.*

The left-hand rule for coils can be used to determine the polarity of a coil. A coil has poles just like a bar magnet. When a coil is wrapped with the left hand with the fingers in the direction of the current flow, the thumb points in the direction of the north pole. These rules apply only to DC, as the changing polarities of AC reverse the direction of the magnetic flux twice for every cycle. **See Figure 1-7.**

Left-Hand Rule for Coils

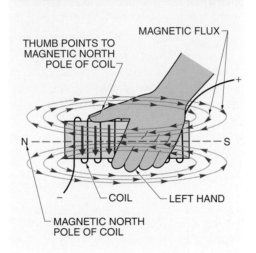

THUMB POINTS TO MAGNETIC NORTH POLE OF COIL

MAGNETIC FLUX

N — — — — — S

+

−

COIL

LEFT HAND

MAGNETIC NORTH POLE OF COIL

Figure 1-7. *The left-hand rule for coils states that when the fingers are wrapped in the direction of current flow, the thumb points in the direction of the north pole of the coil.*

Measurement Units

When discussing electrical properties of circuits and devices, the common electrical measurements of voltage, current, and resistance are universally understood. The units of measure are volts, amperes, and ohms. With prefixes like milli, micro, kilo, and mega, these units are used to describe any level of these units. When discussing magnetic systems, common quantities to be measured include magnetomotive force, magnetic flux, reluctance, magnetic field intensity, magnetic flux density, and permeability. Many of these properties are analogous to electrical units.

Magnetic Quantities. *Magnetomotive force* is the force that produces magnetic lines of flux in a circuit as a result of current in a conductor. Magnetomotive force is analogous to voltage in an electric circuit. **See Figure 1-8.** Magnetic flux is the imaginary lines of force that make up the total quantity of an electromagnetic field. Magnetic flux is analogous to current in an electric circuit. *Reluctance* is the opposition to magnetic flux in a magnetic circuit in a given volume of space or material. Reluctance is analogous to resistance in an electric circuit. *Magnetic field intensity* is the amount of magnetomotive force distributed over the length of a magnet. Magnetic field intensity is also known as magnetizing force. *Magnetic flux density* is the concentration of the magnetic flux in a given area. *Permeability* is a measure of the ability of a material to conduct magnetic flux. Permeability is analogous to specific conductivity in an electric circuit. Greater permeability means that it is easier to conduct magnetic flux.

Measurement Quantities

Magnetic Quantity	Symbol	Electrical Equivalent
Magnetomotive force (field force)	mmf	Voltage
Magnetic flux (field flux)	ϕ	Current
Reluctance	$\mathfrak{R}$	Resistance
Field intensity	H	–
Flux density	B	–
Permeability	μ	Specific conductivity

Figure 1-8. *Many magnetic quantities are analogous to familiar electrical quantities.*

Measurement Systems. There are three complete, separate sets of measurement systems for magnetic quantities. There are two separate metric systems in common use. The most common was originally known as the meter-kilogram-second (MKS) system, for the basic units used. This has been adopted as the official international standard and is now called the SI (Système International d'Unités). Another common metric system is the centimeter-gram-second (CGS) system. In addition, in the United States, the English system is normally used for measurements.

All three of these systems, the SI, CGS, and English, have different units for measuring magnetic units. For example, the unit of magnetic flux density (B) is the tesla in the SI system, the gauss in the CGS system, and lines per square inch in the English system. **See Figure 1-9.** In the English system, the magnetomotive force (mmf) is defined as the product of the number of turns and the current through a coil and is calculated as follows:

$$mmf = N \times I$$

where

mmf = magnetomotive force (in amp-turns)

N = number of turns

I = current (in A)

For example, the magnetomotive force in a coil of 1000 turns with a current of 1.5 A is calculated as follows:

$$mmf = N \times I$$
$$mmf = 1000 \times 1.5$$
$$mmf = \textbf{1500 amp-turns}$$

Electromagnets

An *electromagnet* is a magnet whose magnetic energy is produced by the flow of electric current. Some electromagnets are so large and powerful that they can lift tons of scrap metal at one time. Other electromagnets used in electrical and electronic circuits, such as those found in solenoids and relays, are very small. An electromagnet consists of an iron core inserted into a coil. The iron core concentrates the magnetic flux produced by the coil. With the core in place and the coil energized, the polarity of the magnet can be determined by the left-hand rule for coils.

The advantages of electromagnets are that they can be made stronger than permanent magnets and that magnetic strength can be easily controlled by regulating the electric current. The main characteristics of an electromagnet include the following:

- When electricity flows through a conductor, an electromagnetic field is created around that conductor.

- An electromagnetic field is stronger close to the conductor and weaker further away.

- The strength of the electromagnetic field and the current are directly related: more current, the stronger the electromagnetic field; less current, the weaker the electromagnetic field.

DEFINITION

*An **electromagnet** is a magnet whose magnetic energy is produced by the flow of electric current.*

Measurement Units

Quantity	Symbol	Unit of Measurement		
		SI	CGS	English
Magnetomotive force (field force)	mmf	Amp-turn	Gilbert (Gb)	Amp-turn
Magnetic flux (field flux)	φ	Weber (Wb)	Maxwell (Mx)	Line
Reluctance	ℜ	Amp-turns per weber	Gilberts per maxwell	Amp-turns per line
Field intensity	H	Amp-turns per meter	Oersted (Oe)	Amp-turns per inch
Flux density	B	Tesla (T)	Gauss (G)	Lines per square inch
Permeability	μ	Tesla-meters per amp-turn	Gauss per oersted	Lines per inch-amp-turn

Figure 1-9. Three common magnetic systems of measurement are used.

- The direction of the electromagnetic field is determined by the direction of the current flowing though the conductor.
- The more permeable the core, the greater the concentration of magnetic flux.

Permeability

Permeability is a measure of the ability of a material to conduct magnetic flux. The symbol for permeability is the Greek letter mu (μ). Most often, permeability is actually given as relative permeability and compared to the ability of a standard material (usually vacuum or air) to conduct the magnetic flux. The reference permeability of a vacuum is 1. **See Figure 1-10.**

DEFINITION

Saturation *is the condition where a magnetic core has substantially all the magnetic domains aligned with the field and any increases in current no longer result in a stronger electromagnet.*

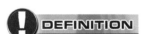

Typical Permeability Values

Material	Relative Permeability
Copper	0.999991
Vacuum	1
Air	1.0000004
Nickel	400 to 1000
Ferrite	2300 to 5000
Silicon steel	5000 to 10,000
Pure iron	6000 to 8000
Supermalloy	800,000

Figure 1-10. *Permeability values refer to the ability of a material to conduct magnetic flux.*

Permeability is analogous to specific conductivity in an electrical circuit. A material or device with a high conductivity can conduct more current than a material or device with a low conductivity. Similarly, a material with a high permeability can conduct more magnetic flux than a material with a low permeability. A material with a permeability of 1 has a magnetic flux density the same as vacuum when the same magnetic field strength is applied. A material with a permeability of 5000 has a magnetic flux density 5000 times greater than vacuum, when operated in its normal range.

All materials have their permeability near 1 except ferromagnetic materials. Iron is relatively inexpensive and has a much higher permeability than air or copper. Therefore, iron and iron alloys are often used as the core of electromagnets and transformers. There are also many specialized alloys used in transformer cores. The alloys are optimized for specific applications and can be very expensive. High permeability means that a very low input results in a very high output. For example, supermalloy has a permeability of 800,000. This is used for very low-level signal transformers, low-level magnetic preamplifiers, and precision current transformers.

A significant difference between permeability and conductivity is that permeability is nonlinear. When the voltage is increased in an electric circuit, the current increases proportionally, as long as the circuit components can handle the increase. When the mmf is increased in a magnetic circuit, the magnetic flux increases, but not proportionally. A magnetic circuit can reach a point of saturation.

TECH FACT

When a magnetic circuit reaches a point of saturation, essentially all the magnetic domains are aligned. Further increases in flux density are not possible.

Saturation

As the current flow in a coil is increased, the strength of the magnetic field (field intensity) increases. As the field intensity increases, the magnetic flux density increases. In other words, at low current, only a relatively small number of magnetic domains in the core are aligned with the magnetic field. At higher and higher currents, more and more of the magnetic domains align with the magnetic field. *Saturation* is the condition where a magnetic core has substantially all the magnetic domains aligned with the field and any increases in current no longer result in a stronger electromagnet.

B-H Curves. Saturation is often shown on a graph of magnetic field intensity (H) and magnetic flux density (B). **See Figure 1-11.** This graph is called a B-H curve, or normal magnetization curve, and shows the relationship between the applied magnetomotive force and the amount of magnetic flux carried by the particular materials. Each of these curves has a knee where increases in the field strength no longer have much effect on the magnetic flux density. As more magnetic flux gets forced into the same area, fewer magnetic domains are available to be aligned. Therefore, it takes more and more force (H) to get smaller and smaller increases in the amounts of magnetic flux (B). When this happens, the only way to increase the strength of an electromagnet or the efficiency of a transformer is to increase the size of the core or substitute a core with a higher permeability. It should be noted that the exact shape of these curves depends strongly on the purity, heat treatment, and other processing of the materials.

Transformers are normally operated in the region of the curve where the relationship is linear. If transformers are operated at the point of saturation, waveform distortion can occur.

Waveform Distortion. When a transformer core approaches saturation, a larger and larger amount of magnetomotive force is required to deliver increases in magnetic flux. The magnetomotive force is proportional to the current through the magnetizing coil. Therefore, larger and larger amounts of magnetomotive force require proportionally larger and larger increases in the coil current. In other words, a small increase in applied voltage will not produce the required increase in CEMF. The primary current increases disproportionally to the rise in voltage. **See Figure 1-12.** The coil current increases dramatically at the peaks in order to maintain the shape of the magnetomotive force waveform. Proper transformer design and application allow the transformer to operate far from the saturation point to avoid waveform distortion.

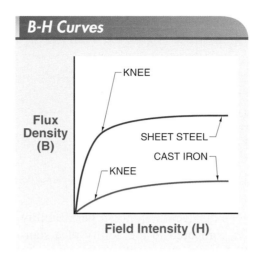

Figure 1-11. *A B-H curve shows the relationship between the applied magnetic force and the amount of magnetic flux carried by the particular materials.*

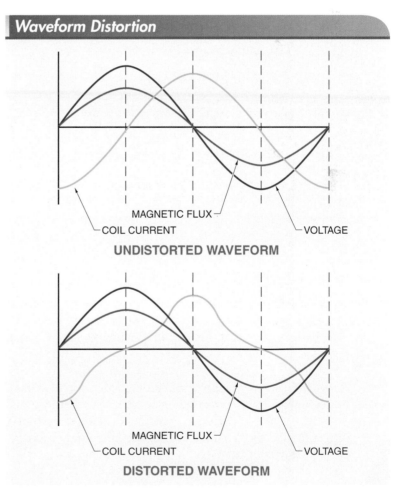

Figure 1-12. *A transformer or electromagnet operated near the saturation point causes a distorted current waveform.*

SUMMARY

- Magnetism is a force that acts at a distance and is caused by a magnetic field.

- Common ferromagnetic materials are iron, nickel, and cobalt.

- The molecular theory of magnetism states that a magnet consists of organized molecular domains.

- Magnetic flux leaves the north pole and enters the south pole of a magnet or magnetic field.

- The left-hand rules are used to determine the direction of magnetic flux around a conductor and the polarity of a coil.

- An electromagnetic field is steady in a DC circuit and varies continuously in an AC circuit.

- Many magnetic units of measure are analogous to familiar electrical units of measure.

- Saturation is the condition where a magnetic core has substantially all the magnetic domains aligned with the field and any increases in current no longer result in a stronger electromagnet.

DEFINITIONS . . .

- *Magnetism* is a force that acts at a distance and is caused by a magnetic field.

- A *magnet* is a substance that produces a magnetic field that attracts ferromagnetic materials.

- A *magnetic field* is a force produced by a magnet and interacts with other magnets or other magnetic fields.

- A *ferromagnetic material* is a material that is easily magnetized.

- A *permanent magnet* is a magnet that can hold its magnetism for a long period of time.

- A *temporary magnet* is a magnet that retains only trace amounts of magnetism after a magnetizing force has been removed.

- *Retentivity* is a measure of the ability of a magnet to retain magnetism after the magnetizing force has been removed.

- *Magnetic flux*, or *field flux*, is the imaginary lines of force that make up the total quantity of an electromagnetic field.

- *Electromagnetism* is magnetism produced when electricity passes through a conductor.

. . . DEFINITIONS

- ***Magnetomotive force*** is the force that produces magnetic lines of flux in a circuit as a result of current in a conductor.

- ***Reluctance*** is the opposition to magnetic flux in a magnetic circuit in a given volume of space or material.

- ***Magnetic field intensity*** is the amount of magnetomotive force distributed over the length of a magnet.

- ***Magnetic flux density*** is the concentration of the magnetic flux in a given area.

- ***Permeability*** is a measure of the ability of a material to conduct magnetic flux.

- An ***electromagnet*** is a magnet whose energy is produced by the flow of electric current.

- ***Saturation*** is the condition where a magnetic core has substantially all the magnetic domains aligned with the field and any increases in current no longer result in a stronger electromagnet.

❓ REVIEW QUESTIONS

1. Describe the molecular theory of magnetism.

2. What are the three most common naturally occurring ferro magnetic materials?

3. Describe the direction of magnetic flux around a magnet.

4. Describe the left-hand rule for conductors.

5. Describe the left-hand rule for coils.

6. Describe the differences between an electromagnetic field around a conductor produced by DC and one produced by AC.

7. What are the three magnetic units of measure that are analogous to voltage, current, and resistance?

8. What is an electromagnet?

9. What is magnetic saturation?

10. What is a B-H curve and what is the significance of the knee in the curve?

11. Why does saturation cause a distorted current waveform?

Operating Principles

Transformers use the principle of mutual induction to transfer power from one coil to another. The turns ratio represents the ratio of the voltage in the primary to the voltage in the secondary. All transformers have losses from wire resistance, hysteresis, flux loss, and eddy currents. Transformer design has a significant effect on power losses and the temperature rating.

INTRODUCTION TO TRANSFORMERS

A *transformer* is an electric device that uses electromagnetism to change AC voltage from one level to another or to isolate one voltage from another through the process of mutual induction. A transformer typically consists of two separate coils with different numbers of turns of conductor wound around the same closed laminated iron core. **See Figure 2-1.** The primary winding is the coil in a transformer that is energized by the source. The secondary winding is the coil in a transformer

that is connected to the load. The primary circuit in a transformer can be the high-voltage or the low-voltage circuit, depending on whether it is a step-up or a step-down transformer. The high-voltage leads are marked with the letter H and the low-voltage leads are marked with the letter X. The secondary neutral is often labeled X0.

DEFINITION

A transformer is an electric device that uses electromagnetism to change voltage from one level to another or to isolate one voltage from another through the process of mutual induction.

Transformer Windings

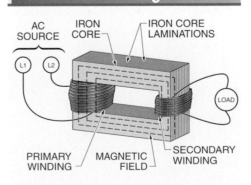

Figure 2-1. The primary winding is connected to the AC input and the secondary winding is connected to the load.

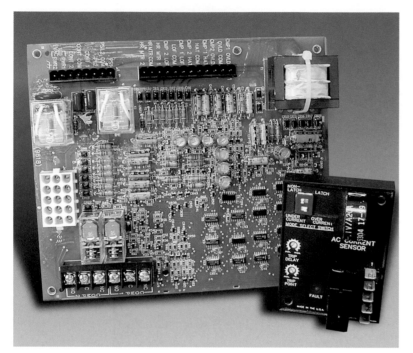

In a typical heavy industrial facility, the electricity may be delivered directly from a transmission substation to an outside transformer vault. Service-entrance conductors are routed from the outside transformer vault through an outdoor busway to a metered switchboard. Power is then fed through circuit breakers in the panelboard and routed through busways to power distribution panels and busways with plug-in sections to the points of use. **See Figure 2-2.** Depending on customer needs, the power distribution system delivers power at standard voltage levels and fixed current ratings to set points such as receptacles.

Ametek®

Small transformers are often used in circuit boards to raise or lower the voltage or for circuit isolation.

TECH FACT

Alternating current produced at generating plants is transformed to a higher voltage to allow efficient transmission of electrical power between power stations and end users.

Power Distribution

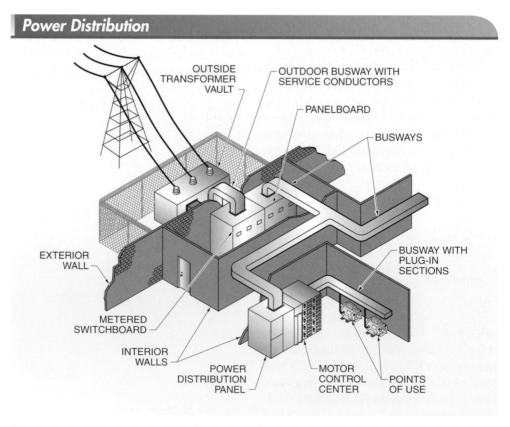

OUTSIDE TRANSFORMER VAULT

OUTDOOR BUSWAY WITH SERVICE CONDUCTORS

PANELBOARD

BUSWAYS

BUSWAY WITH PLUG-IN SECTIONS

EXTERIOR WALL

METERED SWITCHBOARD

INTERIOR WALLS

POWER DISTRIBUTION PANEL

MOTOR CONTROL CENTER

POINTS OF USE

Figure 2-2. Transformers are used to reduce the transmission line voltage to a level usable by the customer.

Induction

Inductance is the property of a device or circuit that causes it to store energy in an electromagnetic field. *Induction* is the ability of a device or circuit to generate reactance to oppose a changing current (self-induction) or the ability to generate a current (mutual induction) in a nearby circuit. The current flowing in the coil produces a field that expands out of and surrounds the conductor. Energy is stored in that field. When the source voltage goes from peak to zero, the energy in the field is returned to the coil and converted back to electrical energy. The energy actually opposes the changes in the source voltage.

The three requirements for induction are a conductor, a magnetic field, and relative motion between the conductor and the magnetic field. In a transformer, the conductor is the wire making up the coil. The AC power flowing through a conductor generates an expanding and collapsing magnetic field. The expanding and collapsing magnetic field flows through the laminated core and provides the relative motion between the conductor in the secondary and the magnetic field.

The core is constructed of layers of a material that is low in reluctance and offers little opposition to the magnetic lines of flux. The closed, laminated core gives a path of low reluctance for the flux to flow, which aligns the flux and allows the maximum number of conductors to be cut. This induces a voltage in the secondary. The total power is the same in the primary and the secondary circuits, except for some losses in the transformer. This means that when the voltage in the secondary is higher than in the primary, the current is proportionally lower in the secondary.

Self-Induction. *Self-induction* is the ability of an inductor in a circuit to generate inductive reactance, which opposes change in the circuit. When an AC source voltage rises and the magnetic flux expands around the circuit conductors, an opposing voltage, or countervoltage, is induced in the circuit. The magnitude of the induced voltage is determined by the rate of change of the current. Lenz's law states that the polarity of the induced voltage is such that it produces a current whose magnetic field opposes the change that produced it. **See Figure 2-3.** The induced magnetic field in any inductor acts to oppose any change in current.

The countervoltage limits the circuit current, as the circuit current is determined by the impedance and the difference between the source and the countervoltage. As the AC source voltage falls back to zero and the field collapses back into the circuit, the countervoltage acts to prevent the current from falling. This shows that the first 90° of a cycle is spent charging the inductor. The electrical energy is converted into magnetic energy in the inductor. When the voltage peaks, the current is at the maximum, the field stops expanding, and all the energy is stored in the magnetic field. When the source voltage starts to drop from peak, the magnetic field starts to collapse back into the inductor and aids the current provided by the source. This makes the current lag 90° behind the source.

When the AC frequency and the coil inductance are known, inductive reactance is calculated as follows:

$$X_L = 2\pi f L$$

where
X_L = inductive reactance (in Ω)
π = 3.14
f = frequency (in Hz)
L = inductance (in H)

The total opposition or impedance is calculated as follows:

$$Z = \sqrt{R^2 + X_L^2}$$

where
Z = impedance (in Ω)
R = resistance (in Ω)
X_L = inductive reactance (in Ω)

DEFINITION

Inductance is the property of a device or circuit that causes it to store charge in an electromagnetic field.

Induction is the ability of a device or circuit to generate reactance to oppose a changing current (self-induction) or the ability to generate a current in a nearby circuit (mutual induction).

Self-induction is the ability of an inductor in a circuit to generate inductive reactance, which opposes change in the circuit.

Self-Induction

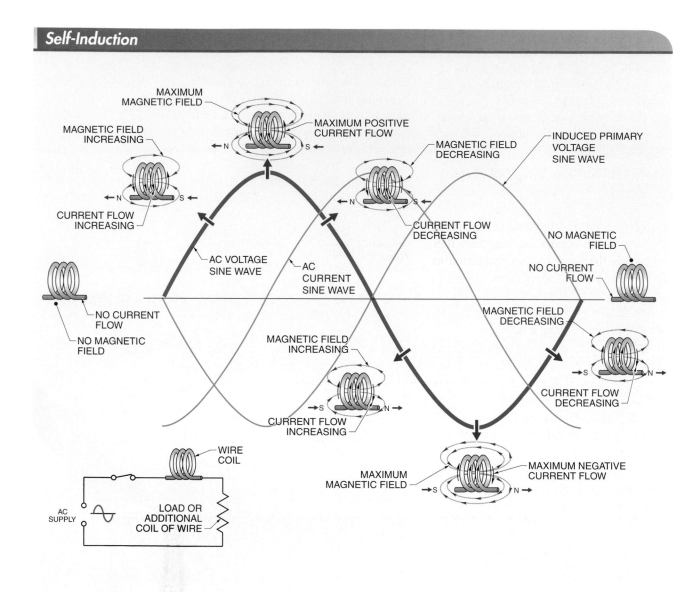

Figure 2-3. *Self-induction in a coil opposes changes to the current.*

A transformer primary is a coil and the opposition induced in that coil can be calculated. If the winding in the coil is 500′ of AWG #22 copper wire, the wire has a resistance of approximately 8 Ω. If the coil has an inductance of 0.5 H, the inductive reactance and total impedance are calculated as follows:

$$X_L = 2\pi f L$$
$$X_L = 2\pi \times 60 \times 0.05$$
$$X_L = \mathbf{188.5}\ \mathbf{\Omega}$$

$$Z = \sqrt{R^2 + X_L^{\,2}}$$
$$Z = \sqrt{8^2 + 188.5^2}$$
$$Z = \sqrt{64 + 35,532}$$
$$Z = \mathbf{188.7}\ \mathbf{\Omega}$$

If the wire is not wound into a coil, the resistance of the wire alone is 8 Ω. Because a coil has a higher impendance than a wire, the total impedance when the wire is wound into a coil is 188.7 Ω. The current capacity in the wire alone is about 23.6 times higher than the current capacity through the coil because of the increased impedance from the inductive reactance of the coil.

The comparison of the coil and the straight conductor shows how a transformer primary can be connected to a source and act to limit the current.

The four factors that determine or control the inductance of a coil are the cross section of the core, the number of turns, the type of the core, and the length of the coil. Many people use the acronym CoNTroL to remember these factors, where the capitalized letters represent the first letter of each of the factors.

Mutual Induction. *Mutual induction* is the ability of an inductor in one circuit to induce a voltage in another circuit. When a transformer primary has alternating current flowing in the conductor, magnetic flux surrounds the conductor in proportion to the amount of the current. The expanding and contracting flux cuts the conductors in the secondary and induces a voltage in the secondary. **See Figure 2-4.**

When an AC source is applied to the primary of a transformer, there is a countervoltage induced in the primary coil that is opposite and nearly equal to the applied voltage. There is a very small difference between the applied and the induced voltages that allows just enough current to magnetize the primary core. The *exciting current,* or *magnetizing current,* is the no-load current through a primary core. The exciting current causes a magnetic field that cuts across the secondary and induces a voltage in the secondary. Exciting current actually has two components. The first component is the true power no-load current (in kW) that magnetizes the core. The second component is the reactive power (in kVAR) that builds the field. For very small transformers, the exciting current may be as high as 10% of the maximum current. For very large transformers, the exciting current may be less than 1% of the maximum current.

When a load is connected across the secondary, the induced voltage causes a current to flow in the secondary. The current causes a magnetic field with polarity opposite to the field in the primary in proportion to the relative field strengths. The secondary magnetic field tends to neutralize and reduce the magnetic field in the primary because the lines of flux oppose each other. The reduced

magnetic field reduces the inductance and allows more current to flow in the primary. The increased current in the primary produces a stronger magnetic field that induces an increased voltage in the secondary. The increased voltage in the secondary allows more current to flow. This continues until the load is drawing the needed current.

While conduit can be used for the grounding of a circuit, installing a separate equipment grounding conductor is always recommended.

Mutual induction is the ability of an inductor in one circuit to induce a voltage in another circuit.

The exciting current, or magnetizing current, is the no-load current through a primary core.

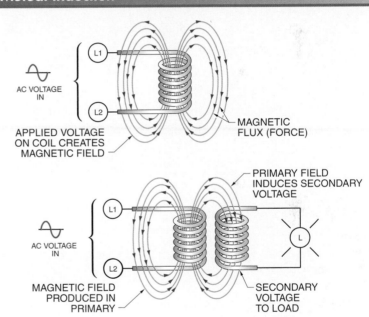

Figure 2-4. *Mutual induction is the process where one coil induces current flow in another coil.*

Turns Ratio

The *turns ratio,* or the *turns-to-turns ratio,* is the ratio of the number of the turns in the primary to the number of turns in the secondary. The turns ratio is written with two numbers, such as 2:1 or 2 to 1. The first number represents the relative number of turns in the primary and the second number represents the relative number of turns in the secondary. A *step-up transformer* is a

The turns ratio, or the turns-to-turns ratio, is the ratio of the number of turns in the primary to the number of turns in the secondary.

*Volts per turn (V/turn)
is the voltage dropped
across each turn of
a coil or the voltage
induced into each turn
of the secondary coil.*

transformer with the source connected to the winding with the fewest turns and the load connected to the winding with the most turns. A *step-down transformer* is a transformer with the source connected to the winding with the most turns and the load connected to the winding with the fewest turns. **See Figure 2-5.**

The turns ratio between the two coils determines if the device is a step-up or step-down transformer. For example, if the coil connected to the source has 500 turns and the coil connected to the load has 1000 turns, the device is a step-up transformer. The turns ratio is 1:2 and the flux from each turn in the primary cuts two turns in the secondary. If the source connected to the primary is 120 V, the secondary voltage is calculated as follows:

$$\frac{N_P}{N_S} = \frac{E_P}{E_S}$$

$$\frac{500}{1000} = \frac{120}{E_S}$$

$$500 \times E_S = 120 \times 1000$$

$$E_S = 120 \times \frac{1000}{500}$$

$$E_S = \mathbf{240\,V}$$

If a 240 V source needs to be stepped down to 120 V, the coils could be reversed. The 240 V coils are the primary and the 120 V coils are the secondary. However, the voltage rating of the coil must never be exceeded. A transformer with a 2:1 ratio with a 240 V primary and a 120 V secondary should not be connected to a 480 V line to build a 240 V secondary. The turns ratio is correct, but the inductance of the coils is too low to provide the required current limits in the coil due to the lower reactance.

Turns Ratio

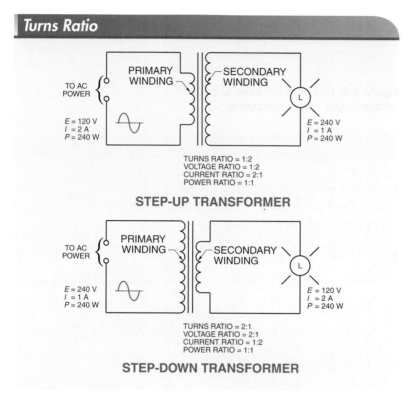

Figure 2-5. The turns ratio determines whether a transformer is a step-up or step-down transformer.

DEFINITION

*A **step-up transformer**
is a transformer with the
source connected to the
winding with the fewest
turns and the load con-
nected to the winding
with the most turns.*

*A **step-down transformer**
is a transformer with the
source connected to the
winding with the most
turns and the load con-
nected to the winding
with the fewest turns.*

For both step-up and step-down transformers, the power rating is always the same on the primary and secondary sides.

The turns ratio can be used to calculate the secondary voltage and current as follows:

$$\frac{N_P}{N_S} = \frac{E_P}{E_S} = \frac{I_S}{I_P}$$

where

N_P = number of turns in primary

N_S = number of turns in secondary

E_P = voltage in primary (in V)

E_S = voltage in secondary (in V)

I_P = current in primary (in A)

I_S = current in secondary (in A)

TECH FACT

When installing (or replacing) fuses, install so that the fuse label is displayed facing out and can be clearly read.

Volts per Turn. The turns ratio can be used to explain a related concept called volts per turn. *Volts per turn (V/turn)* is the voltage dropped across each turn of a coil or the voltage induced into each turn of the secondary coil. Each transformer has a design value for the volts per turn. For example, if a transformer primary has 120 turns with a source of 120 V, it has 1 V/turn. The secondary coil has the same volts per turn value. If the secondary coil has 24 turns, the voltage across the secondary is 24 V. Therefore, a transformer with a turns ratio of 120:24, volts per turn of 1, and a primary voltage of 120 V has a secondary voltage of 24 V.

Coil Taps. A *coil tap* is an extra electrical connection on a transformer coil that allows a varying number of turns of a coil to be part of the circuit. **See Figure 2-6.** Coil taps are sometimes necessary depending on the location of the service in reference to the substation that is providing power. At the end of the distribution line a long distance from the source, the voltage will sometimes be below normal. In order to get the proper voltage to the equipment in the plant, taps on the transformer primary are offered. These taps allow the turns ratio of a transformer to be modified in the field to compensate for a low primary voltage.

For example, if a primary coil is rated at 7200 VAC and has 1620 turns, what is the volts per turn value of the transformer? The transformer drops 7200 VAC across 1620 turns. The volts per turn is calculated as follows:

$$V / turn = \frac{E}{N_P}$$

where

$V / turn$ = volts per turn

E = voltage (in V)

turns = number of turns in primary

$$V / turn = \frac{E}{N_P}$$

$$V / turn = \frac{7200}{1620}$$

$$V / turn = \mathbf{4.444}$$

What is the volts per turn value of the transformer if the voltage is 5% low (6840 VAC)?

$$V / turn = \frac{E}{N_P}$$

$$V / turn = \frac{6840}{1620}$$

$$V / turn = \mathbf{4.222}$$

Coil Taps

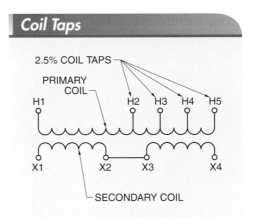

Figure 2-6. Coil taps are used to adjust the voltage output from a transformer.

This shows that when the primary voltage is reduced by 5%, the secondary voltage is reduced by the same percent. By changing taps and removing turns from the primary circuit, the volts per turn can be brought back up to a level that gives us the proper secondary voltage. The number of turns in the primary required to drop 4.444 V/turn at 6840 VAC is calculated as follows:

$$N_P = \frac{E}{V / turn}$$

where

N_P = number of turns in primary

E = voltage (in V)

$V / turn$ = volts per turn

$$N_P = \frac{E}{V / turn}$$

$$N_P = \frac{6840}{4.444}$$

$$N_P = \mathbf{1539 \ turns}$$

This type of transformer is normally furnished with coil taps at multiples of 2.5% above normal (AN) and 2.5% below normal (BN). For a primary with 1620 turns, 2.5% represents about 40 turns (1620 × 0.025 = 40.5). Moving the connection by two tap locations changes the number of turns in the primary coil by about 80 turns. The primary is changed from 1620 turns to 1540 turns. The turns ratio is changed so that the transformer can compensate for the low voltage and ensure that the secondary is at the rated voltage.

*A **coil tap** is an extra electrical connection on a transformer coil that allows a varying number of turns of a coil to be part of a circuit.*

Transformer Losses

Transformers, like all devices, are not perfect. The power output from a transformer is always slightly less than the power input to a transformer. These power losses end up as heat that must be removed from the transformer. The four main types of loss are resistive loss, eddy currents, hysteresis, and flux loss.

Resistive Loss. *Resistive loss,* or *I^2R loss,* or *copper loss,* is the power loss in a transformer caused by the resistance of the copper wire used to make the windings. Since higher frequencies cause the electrons to travel more toward the outer circumference of the conductor (skin effect), electrical disturbances called harmonics have the effect of reducing the wire size and increasing resistive loss. These losses are the same as the power losses in any conductor and are calculated as follows:

$$P = I^2R$$

where
P = power (in W)
I = current (in A)
R = resistance (in Ω)

For example, if a transformer primary is wound with 100′ of #12 copper wire that carries 15 A, what is the resistive loss in that coil? The resistance of #12 copper wire is 1.588 Ω/1000′ at room temperature. Therefore, the resistance of 100′ of the wire is 0.1588 Ω.

$$P = I^2R$$
$$P = 15^2 \times 0.1588$$
$$P = 225 \times 0.1588$$
$$P = \mathbf{35.7\ W}$$

The transformer primary wiring consumes 35.7 W of power that is wasted as heat. If the transformer is not cooled properly, this heat increases the temperature of the transformer and the wires. This increased temperature causes an increase in the wire resistance and in the voltage dropped across the conductor. This loss varies with the current and is always present in the primary when it is energized. The secondary sees very little loss of this type when unloaded.

Eddy Current Loss. *Eddy current loss* is power loss in a transformer or motor due to currents induced in the metal field structure from the changing magnetic field. Any conductor that is in a moving magnetic field has a voltage and current induced in it. The iron core offers a low reluctance to the magnetic flux for mutual induction. The magnetic flux induces current at right angles to the flux. This means that current is induced across the core. This current causes heating in the core. At higher frequencies than the 60 Hz line frequency, heat produced by eddy currents increases as the square of the frequency. For example, the third harmonic (180 Hz) has nine (3^2) times the heating effect of the fundamental (60 Hz) frequency.

Constructing the core of thin sheets of iron laminated together can minimize this loss. **See Figure 2-7.** Each sheet is coated with an insulating varnish that forces these currents to only flow within individual laminations. This reduces the overall eddy currents in the entire core. These thin sheets are manufactured from silicon-iron or nickel-iron alloys that can be magnetized more readily than pure iron. The use of alloy cores also improves the age resistance of the core. The sheets are often made from 29-gauge alloy, which is only 0.014″ thick.

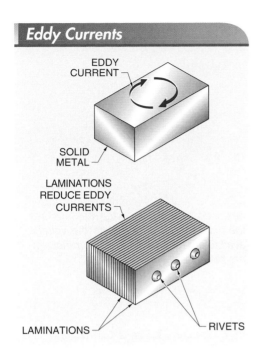

Eddy Currents

EDDY CURRENT

SOLID METAL

LAMINATIONS REDUCE EDDY CURRENTS

LAMINATIONS — — RIVETS

Figure 2-7. *Laminations in the core material reduce the amount of eddy current loss.*

Hysteresis Loss. *Hysteresis* is the property of ferromagnetic materials where the magnetic induction of a coil lags the magnetic field that is charging the coil. Magnetic domains are small sections of a magnetic material that act together when subject to an applied magnetic field. Magnetic domains have magnetic properties and move in the iron when influenced by a magnetic field. When the iron is subjected to a magnetic field in one polarity, the magnetic domains will be forced into alignment with the field. When the polarity changes twice each cycle, power is consumed by this realignment and this reduces the efficiency of the transformer. This movement of the molecules produces friction in the iron and thus heat is a result. The presence of harmonics may cause the current to reverse direction more often, resulting in greater hysteresis loss. Hysteresis is reduced through the use of highly permeable magnetic core material.

Flux Loss. *Flux loss* is a power loss that occurs in a transformer when some of the lines of flux from the primary do not travel through the core to the secondary. There

are two main reasons for the lines of flux to travel through air instead of through the core. First, the iron core can become saturated so that the core cannot accept any more lines of flux. The lines of flux then travel through the air and are not cut by the secondary. Second, the ratio of the reluctance of the air and the core in the unsaturated region is typically about 10,000:1. This means that for every 10,000 lines of flux through the core, there is 1 line of flux through the air. Flux loss is generally small in a well-designed transformer.

Transformers are often placed on concrete pads. Barriers are then put in place to protect the transformer from damage.

Transformer Mutual Inductance

The *mutual inductance,* or *coefficient of coupling,* of a transformer is a measure of the efficiency by which power is transferred from the primary to the secondary coils. **See Figure 2-8.** If the power transfer is perfect, the coefficient of coupling is 1. If there is no power transfer, the coefficient of coupling is 0. The coefficient of coupling depends on the design of the transformer. The most important factor is the position of each coil with respect to the other. If the coils are wound over one another and each line of flux from the primary cuts a turn in the secondary, then the coefficient of coupling is very close to 1. If any flux is lost, then the coefficient of coupling is less than 1. The coefficient of coupling of typical transformers ranges from 0.95 to 0.99, depending on design and purpose.

DEFINITION

Hysteresis is the property of ferromagnetic materials where the magnetic induction of a coil lags the magnetic field that is charging the coil.

DEFINITION

Flux loss is a power loss that occurs in a transformer when some of the lines of flux from the primary do not travel through the core to the secondary.

*The **mutual inductance**, or **coefficient of coupling**, of a transformer is a measure of the efficiency by which power is transferred from the primary to the secondary coils.*

Mutual Inductance

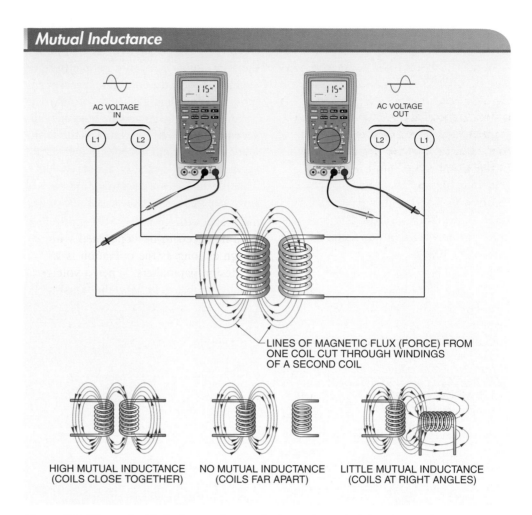

LINES OF MAGNETIC FLUX (FORCE) FROM ONE COIL CUT THROUGH WINDINGS OF A SECOND COIL

HIGH MUTUAL INDUCTANCE (COILS CLOSE TOGETHER)

NO MUTUAL INDUCTANCE (COILS FAR APART)

LITTLE MUTUAL INDUCTANCE (COILS AT RIGHT ANGLES)

Figure 2-8. The mutual inductance of two coils depends on their location and orientation.

Power Factor

A transformer, like a motor, is a reactive generator. In accordance with Lenz's law, this generator induces a countervoltage that opposes the source voltage. When there is no load on the secondary, the exciting current creates a small magnetic field that induces a potential in the secondary. When there is no load to consume power, the circuit is almost purely reactive. This means the countervoltage is at its maximum, the power factor is low, and the current in the primary and secondary are 180° out of phase.

Since the current in the secondary is 180° out of phase with the current in the primary, the magnetic field in the secondary is also 180° out of phase with the magnetic field in the primary. Magnetic lines of flux of opposite polarities that occupy the same space cancel each other. As the lines cancel one another, the ability to store energy in the magnetic field (inductance) is reduced. The reduced field size means that the inductance is lowered and therefore the inductive reactance is lowered. As the reactance is lowered, the relative importance of resistive load elements increases. The phase angle between the current in the two coils decreases and the power factor increases.

 TECH FACT

Magnetic lines of flux of opposite polarity cancel each other when they occupy the same space.

In other words, a transformer operating under no-load conditions has a low power factor because the circuit is almost purely reactive. As the load on a transformer increases, the reactance decreases and the power factor increases. At full load, the power factor approaches 1. Loads with a low power factor draw considerably more current than loads with a power factor near unity. **See Figure 2-9.** For example, a load that draws 35 kW with a power factor of 100% draws 76 A. A load that draws the same power with a power factor of 70% draws over 108 A.

Transformer Impedance

The induced countervoltage in a transformer is 180° out of phase with the source. The difference between this countervoltage and the source is the voltage dropped across the coil to provide flux to excite the coil. If the circuit were purely resistive without the inductive reactance, the low nominal resistance would allow very high current to flow, which could destroy the coil. Therefore, the transformer coil impedance is very important to transformer operation.

A single-phase transformer is designed with enough turns in the coils to provide the proper voltage at the output. The user does not need to be concerned with the design as long as the operation is within the design parameters. When a voltage is induced in the secondary, the transformer provides the rated voltage and current. As the load increases from zero to full load, the voltage drop increases.

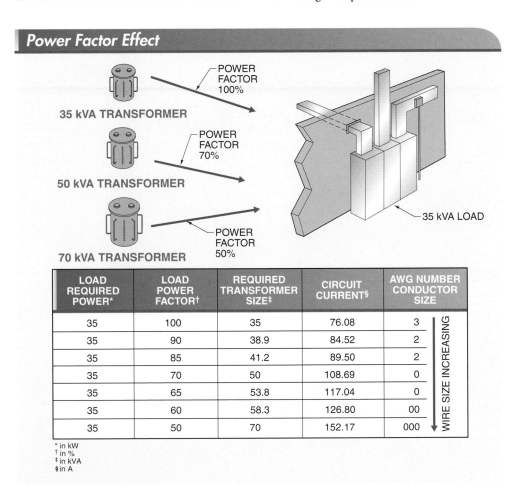

Power Factor Effect

35 kVA TRANSFORMER — POWER FACTOR 100%

50 kVA TRANSFORMER — POWER FACTOR 70%

70 kVA TRANSFORMER — POWER FACTOR 50%

35 kVA LOAD

LOAD REQUIRED POWER*	LOAD POWER FACTOR†	REQUIRED TRANSFORMER SIZE‡	CIRCUIT CURRENT§	AWG NUMBER CONDUCTOR SIZE	
35	100	35	76.08	3	WIRE SIZE INCREASING
35	90	38.9	84.52	2	
35	85	41.2	89.50	2	
35	70	50	108.69	0	
35	65	53.8	117.04	0	
35	60	58.3	126.80	00	
35	50	70	152.17	000	

* in kW
† in %
‡ in kVA
§ in A

Figure 2-9. A load with a lower power factor requires more current to deliver the power than the same load with a higher power factor.

In a 3-phase transformer bank, comprised of three individual single-phase units, the impedance is very important. At no load, the output voltage is equal to the rating of the coils. As a load is placed on the system, the situation changes. All three transformers in the bank should have the same impedance to ensure that the voltage drop across each coil is the same. **See Figure 2-10.** If the three transformers in the bank have different impedances, the voltage drop across the coil with the highest impedance is higher than the voltage drop across the other coils. When the transformers have unequal impedances, the voltage available from the transformer is different for each phase as the load increases from zero to full load.

It is common for delta systems to use two transformers of one kVA rating, and the third transformer with a higher rating. **See Figure 2-11.** This is used on the 120/240 3-phase 4-wire systems with a high leg to ground. The large transformer is used for the single-phase loads, while also providing power for the 3-phase loads. The kVA ratings are not equal, but the impedance of each is as close as possible.

TECH FACT

For every 10°C rise in temperature above the transformer's rated limit, the life of the transformer will be reduced by about 50%.

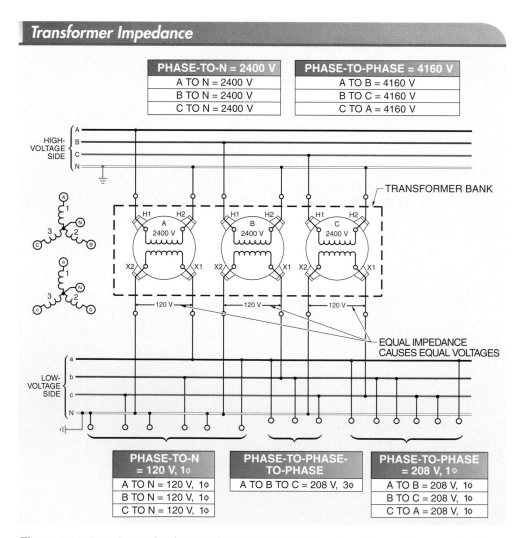

Figure 2-10. Transformer banks must have equal impedance in order to deliver equal voltage.

Delta High Leg to Ground

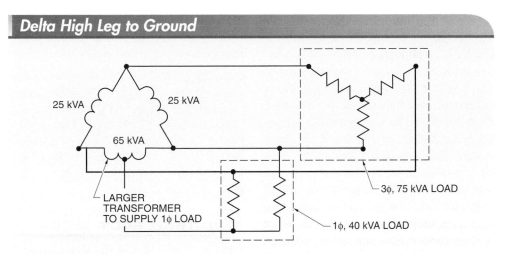

Figure 2-11. Single-phase loads on a delta system add extra load. One transformer must be larger to balance the load.

Short Circuit Stress

An energized coil on a transformer creates an electromagnet. When a current flows through a conductor that has been wound into a coil, the various turns attract each other at all points because the current is flowing in the same direction in all turns. Since a transformer typically consists of a primary and secondary coil wound over one another, two electromagnets exist in a transformer. The currents in these two windings are opposite in direction and there is a force of repulsion between the coils at all times. Under normal circumstances, these forces are small. However, these forces are multiplied many times over during a short circuit.

There is a tendency for the windings to deform in response to the stress applied by the electromagnets during a short circuit. This can cause the coils to move, or telescope, in opposite directions. Once movement of the coils starts, the transformer is often damaged or destroyed. To minimize the probability of telescoping, the primary and secondary windings are electrically centered with respect to each other on the core leg. This means that the coils are designed so that the electrical center of the two coils is in an identical position. The core and coil assembly of a 3-phase core-type transformer has the coils supported at both ends by support blocks with resilient pads. The support blocks are mounted on the frame, which supports the entire core and coil assembly. The core and coil assembly of a single-phase shell-type transformer has a permanently welded core clamp that cradles the core and minimizes the stress.

Telescoping problems become even more complex when transformers are specified with taps. Taps may appear on either or both of the windings. Most commonly, taps are built into the primary or high-voltage winding. When the customer connects to a tap point other than the standard ratio, the electrical center of one of the windings is shifted from the center of the companion winding. This increases the probability of telescoping under fault conditions.

TRANSFORMER DESIGN

Two key transformer design parameters are the type of core and the type of cooling. The type and shape of the core influence the efficiency of the transformer. Since all transformers have power losses, transformer cooling is part of the power rating. Common types of cooling include dry and oil-immersed transformers, with or without forced-air ventilation.

Cores

The design of transformer cores has an effect on the efficiency of the transformer. Cores are constructed of legs and yokes. The vertical legs support the coils and upper and lower yokes connect the legs. The ends of the laminations used to construct the core are often cut at a 45° angle instead of square. **See Figure 2-12.** This allows the lamination layers of the legs and yokes to overlap slightly at the corner. This helps improve the magnetic conduction path through the core.

Three common types of transformers are core-type, shell-type, and toroidal transformers.

Core-Type. A core-type transformer has windings placed around each leg of the core material. **See Figure 2-13.** A thick layer of insulating material is wrapped around the legs to prevent electrical contact between the wire of the coil and the iron of the leg. Three-phase transformers typically use a 3-leg design. Single-phase transformers typically use a 2-leg design. Core-type transformers are generally less expensive than other types as less iron is used and the enclosure is smaller.

Core Construction

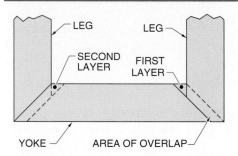

Figure 2-12. The laminations used to construct cores are often cut at 45° angles to allow overlap to improve the magnetic conduction path.

Core-Type Transformers

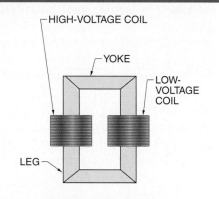

SINGLE-PHASE

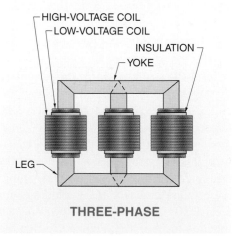

THREE-PHASE

Figure 2-13. A core-type transformer has windings placed around each leg of the core material.

For a step-down transformer, the secondary coil is wound first and placed closest to the core iron with a layer of insulating material between them. The primary coil is then wound and placed over the low-voltage coil with a layer of insulating material between them. This construction places the conductors energized at the high voltage at a greater physical distance away from the iron, which is normally grounded. For a step-up transformer, the primary is wound first and the secondary is wound on top.

The core is electrically interconnected with core clamps, steel structure, and enclosing case, all of which are connected with a lead to the plant or system ground.

Shell-Type. A shell-type transformer has extra legs and a metal body surrounding the core. The extra conductive material helps keep the core leakage flux to a minimum. Common shell-type transformers include 3-, 5-, and 7-leg cores. **See Figure 2-14.** Three-phase transformers typically use a 5-leg or 7-leg design. The two auxiliary legs of the 7-leg design provide for symmetry among the three magnetic circuits of the core. However, the extra legs of a shell-type transformer result in increased capacitance between the primary and the secondary. This design allows more uniform distribution of flux between the various legs of the core. The more uniform distribution of flux results in reduced harmonics. Single-phase transformers typically use a 3-leg design.

Toroidal. A toroidal core transformer has a doughnut-shaped core wrapped with copper wire wound all around the core. **See Figure 2-15.** The toroidal core is constructed by taking a long strip of magnetic material and rolling it tightly into shape like a spring. Since the core is constructed of one piece of magnetic material, this design offers the best possible coupling between the primary and the secondary. The primary and secondary wires are wrapped through the center hole, around the core, and back through the center hole. The wire can be wrapped to provide complete coverage of the core and completely keep the magnetic field confined to the core to improve efficiency. Toroidal transformers are also quieter than other transformers. Toroidal transformers are most commonly used in relatively high-frequency applications, such as audio and RF components.

Shell-Type Transformers

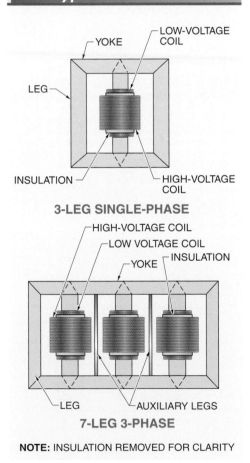

3-LEG SINGLE-PHASE

7-LEG 3-PHASE

NOTE: INSULATION REMOVED FOR CLARITY

Figure 2-14. A shell-type transformer has more legs than windings.

Toroidal Transformers

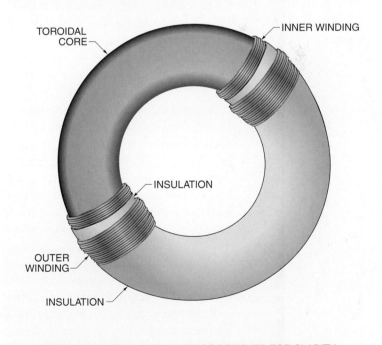

NOTE: INSULATION AND WINDINGS REMOVED FOR CLARITY

Figure 2-15. A toroidal transformer has a doughnut-shaped core with the windings wrapped around the core.

Cooling

Part of the rating of a transformer is determined by the dissipation of the heat losses generated in the transformer and by the temperature rise of the transformer when delivering a certain load. Because of the close concentration of the conductor in the windings, heat is a factor in transformer installation and design. If excessive heat is permitted to remain in or around the transformer, the insulation on the conductors that comprise the windings is subjected to damage and perhaps failure. Any means that can economically dissipate the heat generated in the coils and magnetic circuit more readily permits an increase in the rating of a transformer having a definite physical size or permits a decrease in the dimensions of a unit having a definite kVA rating. Transformers may be designated as dry-type, where air is used to cool the coils, or liquid-immersed, where the coils are immersed in oil.

Coolants. While dry-type transformers are cooled by air, sometimes additional cooling is needed to reach the desired load rating. This is accomplished by immersing the coils in a liquid coolant. The oils typically used as coolants also act as electrical insulators so they have a very high dielectric strength.

Mineral oil is a common coolant used in transformers. Mineral oil has a higher dielectric strength than air, so it acts as an insulator. It also has a higher heat capacity than air, so it acts as a coolant. However, mineral oil is somewhat flammable. Askarels are a class of oils that were developed to replace mineral oils in locations where flammability was a concern. Askarels are considered to be nonflammable. However, they can decompose when exposed to heat to form hydrochloric acid and toxic chemicals like dioxins. In addition, some types of askarels contain polychlorinated biphenyls (PCBs). The use of askarels in new transformers in the U.S. was banned in 1977. Many older transformers in the field still contain poisonous askarels. Extreme care must be taken when working on older oil-filled transformers.

High-temperature hydrocarbons and synthetic esters are seeing increased use as transformer coolants. Both types of materials are very stable at relatively high temperatures, but high cost somewhat limits their use to only the most critical applications. Silicones and halogenated fluids have been used in the past, but are seldom used now because both exhibit biological persistence in the environment if spilled.

Cooling Class. Historically, the cooling class of a transformer was given by a 2- to 4-letter designation indicating air (A), water (W), mineral insulating oil (O), or a synthetic nonflammable insulating liquid (L) and whether natural convection (N) or forced convection (F) was used. The cooling class of dry-type transformers is defined in IEEE C57.94-1982 (R-1987). **See Figure 2-16.**

Dry-Type Cooling Classes

Cooling Class	IEEE Designation
Ventilated self-cooled	AA
Ventilated forced-air cooled	AFA
Ventilated self-cooled/forced-air cooled	AA/FA
Nonventilated self-cooled	ANV
Sealed self-cooled	GA

Definitions
Ventilated – Ambient air may circulate and cool the core and windings
Nonventilated – No circulation of external air through the core and windings
Sealed – Self-cooled transformer with hermetically sealed tank
Self-cooled – Cooled by natural circulation of air
Forced-air cooled – Cooling by forced circulation of air
Self-cooled/forced-air cooled – Cooling by natural circulation of air with cooling by forced circulation of air

Figure 2-16. The cooling class of dry-type transformers tells whether the transformer is ventilated and whether the transformer is self-cooled or forced-air cooled.

The cooling class of liquid-immersed transformers historically had a similar designation. The cooling class for liquid-immersed transformers has recently been replaced. The cooling class of liquid-immersed transformers is now defined in IEEE C57.12.00-2000. This standard provides a 4-letter designation that indicates specific criteria relative to the type of oil, how the oil is circulated, what is used to cool the oil, and how the oil is cooled externally. **See Figure 2-17.**

Temperature Limits. The *temperature rise* in a transformer is the difference between the hot-spot maximum core temperature at full load and the temperature when not operating. The temperature rise must be limited so that the local temperature does not exceed the insulation rating. The amount of transformer temperature rise depends on the transformer design, ambient conditions,

and the load. Common design factors that influence the amount of temperature rise include the diameter of the winding wire and the size and design of the core.

There are several common transformer ratings based on the allowed temperature rise. **See Figure 2-18.** For example, Class 105 (A) insulating materials are limited to use in transformers that are designed to have a continuous full-load temperature rise not exceeding 55°C over a 40°C ambient temperature. The standards allow for a hottest spot temperature of 10°C over the normal temperature rise. Therefore, a Class 105 (A) transformer under full load has an average conductor temperature of 95°C when operating in a 40°C ambient temperature and a maximum conductor hottest spot temperature of 105°C. Other classes have different values for the allowable temperature rise and the hotspot temperature rise.

*The **temperature rise** in a transformer is the difference between the hot-spot maximum core temperature at full load and the temperature when not operating.*

Liquid-Immersed Cooling Classes

	Term	Designation
First letter	K	Insulating liquid with fire point > 300°C
	L	Insulating liquid with fire point ≤ 300°C
	O	Mineral oil or synthetic insulating liquid with fire point ≤ 300°C
Second letter	D	Forced circulation through cooling equipment; directed flow in main windings
	F	Forced circulation through cooling equipment; directed flow in natural convection in windings
	N	Natural convection through cooling equipment and through windings
Third letter	A	Air
	W	Water
Fourth letter	F	Forced convection
	N	Natural convection

ANSI Designations	CSA Designations
OA	ONAN
FA	ONAF
OA/FA/FA	ONAN/ONAF/ONAF
OA/FA/FOA	ONAN/ONAF/OFAF
OA/FOA	ONAN/ODAF
OA/FOA/FOA	ONAN/ODAF/ODAF
FOA	OFAF or ODAF
FOW	OFWF or ODWF

Figure 2-17. *The cooling class of liquid-immersed transformers describes the type of oil, how the oil is circulated, what is used to cool the oil, and how the oil is cooled externally.*

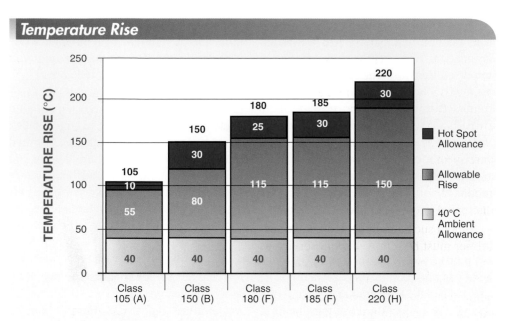

Temperature Rise

Figure 2-18. The temperature rating of a transformer is determined by the allowable temperature rise and the hot spot allowance temperature rise.

TRANSFORMER ACCESSORIES AND FITTINGS

The accessories found on power transformers enable safe operation of the transformer. Common accessories found on power transformers include explosion vents, gas relays, bushings, level gauges, thermometers, and liquid-handling and sampling valves.

Explosion Vents

High pressures are possible when an electrical fault occurs under oil. The pressure could burst the steel tank of a transformer. An *explosion vent* is a pipe, 4″ in diameter or greater, that extends a few feet above the cover of a transformer and is curved toward the ground at the outlet end of the pipe. A diaphragm fitted at the curved end breaks at a relatively low pressure to release the forces from within the transformer. The diaphragm may be glass or thin phenolic sheeting (0.16″).

Gas Relays

The chemical and electrical phenomena associated with faults in oil-filled transformers generate gas. A volume of gas is frequently formed in the early stages of a fault. For this reason, a protective device is needed to draw attention to the fault in its early stage, allowing action to be taken to prevent damage.

The boiling point of transformer oil is typically in the range of 280°C to 400°C. At the upper end of the boiling point range, traces of decomposition begin to appear. As the temperature rises higher, breakdown continues. At arcing temperature, the oil is completely broken down into carbon and simple gases that are formed from the elements in the oil.

 TECH FACT

If a transformer is deenergized in a humid area, condensation may form on the coils. A common failure in transformers is due to moisture in the insulation. To prevent this type of failure, hot air can be used to dry out the coils before energizing. Alternatively, heat can be applied during the down time so that condensation never appears.

Electrical insulating oils decompose when subjected to an electric arc during silent discharge in regions of high electric stress, or when overheating occurs from high resistance joints, connections, or splices. The products of the decomposition are carbon, water or steam, carbon dioxide, and a number of flammable gases such as carbon monoxide, hydrogen, methane, and acetylene. The composition of gases produced varies with the oil, fault, power dissipated in the fault, and other factors.

Should such a fault develop, the transformer must be removed from service as quickly as possible to prevent damage. A thin diaphragm in the gas relay moves when acted upon by the pressure wave. A switch connected to this diaphragm energizes relays to switch the transformer OFF the load. However, gradual overheating of any part, such as a hot joint, while not causing a pressure wave, can ultimately result in failure of the transformer. This local overheating decomposes or cracks the oil, forming gases that rise to the top of the tank. The gases accumulate in a dome-like section of the relay in which a float is riding on the oil. The gas displaces the oil, dropping the level of the oil. Because the float rides the surface of the oil, the float also drops. When the float drops, it operates a switch that engages an alarm circuit. Upon receiving an alarm, the condition may be investigated before extensive damage results.

Bushings

Electrical power circuits must be insulated by bushings where they enter the transformer tank. **See Figure 2-19.** In addition, the bushings and entrance must be oil-tight and weatherproof. The bushing is normally composed of an outer porcelain body. At high voltages, additional insulation in the form of oil and molded paper is used within the porcelain. The four types of bushings used on transformers as main lead entrances include solid porcelain, cable or pothead terminators, oil-filled, and condenser bushings at the transition from underground to overhead service. A *pothead* is a transition device between underground cable and overhead lines.

Bushings

OIL-FILLED BUSHING

Figure 2-19. Bushings provide electrical resistance that insulates power lines from the transformer enclosure.

Low-voltage transformers with separate leads normally have solid porcelain bushings. These bushings consist of high-grade porcelain cylinders through which the connections pass. The outside surface may be plain or have a series of corrugations or skirts to increase the surface leakage path to the metal case. When the conductors are brought to the transformers in lead-covered cables, the leads often enter through cable terminators. These terminators are similar to ordinary potheads and are attached to the transformer with the bushings inside the case.

High-voltage bushings are either oil-filled or condenser bushings. **See Figure 2-20.** Oil-filled bushings have a central con-

DEFINITION

*A **pothead** is a transition device between underground cable and overhead lines.*

ducting rod or tube through which the conductor passes. Around this, a series of insulating barriers are held apart by spacers. The barriers and spacers are enclosed in a skirted porcelain shell that is filled with oil. Condenser bushings are similar except that the central rod is wound with alternating layers of insulation and foil. This results in a path from the conductor to the case consisting of a series of condensers. The layers are designed to provide an approximately equal voltage drop between each condenser. In some condenser bushings, the whole bushing is enclosed in a skirted porcelain shell. Other types have only the exposed part of the bushing enclosed in the porcelain shell.

Level Gauges

A level gauge is fitted to the transformer to ensure that the correct liquid level exists in the transformer. The level gauge is located on the transformer tank of small

transformers that have no expansion tank. The level gauge is located on the expansion tank of larger transformers. A level gauge has a mark at the correct level for oil at 25°C. This ensures that, should the temperature rise, the unit does not overflow and a dangerously low level cannot be reached at low temperature conditions. Contacts are provided on large transformers to signal an alarm when the oil level drops to a dangerous level.

Thermometers

A thermometer is fitted to a transformer to measure the temperature of the top of the oil. In small transformers, this may be in the form of a liquid-filled thermometer mounted at the top of the tank. In large units, the thermometer is normally a gas-filled type that has the bulb fitted into a well in the cover. An indicator is located at the bottom of the tank at eye level. Contacts are provided to signal an alarm at high temperatures.

High-Voltage Bushings

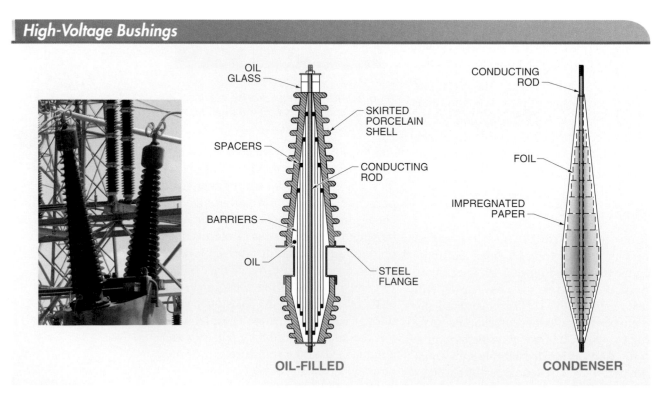

Figure 2-20. *High-voltage bushings are designed to protect the transformer from the high voltage present in the surrounding wires.*

Liquid-Handling and Sampling Valves

A liquid-handling valve is fitted at the bottom of the transformer tank to add or remove liquid from the tank. The valve is normally connected to a sump to ensure that all liquid is removed. Adjacent to the liquid-handling valve is a ½″ sampling valve, which also leads to the sump. The sampling valve is a needle valve used to draw a sample of liquid for test purposes. By taking liquid from the lowest point in the tank, any free water should appear in the sample and give warning of contamination of the liquid.

Nameplates

A transformer, like all other electrical equipment, is designed and manufactured to function in an electrical system at a specified voltage, frequency, load, etc. The nameplate data is carefully checked to ensure that the transformer is installed and maintained according to the data. **See Figure 2-21.** The nameplate should never be defaced or removed. The five sections of a nameplate include general information, the physical terminal arrangement, a schematic diagram, a vector reference, and a voltage ratio table.

General. This section includes information on transformer serial number, type, rating, nominal voltages, impedance, tap changer range, oil capacity, weights, etc.

Physical Terminal Arrangement. Normally as a block drawing, the physical locations of all terminals are given with reference to some obvious feature of outside construction. This enables all connections to be positively identified.

Schematic Diagram. A complete winding schematic diagram shows all internal connections, coil tap numbers, internal selector switches, and main terminal markings. The combined use of the physical terminal arrangement drawing and the schematic diagram enables the user to determine exactly how each bushing is connected internally. Also included on the schematic diagram are all auxiliary components such as current and potential transformers showing their actual electrical positions and the ratios.

Vector Reference. On all 3-phase units, primary and secondary voltage reactors are drawn to indicate the phase angle between the primary and secondary voltages. The vector diagrams are labeled with the terminal designations used on the schematic diagram.

Nameplates

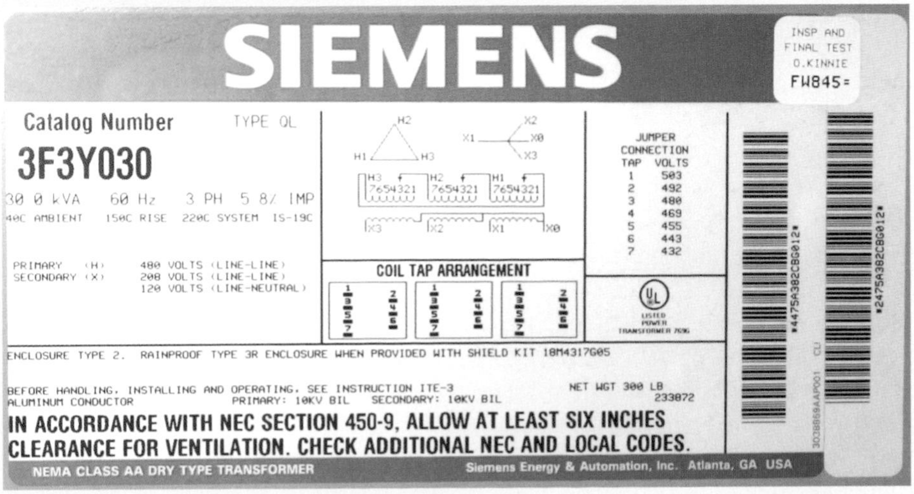

Figure 2-21. *A nameplate describes the operating parameters of a transformer.*

Voltage Ratio Table. One or more tables are given for transformers with a tap changer. The table lists the tap indicator number, the actual coil numbers, and how they are connected. The voltage ratio is given for each tap position and, in some instances, the full-load current at that tap position. On many nameplates, to avoid the use of voltage ratios, which would mean small decimal numbers, the nominal voltage of the winding without the tap is taken and the voltage required to produce this nominal value is listed for each tap. Both methods give the same information.

Sound Levels

All transformers hum when energized. The hum is due to a property of magnetic materials called magnetostriction. A changing magnetic field causes small changes in the size of the metal parts in the direction of the magnetic field. The change in size results in vibrations within the laminated steel core structure. The vibrations generate a humming sound. The hum has a fundamental frequency of twice the applied frequency. The sound volume is determined by the transformer design, construction characteristics, and the methods used in installation. Toroidal transformers typically have the lowest volume of all transformer types. Electrical maintenance personnel should monitor transformer sound levels. Any noticeable change in sound level could be the result of loosening clamp hardware, vibration isolators, or overexcitation and should be investigated

The level of sound produced by a transformer is measured in decibels (dB). A *decibel (dB)* is the unit used to measure the intensity of sound. The decibel number represents a ratio of the level of sound to a reference level (usually 0 dB). The noises in and around most indoor locations (ambient sound level) normally mask transformer sounds. Tables are available for the average ambient sound levels in areas where transformer noise could be a problem. **See Figure 2-22.**

DEFINITION

*A **decibel (dB)** is the unit used to measure the intensity level of sound.*

Ambient Sound Levels

Decibels	Example	Loudness
180	Rocket engine	Deafening
160	Jet engine	
150	Explosion	
140	Loud rock music	
130	Air raid siren	Very loud
120	Thunder	
110	Chain saw	
100	Subway	
90	Heavy truck traffic	
80	Vacuum cleaner	Loud
70	Busy street	
60	Hair dryer	Moderate
50	Normal conversation	
40	Running refrigerator	
30	Quiet conversation	Faint
20	Quiet living room	
10	Whisper	Very faint
0	Intolerably quiet	

Figure 2-22. The ambient sound level must be considered when selecting a transformer.

The sound level produced by a transformer is normally not a problem when transformers are installed in substations, in vaults, or outdoors. There are specific critical locations where sound is an important factor. Test procedures have been established so that transformer manufacturers can publish the sound level ratings of their transformers.

Sound levels are based on the kVA rating of the transformer. **See Figure 2-23.** For example, if a 150 kVA distribution transformer with a 50 dB rating is installed in a factory that has an ambient sound level of 85 dB, the sound of the transformer would not be heard above the ambient sound. If this same transformer were installed in an apartment building where the ambient sound level is 30 dB, the transformer sound would be noticeable and would be considered objectionable.

In other areas such as schools, churches, and hospitals where the ambient sound level is very low, special precautions must be taken to select a transformer with a low sound rating and to locate and install the transformer to keep the sound level at a minimum.

Transformer Sound Levels

kVA Rating	Sound Level*
3–9	40
10–50	45
51–150	50
151–300	55
301–500	60
501–700	62
701–1000	64
1001–1500	65
1501–2000	66
2001–3000	68

*in dB

Figure 2-23. The transformer sound level is based on the kVA rating.

TECH FACT

Sound levels are measured 1' from the outside edge of the transformer.

POWER RATINGS

Transformers come in many sizes and types. Transformers are rated by their capacity in volt amps (VA). The smallest transformers may be rated at only a few volt amps, while large utility distribution transformers are rated at thousands of volt amps, or kilovolt amps (kVA).

True Power

True power is the power, in watts or kilowatts, used by motors, lights, and other devices to produce useful work or heat energy. True power is the resistive part of the circuit that performs the work. True power can be produced only when current and voltage are both positive or both negative. **See Figure 2-24.** If the current and voltage are out of phase, some of the current does not produce useful work. Purely resistive circuits exist only in theory. In real situations, purely resistive circuits do not exist because the circuit conductors themselves produce some amount of inductance and capacitance.

Many loads are mainly resistive and make only very small inductive and capacitive contributions. These loads draw true

power and are rated in watts (W). Therefore, it is appropriate to label primarily resistive loads in watts or kilowatts and ignore the reactive component. Examples of resistive loads are incandescent light bulbs, water heaters, unit heaters, hair dryers, and cooking ranges. Resistive loads have very small reactive components. Therefore, an apparent power rating (VA) would not be appropriate for these loads.

True Power

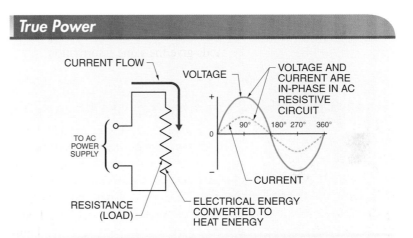

Figure 2-24. True power is the power, in watts or kilowatts, used by motors, lights, and other devices to produce useful work or heat energy.

Reactive Power

Reactive power is the power, in VAR or kVAR, stored and released by inductors and capacitors. Reactive power is measured in volts-amps-reactive (VAR). Reactive power is the power that flows back into the source from the inductors and capacitors. It is this opposing power that affects the power factor of a circuit.

In a circuit with reactive components, the voltage and current are out of phase. For inductive circuits, the current lags the voltage. **See Figure 2-25.** True power is being consumed by the resistive load at those times when the voltage and current are in the same direction (both positive or both negative). Reactive power is flowing through the inductor or capacitor at the times when the voltage and current are not in the same direction (one positive and one negative).

Reactive Power

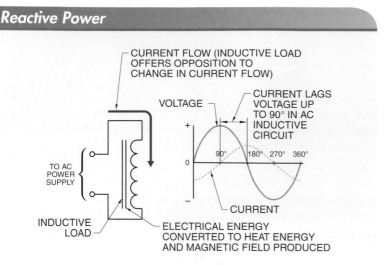

Figure 2-25. *Reactive power is the power, in VAR or kVAR, stored and released by inductors and capacitors.*

Reactive power is returned to the source without being consumed. However, current flows through the circuit to supply the reactive power. Wires, components, and devices must be sized to allow for the increased current flow from the reactive power.

Apparent Power

Apparent power is power, in VA or kVA, that is the sum of true power and reactive power. Apparent power is the product of the total current and voltage in a circuit. Many loads include reactive components. Apparent power, in VA, is made up of the true power consumed by resistive loads and the reactive power flowing through

capacitive and inductive loads. A common type of inductive load is an inductive motor. Inductive loads have the current lagging the voltage. In addition, many circuits contain harmonics. Harmonics contribute to apparent power.

Power Factor

Power factor is a ratio of the true power (W) to apparent power (VA). It is ideal to design a system with a 95% power factor. An increase in reactive power (VAR) causes the power factor to decrease. The decreased power factor means more wasted power used to perform the work. A utility company often penalizes customers for poor power factor because of the wasted power.

Transformer Ratings

Transformers are rated in kVA because apparent power represents the total power (current × voltage) the transformer can supply. The single-phase total full-load current is calculated by dividing the power by the voltage. For example, a 25 kVA transformer can deliver 104 A to a 1φ load at 240 V (25,000 ÷ 240 = 104 A). The technician must ensure that the load connected to the transformer does not exceed the full current rating of the transformer.

The actual power consumed by the load is in watts (W) or kilowatts (kW). The power factor of a circuit can be calculated by dividing the kVA supplied by the transformer by the kW consumed by the load.

SUMMARY

- Transformers use the principle of mutual induction.

- Power transmission systems step up and step down voltages.

- The turns ratio determines the relative voltages of the primary and secondary coils.

- Transformer losses include resistive, hysteresis, eddy current, and flux leakage losses.

- The core design and cooling class have significant effects on the power rating of a transformer.

- Most large utility transformers contain gas relays, pressure vents, temperature gauges, and level gauges.

- A nameplate gives information about the transformer type, rating, nominal voltages, terminal arrangements, and other information needed in the field.

- Transformer sound levels are specified in international standards based on the kVA rating.

DEFINITIONS . . .

- A *transformer* is an electric device that uses electromagnetism to change voltage from one level to another or to isolate one voltage from another through the process of mutual induction.

- *Inductance* is the property of a device or circuit that causes it to store charge in an electromagnetic field.

- *Induction* is the ability of a device or circuit to generate reactance to oppose a changing current (self-induction) or the ability to generate a current in a nearby circuit (mutual induction).

- *Self-induction* is the ability of an inductor in a circuit to generate inductive reactance, which opposes change in the circuit.

- *Mutual induction* is the ability of an inductor in one circuit to induce a voltage in another circuit.

- The *exciting current*, or *magnetizing current*, is the no-load current through a primary core.

- The *turns ratio*, or the *turns-to-turns ratio*, is the number of the turns in the primary to the number of turns in the secondary.

- A *step-down transformer* is a transformer with the source connected to the winding with the most turns and the load connected to the winding with the fewest turns.

- A *step-up transformer* is a transformer with the source connected to the winding with the fewest turns and the load connected to the winding with the most turns.

- *Volts per turn (V/turn)* is the voltage dropped across each turn of a coil or the voltage induced into each turn of the secondary coil.

- A *coil tap* is an extra electrical connection on a transformer coil that allows a varying number of turns of a coil to be part of a circuit.

- *Resistive loss,* or *I^2R loss,* or *copper loss,* is the power loss in a transformer caused by the resistance of the copper wire used to make the windings.

- *Eddy current loss* is power loss in a transformer or motor due to currents induced in the metal field structure from the changing magnetic field.

- *Hysteresis* is the property of ferromagnetic materials where the magnetic induction of a coil lags the magnetic field that is charging the coil.

- *Flux loss* is a power loss that occurs in a transformer when some of the lines of flux from the primary do not travel through the core to the secondary.

- The *mutual inductance*, or *coefficient of coupling*, of a transformer is a measure of the efficiency by which power is transferred from the primary to the secondary coils.

- The *temperature rise* in a transformer is the difference between the hot-spot maximum core temperature at full load and the temperature when not operating.

- An *explosion vent* is a pipe, 4″ in diameter or greater, that extends a few feet above the cover of a transformer and is curved toward the ground at the outlet end of the pipe.

- A *pothead* is a transition device between underground cable and overhead lines.

- A *decibel (dB)* is the unit used to measure the intensity level of sound.

REVIEW QUESTIONS

1. Explain how transformers use the principle of induction in their operation.

2. What are the three requirements for induction?

3. What is the impedance of a transformer coil that has 600′ of #22 copper wire with a resistance of 10 Ω and an inductance of 0.06 H on 60 Hz?

4. What is the meaning of the word "inductance"?

5. Can a transformer with a 2:1 ratio that has a 240 V primary and a 120 V secondary be connected to a 480 V line to build a 240 V secondary?

6. Explain how placing a load on the secondary of a transformer increases the current flow in the primary.

7. List four losses found in all transformers.

8. Why are most dry-type transformers wound with the high-voltage coil on the outside of the low-voltage coil?

9. What is telescoping in a transformer and how is this effect minimized?

10. What is the purpose of taps in a transformer?

11. When is a transformer closest to unity power factor?

12. What are askarels?

Electrical Safety

There are many dangers associated with electrical power. A technician must be knowledgeable about these dangers in order to work safely. Codes and standards establish the safety requirements. Electrical safety rules must be followed when working with electrical equipment to help prevent injuries from electrical energy sources. Safe work habits, proper personal protective equipment, and proper procedures minimize the possibility of personal injury.

equipment is built and installed safely and every effort is made to protect people from electrical shock. Electrical safety has been advanced by the efforts of the Occupational Safety and Health Administration (OSHA), National Fire Protection Association (NFPA), and state safety laws.

CODES AND STANDARDS

Standards and national, state, and local codes are used to protect people and property from electrical dangers. A code is a regulation or legal minimum requirement. A standard is an accepted reference or practice. Codes and standards ensure electrical

NFPA 70E is the primary standard used to protect technicians from the hazards of electrical energy.

Occupational Safety and Health Administration (OSHA)

The *Occupational Safety and Health Administration (OSHA)* is a federal agency that requires all employers to provide a safe environment for their employees. All work areas must be free from hazards likely to cause serious harm. The provisions of this act are enforced by federal inspection.

OSHA has developed color codes to help ensure a safe environment. **See Figure 3-1.** The color codes help quickly identify fire protection equipment, physical hazards, dangerous parts of machines, radiation hazards, and locations of first aid equipment. With few exceptions, OSHA uses the NEC® to help ensure a safe electrical environment.

OSHA Safety Color Codes

Color	Examples
Red	Fire protection equipment and apparatus, portable containers of flammable liquids, emergency stop pushbuttons and switches
Yellow	Caution and for marking physical hazards, waste containers for explosive or combustible materials, caution against starting, using, or moving equipment under repair, identification of the starting point or power source of machinery
Orange	Dangerous parts or machines, safety starter buttons, the exposed parts (edges) of pulleys, gears, rollers, cutting devices, power jaws
Purple	Radiation hazards
Green	Safety, location of first aid equipment (other than fire fighting equipment)

Figure 3-1. *OSHA-mandated color coding is used to provide visual information on the location of safety information and on the dangers that are present in an area.*

National Fire Protection Association (NFPA)

The *National Fire Protection Association (NFPA)* is a national organization that provides guidance in assessing the hazards of the products of combustion. The NFPA sponsors the development of the National Electrical Code® (NEC®) and the NFPA 70E standard.

National Electrical Code® (NEC®). The *National Electrical Code® (NEC®)* is a standard on practices for the installation of electrical products published by the National Fire Protection Association (NFPA). The National Electrical Code® is one of the most widely used and recognized consensus standards in the world. The purpose of the NEC® is to protect people and property from hazards that arise from the use of electricity. Improper procedures when working with electricity can cause injury or death. Many city, county, state, and federal agencies use the NEC® to set requirements for electrical installations. For example, Article 450 of the NEC® covers transformers and has several specific requirements for transformer vaults. **See Figure 3-2.** The NEC® is updated every three years.

NFPA 70E. The National Fire Protection Association standard NFPA 70E, *Standard for Electrical Safety in the Workplace,* addresses "electrical safety requirements for employee workplaces that are necessary for the safeguarding of employees in their pursuit of gainful employment." Per NFPA 70E, "Only qualified persons shall perform testing work on or near live parts operating at 50 V or more."

TECH FACT

According to the NFPA 70E, Standard for Electrical Safety in the Workplace, 2004 Edition, Article 420.10(F), all oil-insulated transformers installed indoors shall be installed in a vault. A door sill or curb that is of sufficient height to confine the oil from the largest transformer within a transformer vault shall be provided, and in no case shall the height be less than 100 mm (4 in.). Doors shall be equipped with locks, and doors shall be kept locked, access being allowed only to qualified persons. Doors typically need a fire rating of 3 hours.

Transformer Vaults

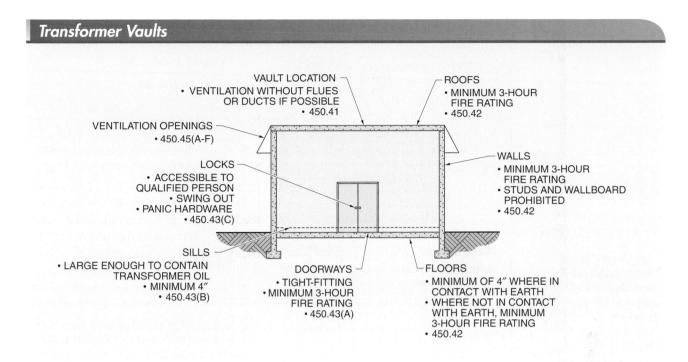

VAULT LOCATION
• VENTILATION WITHOUT FLUES
OR DUCTS IF POSSIBLE
• 450.41

ROOFS
• MINIMUM 3-HOUR
FIRE RATING
• 450.42

VENTILATION OPENINGS
• 450.45(A-F)

WALLS
• MINIMUM 3-HOUR
FIRE RATING
• STUDS AND WALLBOARD
PROHIBITED
• 450.42

LOCKS
• ACCESSIBLE TO
QUALIFIED PERSON
• SWING OUT
• PANIC HARDWARE
• 450.43(C)

SILLS
• LARGE ENOUGH TO CONTAIN
TRANSFORMER OIL
• MINIMUM 4″
• 450.43(B)

DOORWAYS
• TIGHT-FITTING
• MINIMUM 3-HOUR
FIRE RATING
• 450.43(A)

FLOORS
• MINIMUM OF 4″ WHERE IN
CONTACT WITH EARTH
• WHERE NOT IN CONTACT
WITH EARTH, MINIMUM
3-HOUR FIRE RATING
• 450.42

Figure 3-2. There are several NEC® requirements for transformer vaults to ensure that hazards from explosion or fire are not transmitted to adjacent buildings or structures.

American National Standards Institute (ANSI)

The *American National Standards Institute (ANSI)* is a U.S. national organization that helps identify industrial and public needs for standards. ANSI coordinates and encourages activities in national standards development. **See Figure 3-3.**

Canadian Standards Association (CSA)

The *Canadian Standards Association (CSA)* is a Canadian nonprofit membership association for standards development, sale of publications, training, and membership services. CSA International is a related organization for product testing and certification. CSA publishes the Canadian Electric Code.

Underwriters Laboratories Inc. (UL)

Underwriters Laboratories Inc. (UL) is an independent organization that tests equipment and products to see if they conform to national codes and standards. Equipment tested and approved by UL receives the UL label. UL-approved equipment and products are listed in its annual publication.

International Electrotechnical Commission (IEC)

The *International Electrotechnical Commission (IEC)* is an international organization that develops international safety standards for electrical equipment. IEC standards reduce safety hazards that can occur from unpredictable circumstances when using electrical test equipment such as digital multimeters.

National Electrical Manufacturers Association (NEMA)

The *National Electrical Manufacturers Association (NEMA)* is a U.S. national organization that assists with information and standards concerning proper selection, ratings, construction, testing, and performance of electrical equipment. NEMA standards are used as guidelines for the manufacture and use of electrical equipment.

 **DEFINITION**

*The **American National Standards Institute (ANSI)** is a U.S. national organization that helps identify industrial and public needs for standards.*

*The **Canadian Standards Association (CSA)** is a Canadian nonprofit membership association for standards development, sale of publications, training, and membership services.*

***Underwriters Laboratories Inc. (UL)** is an independent organization that tests equipment and products to see if they conform to national codes and standards.*

Codes and Standards

NFPA	**NFPA** National Fire Protection Association 1 Batterymarch Park Quincy, MA 02269 www.nfpa.org	Provides guidance in assessing hazards of products of combustion
		Publishes the National Electrical Code®
		Develops hazardous materials information
ANSI	**ANSI** American National Standards Institute 1819 L St NW Washington, DC 20036 www.ansi.org	Coordinates and encourages activities in national standards development
		Identifies industrial and public needs for standards
		Acts as national coordinator and clearinghouse for consensus standards
CSA	**CSA** Canadian Standards Association 5060 Spectrum Way Suite 100 Mississauga, ON L4W 5N6 www.csa.ca	Tests equipment and products to verify conformance to Canadian national codes and standards
	UL Underwriters Laboratories Inc. 333 Pfingsten Rd Northbrook, IL 60062 www.ul.com	Tests equipment and products to verify conformance to national codes and standards
IEC	**IEC** 3, rue de Varembé P.O. Box 131 CH-1211 GENEVA 20, Switzerland www.iec.ch	Publishes international standards related to electrical, electronic, and related technologies
NEMA	**NEMA** National Electrical Manufacturers Association 1300 N 17th St Suite 1847 Rosslyn, VA 22209 www.nema.org	Assists with information and standards concerning proper selection, ratings, construction, testing, and performance of electrical equipment

Figure 3-3. Many organizations develop standards and test products to ensure worker and product safety.

DEFINITION

*The **International Electrotechnical Commission (IEC)** is an international organization that develops international safety standards for electrical equipment.*

*The **National Electrical Manufacturers Association (NEMA)** is a U.S. national organization that assists with information and standards concerning proper selection, ratings, construction, testing, and performance of electrical equipment.*

*A **qualified person** is a person who is trained in, and has specific knowledge of, the construction and operation of electrical equipment or a specific task, and is trained to recognize and avoid electrical hazards that might be present with respect to the equipment or specific task.*

ELECTRICAL WORK HAZARDS

Electrical work hazards are present whenever technicians are working with electrical power. The primary hazards are electrical shock, arc flash, and arc blast. Therefore, all electrical work must be performed by qualified persons. Safety labels are used to warn workers of hazards in the area. The risks of electrical shock can be reduced by using good grounding practices and using ground fault circuit interrupters. Overvoltage protection is designed into some testing tools to protect technicians from the dangers of unforeseen voltage transients.

Qualified Persons

To prevent an accident, electrical shock, or damage to equipment, all electrical work must be performed by qualified persons. A *qualified person* is a person who is trained in, and has specific knowledge of, the construction and operation of electrical equipment or a specific task, and is trained to recognize and avoid electrical hazards that might be present with respect to the equipment or specific task. NFPA 70E Part II *Safety-Related Work Practices,* Chapter 1 *General,* provides additional information regarding the definition of a qualified person. A qualified person does the following:

- Determines the voltage of energized electrical parts

- Determines the degree and extent of hazards and uses the proper personal protective equipment and job planning to perform work safely on electrical equipment by following all NFPA, OSHA, equipment manufacturer, state, and company safety procedures and practices

- Performs the appropriate task required during an accident or emergency situation

- Understands electrical principles and follows all manufacturer procedures and approach distances specified by the NFPA

- Understands the operation of test equipment and follows all manufacturer procedures

- Informs other technicians and operators of tasks being performed and maintains all required records

Safety Labels

A *safety label* is a label that indicates areas or tasks that can pose a hazard to personnel and/or equipment. Safety labels appear in several ways on equipment and in equipment manuals. Safety labels use signal words to communicate the severity of a potential problem. The three most common signal words are danger, warning, and caution.

SAFETY TIP

A person may be considered a "qualified person" with regard to certain equipment, but not qualified for other equipment.

Danger Signal Word. A *danger signal word* is a word used to indicate an imminently hazardous situation which, if not avoided, results in death or serious injury. The information indicated by a danger signal word indicates the most extreme type of potential situation, and must be followed. The danger symbol is an exclamation mark enclosed in a triangle followed by the word "danger" written boldly in a red box. **See Figure 3-4.**

Warning Signal Word. A *warning signal word* is a word used to indicate a potentially hazardous situation which, if not avoided, could result in death or serious injury. The information indicated by a warning signal word indicates a potentially hazardous situation and must be followed. The warning symbol is an exclamation mark enclosed in a triangle followed by the word "warning" written boldly in an orange box.

DEFINITION

*A **safety label** is a label that indicates areas or tasks that can pose a hazard to personnel and/or equipment.*

*A **danger signal word** is a word used to indicate an imminently hazardous situation which, if not avoided, results in death or serious injury.*

*A **warning signal word** is a word used to indicate a potentially hazardous situation which, if not avoided, could result in death or serious injury.*

Motor drives typically have transformers to reduce voltage to the required level. Technicians working on motor drives must be cautious because of the hazardous voltages present.

*A **caution signal word** is a word used to indicate a potentially hazardous situation which, if not avoided, may result in minor or moderate injury.*

Caution Signal Word. A *caution signal word* is a word used to indicate a potentially hazardous situation which, if not avoided, may result in minor or moderate injury. The information indicated by a caution signal word indicates a potential situation that may cause a problem to people and/or equipment. A caution signal word also warns of problems due to unsafe work practices. The caution symbol is an exclamation mark enclosed in a triangle followed by the word "caution" written boldly in a yellow box.

Other signal words may also appear with danger, warning, and caution signal words used by manufacturers. ANSI Z535.4, *Product Safety Signs and Labels,* provides additional information concerning safety labels. Additional signal words may be used alone or in combination on safety labels.

Signal words and the major message must be understandable to all employees who may be exposed to the hazard.

Safety Labels

Safety Label	Box Color	Symbol	Significance
⚠ DANGER **HAZARDOUS VOLTAGE** • Ground equipment using screw provided. • Do not use metallic conduits as a ground conductor.	red	 ⚠	**DANGER** – Indicates an imminently hazardous situation which, if not avoided, will result in death or serious injury
⚠ WARNING **MEASUREMENT HAZARD** When taking measurements inside the electric panel, make sure that only the test lead tips touch internal metal parts.	orange	 ⚠	**WARNING** – Indicates a potentially hazardous situation which, if not avoided, could result in death or serious injury
⚠ CAUTION **MOTOR OVERHEATING** Use of a thermal sensor in the motor may be required for protection at all speeds and loading conditions. Consult motor manufacturer for thermal capability of motor when operated over desired speed range.	yellow	 ⚠	**CAUTION** – Indicates a potentially hazardous situation which, if not avoided, may result in minor or moderate injury, or damage to equipment. May also be used to alert against unsafe work practices
 **WARNING** Disconnect electrical supply before working on this equipment.	orange	 ⚡	**ELECTRICAL WARNING** – Indicates a high voltage location and conditions that could result in death or serious injury from an electrical shock
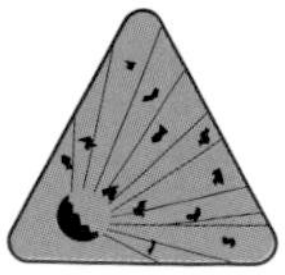 **WARNING** Do not operate the meter around explosive gas, vapor, or dust.	orange	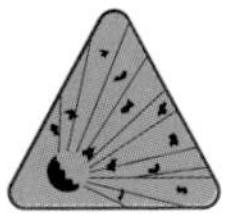 💥	**EXPLOSION WARNING** – Indicates location and conditions where exploding electrical parts may cause death or serious injury

Figure 3-4. *Safety labels are used to indicate a situation with different degrees of likelihood of death or injury to personnel.*

Electrical Warning Signal Word. *Electrical warning signal word* is a word used to indicate a high-voltage location and conditions that could result in death or serious personal injury from an electrical shock if proper precautions are not taken. An electrical warning safety label is usually placed where there is a potential for coming in contact with live electrical wires, terminals, or parts. The electrical warning symbol is a lightning bolt enclosed in a triangle. The safety label may be shown with no words or may be preceded by the word "warning" written boldly.

Explosion Warning Signal Word. *Explosion warning signal word* is a word used to indicate locations and conditions where exploding parts may cause death or serious personal injury if proper precautions and procedures are not followed. The explosion warning symbol is an explosion enclosed in a triangle. The safety label may be shown with no words or may be preceded by the word "warning" written boldly.

Electrical Shock

An *electrical shock* is a shock that results anytime a body becomes part of an electrical circuit. Electrical shock effects vary from a mild sensation, to paralysis, to death. Also, severe burns may occur internally and where current enters and exits the body. The severity of an electrical shock depends on the amount of electric current in milliamps (mA) that flows through the body, the length of time the body is exposed to the current flow, the path the current takes through the body, and the physical size and condition of the body through which the current passes. **See Figure 3-5.**

Prevention is the best medicine for electrical shock. Anyone working on electrical equipment should have respect for all voltages, have knowledge of the principles of electricity, and follow safe work procedures. All technicians should be encouraged to take a basic course in cardiopulmonary resuscitation (CPR) so they can aid a coworker in emergency situations.

Electrical Shock Effects

APPROXIMATE CURRENT*	EFFECT ON BODY†
over 20	Causes severe muscular contractions, paralysis of breathing, heart convulsions
15-20	Painful shock. May be frozen or locked to point of electrical contact until circuit is de-energized
8-15	Painful shock. Removal from contact point by natural reflexes
8 or less	Sensation of shock but probably not painful

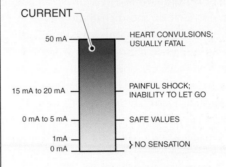

* in mA
† effects vary depending on time, path, amount of exposure, and condition of body

Figure 3-5. Electrical shock results any time a body becomes part of an electrical circuit.

During an electrical shock, the body of a person becomes part of an electrical circuit. The body of a person offers varied resistance to the flow of current. Sweaty hands have less resistance than dry hands. A wet floor has less resistance than a dry floor. The lower the resistance, the greater the current flow. As the current flow increases, the severity of the electrical shock increases.

If a person is receiving an electrical shock, power should be removed as quickly as possible. If power cannot be removed quickly, the victim must be removed from contact with live parts. Action must be taken quickly and cautiously. Delay may be fatal. Individuals must keep themselves from also becoming a casualty while attempting to rescue another person. If the equipment circuit disconnect switch is nearby and can be operated safely, shut OFF the power. Excessive time should not be spent searching for the circuit disconnect. In order to remove the energized part, insulated protective equipment such as a hot stick, rubber gloves, blankets, wood poles, plastic pipes, etc., can be used if such items are accessible.

After the victim is freed from the electrical hazard, help should be called and first aid (CPR, etc.) begun as needed. The injured individual should not be transported unless there is no other option and the injuries require immediate professional attention.

Grounding. *Grounding* is the connection of portions of the distribution system to earth in order to establish a common electrical reference and a low impedance fault path to facilitate operation of overcurrent protective devices. Grounding provides an electrically conductive path designed and intended to carry current under ground fault conditions from the point of a fault on a wiring system to the electrical supply source. Grounding facilitates the operation of overcurrent protection devices.

For systems that are solidly grounded, grounding provides a means to limit the voltage to ground during normal operation and to prevent excessive voltages due to lightning, line surges, or unintentional contact with higher-voltage lines and to stabilize the voltage to ground during normal operation.

The noncurrent-carrying metal parts of a transformer installation are required by the NEC® to be effectively grounded. Conductive materials enclosing conductors or equipment are grounded to prevent a voltage or difference of potential on these materials. Circuits and enclosures are grounded to facilitate overcurrent device operation in case of insulation failure or ground faults.

Grounding is accomplished by connecting the electrical distribution system to a metal underground water pipe, the metal frame of a building, a concrete-encased electrode, or a ground ring. To prevent problems, a grounding path must be as

short as possible and of sufficient ampacity, never be fused or switched, be a permanent part of the electrical circuit, and be continuous and uninterrupted from the electrical circuit to the ground.

A ground is provided at the main service equipment or at the source of a separately derived system (SDS). A *separately derived system (SDS)* is a system that supplies electrical power derived (taken from) transformers, storage batteries, solar photovoltaic systems, or generators. **See Figure 3-6.** The majority of separately derived systems are produced by the secondary of a distribution transformer.

The neutral ground connection must be made at the transformer or at the main service panel only. The neutral ground connection is made by connecting the neutral bus to the ground bus with a main bonding jumper. A *main bonding jumper (MBJ)* is a connection at the service equipment that connects the equipment grounding conductor, the grounding electrode conductor, and the grounded conductor (neutral conductor). The purpose of the main bonding jumper is to bond the neutral and equipment grounding conductor together with the enclosure to create a common reference potential.

Separately Derived Systems

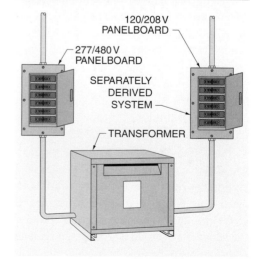

Figure 3-6. *The 120/208V panelboard is a separately derived system.*

An *equipment grounding conductor (EGC)* is an electrical conductor that provides a low-impedance ground path between electrical equipment and enclosures within the distribution system and takes current back to the source. A *grounding electrode conductor (GEC)* is a conductor that connects grounded parts of a power distribution system (equipment grounding conductors, grounded conductors, and all metal parts) to the grounding system.

A *grounded conductor* is a conductor that has been intentionally grounded. The grounded conductor is commonly a neutral conductor. However, not all electrical distribution systems use the grounded conductor as a neutral. For example, corner-grounded delta systems contain a grounded conductor that is not a neutral conductor. Therefore, it is not correct to refer to all grounded conductors as neutral conductors, although that is the case in the majority of electrical distribution systems.

Ground Fault Circuit Interrupters. A *ground fault circuit interrupter (GFCI)* is a device that protects against electrical shock by detecting an imbalance of current in the normal conductor pathways and opening the circuit. When current in the two conductors of an electrical circuit varies by more than 5 mA, a GFCI opens the circuit. A GFCI is rated to trip quickly enough (1/40 of a second) to prevent electrocution. **See Figure 3-7.**

A potentially dangerous ground fault is any amount of current above the level that may deliver a dangerous shock. Any current over 8 mA is considered potentially dangerous—depending on the path the current takes, the physical condition of the person receiving the shock, and the amount of time the person is exposed to the shock. Therefore, GFCIs are required in such places as dwellings, hotels, motels, construction sites, marinas, receptacles near swimming pools and hot tubs, underwater lighting, fountains, and other areas in which a person may experience a ground fault.

➕ **SAFETY TIP**

According to the NFPA 70E, Standard for Electrical Safety in the Workplace, 2004 Edition, Article 410.10(C)(3), alternating current circuits of less than 50 volts shall be grounded where supplied by transformers, if the transformer supply systems exceed 150 volts to ground.

❗ **DEFINITION**

*An **equipment grounding conductor (EGC)** is an electrical conductor that provides a low-impedance ground path between electrical equipment and enclosures within the distribution system and takes current back to the source.*

*A **grounding electrode conductor (GEC)** is a conductor that connects grounded parts of a power distribution system (equipment grounding conductors, grounded conductors, and all metal parts) to the grounding system.*

*A **grounded conductor** is a conductor that has been intentionally grounded.*

Ground Fault Circuit Interrupters

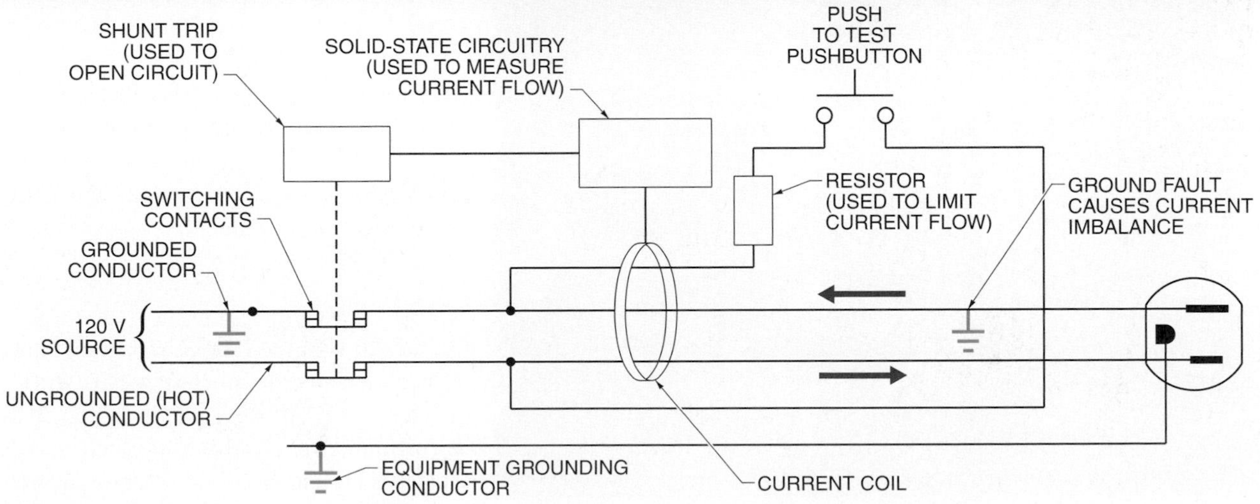

Figure 3-7. A GFCI compares the amount of current in the ungrounded hot conductor with the amount of current in the neutral conductor.

*A **ground fault circuit interrupter (GFCI)** is a device that protects against electrical shock by detecting an imbalance of current in the normal conductor pathways and opening the circuit.*

***Arc flash** is a short circuit through air.*

*An **arc blast** is an explosion that occurs when the surrounding air becomes ionized and conductive.*

A GFCI compares the amount of current in the ungrounded (hot) conductor with the amount of current in the neutral conductor. If the current in the neutral conductor becomes less than the current in the hot conductor, a ground fault condition exists. The amount of current that is missing is returned to the source by some path other than the intended path (fault current).

GFCI protection may be installed at different locations within a circuit. Direct-wired GFCI receptacles provide ground fault protection at the point of installation. GFCI receptacles may also be connected to provide protection at all other receptacles installed downstream on the same circuit. GFCI circuit breakers, when installed in a load center or panelboard, provide GFCI protection and conventional circuit overcurrent protection for all branch-circuit components connected to the circuit breaker.

Plug-in GFCIs provide ground fault protection for devices plugged into them. These plug-in devices are often used by personnel working with power tools in an area that does not include GFCI receptacles.

Portable GFCIs are designed to be easily moved from one location to another. **See Figure 3-8.** Portable GFCIs commonly contain more than one receptacle outlet protected by an electronic circuit module. Portable GFCIs should be inspected and tested before each use. GFCIs have a built-in test circuit to ensure that the ground fault protection is operational.

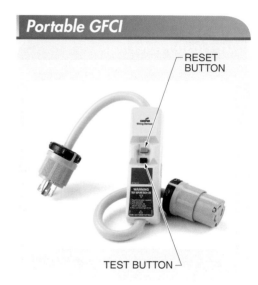

Cooper Wiring Devices

Figure 3-8. *A portable GFCI can be used on a job site to protect workers.*

A GFCI protects against the most common form of electrical shock hazard, the ground fault. A GFCI does not protect against line-to-line contact hazards, such as a technician holding two hot wires or a hot and a neutral wire in each hand. GFCI protection is required in addition to NFPA grounding requirements.

Arc Flash

In addition to the risk of electrical shock, other hazards are present when work is performed on or near energized electrical equipment or conductors. Two serious hazards are electrical arc flashes and electrical arc blasts. *Arc flash* is a short circuit through air. An *arc blast* is an explosion that occurs when the surrounding air becomes

Cooper Wiring Devices

Ground fault circuit interrupters are required in places to protect against a possible ground fault.

ionized and conductive. When insulation or isolation can no longer withstand the applied voltage, an arc flash occurs. An arc flash may occur from phase to ground or from phase to phase. The temperature of the arc flash may reach thousands of degrees and cause an arc blast. The explosion from the arc blast can spread hot gases and melting metal, damage hearing and vision, and send objects flying. **See Figure 3-9.**

OSHA Part 1910.333 gives general requirements for safety-related work practices. OSHA inspectors also carry a copy of NFPA 70E and use it to enforce safety procedures related to arc flash. NFPA 70E gives specific guidelines on actions to be taken to comply with the general OSHA statements. The NEC® requires several types of equipment to be marked to warn qualified persons that a hazard exists. **See Figure 3-10.**

Arc Flash Protection

Figure 3-9. Flash suits protect workers from the thermal energy of arc blasts.

Arc Flash Warning

⚠ **DANGER**

Arc flash and shock hazard.

Follow ALL requirements in NFPA 70E for safe work practices and for Personal Protective Equipment.

Figure 3-10. An arc flash warning label must be placed on any electrical equipment that may remain energized during repair or maintenance.

✚ SAFETY TIP

According to the NFPA 70E, Standard for Electrical Safety in the Workplace, 2004 Edition, Annex K, the temperature of an arc flash can reach 35,000°F. Exposure to these extreme temperatures both burns the skin directly and causes ignition of clothing, which adds to the burn injury. Material and molten metal is expelled away from the arc at speeds exceeding 700 miles per hour, fast enough for shrapnel to completely penetrate the human body.

The NFPA requires facility owners to perform a flash hazard analysis. A flash protection boundary must be established around electrical devices. This boundary is determined by calculations that estimate the maximum energy released and the distance that energy travels before dissipating to a safe level. Technicians working within the boundary must have appropriate personal protective equipment (PPE). The IEEE Standard 1584-2002, *Guide for Performing Arc-Flash Hazard Calculations,* gives procedures for determining the incident energy exposure, flash protection boundary, and level of PPE required. Incident energy is expressed in calories per centimeter squared (cal/cm^2). This is a measure of the heat energy applied to a certain area of an object. The object may be a person. The flash protection boundary must be established at the point where the incident energy has fallen below 1.2 cal/cm^2. It is not safe to assume that similar equipment located in different locations has the same flash protection boundary.

Overvoltage Protection

Voltage surges on a power distribution system can cause a safety hazard. A *voltage surge* is a higher-than-normal voltage that temporarily exists on one or more power lines. Voltage surges vary in voltage level and time present on power lines. One type of voltage surge is a transient voltage. A *transient voltage (voltage spike)* is an unwanted voltage of very short duration in an electrical circuit. Transient voltages typically exist for a very short time, but are often larger in magnitude than voltage surges and are very erratic. Transient voltages occur due to lightning strikes, unfiltered electrical equipment, and power being switched ON and OFF. High transient voltages may reach several thousand volts. A transient voltage on a 120 V power line can reach 1000 V (1 kV) or more.

High transient voltages exist close to a lightning strike or when large (high-current) loads are switched OFF. For example, when a large motor (100 HP) is turned OFF, a transient voltage can move down the power distribution system. If a DMM is connected to a system when a high transient voltage occurs, an arc can be created inside the DMM. Once started, the arc can cause a high-current short in the power distribution system even after

the original high transient voltage is gone. The high-current short can turn into an arc blast. **See Figure 3-11.**

The amount of current drawn and potential damage caused depends on the specific location of the power distribution system. All power distribution systems have current limits set by fuses and circuit breakers along the system. The current rating (size) of fuses and circuit breakers decreases farther away from the main distribution panel. The farther away from the main distribution panel, the less likely the high transient voltage is to cause damage.

Overvoltage Installation Categories. The IEC 1010-1 standard defines four overvoltage installation categories in which a DMM may be used (Category I—Category IV). These categories are typically abbreviated as CAT I, CAT II, CAT III, and CAT IV. **See Figure 3-12.** They determine the magnitude of transient voltage a DMM or other electrical appliance needs to withstand when used on the power distribution system. For example, a DMM or other electrical appliance used in a CAT III environment must withstand a 6000 V transient voltage without causing an arc. If the DMM or other appliance is operated on voltages above 600 V, then the DMM must withstand an 8000 V transient voltage.

Transient Voltages

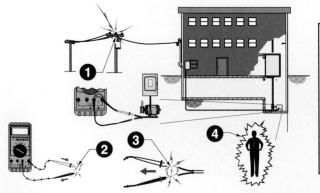

1. LIGHTNING STRIKE OR LARGE LOAD SWITCHING CAUSES A TRANSIENT VOLTAGE ON POWER LINE, CREATING ARC BETWEEN DMM INPUT TERMINALS

2. HIGH CURRENT FLOWS IN CLOSED CIRCUIT; ARC STARTS AT PROBE TIPS

3. WHEN PROBES ARE PULLED IN REACTION TO LOUD NOISE, ARCS ARE DRAWN TO TERMINALS

4. IF ARCS ARE JOINED, RESULTING HIGH-ENERGY ARC CAN CREATE A LIFE-THREATENING SITUATION FOR USER

Figure 3-11. When taking measurements in an electrical circuit, transient voltages can cause electrical shock and/or damage to equipment.

IEC 1010 Overvoltage Installation Categories

CATEGORY	IN BRIEF	EXAMPLES
CAT I	Electronic	• Protected electronic equipment • Equipment connected to (source) circuits in which measures are taken to limit transient overvoltage to an appropriately low level • Any high-voltage, low-energy source derived from a high-winding-resistance transformer, such as the high-voltage section of a copier
CAT II	1φ receptacle-connected loads	• Appliances, portable tools, and other household and similar loads • Outlets and long branch circuits • Outlets at more than 30′ (10 m) from CAT III source • Outlets at more than 60′ (20 m) from CAT IV source
CAT III	3φ distribution, including 1φ commercial lighting	• Equipment in fixed installations, such as switchgear and polyphase motors • Bus and feeder in industrial plants • Feeders and short branch circuits and distribution panel devices • Lighting systems in larger buildings • Appliance outlets with short connections to service entrance
CAT IV	3φ at utility connection, any outdoor conductors	• Refers to the origin of installation, where low-voltage connection is made to utility power • Electric meters, primary overcurrent protection equipment • Outside and service entrance, service drop from pole to building, run between meter and panel • Overhead line to detached building

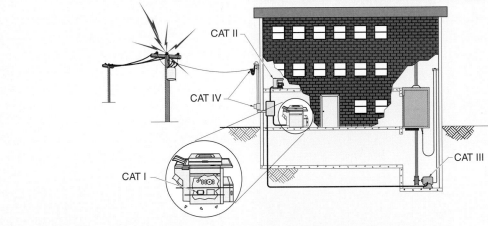

Figure 3-12. The applications in which a DMM may be used are classified by the IEC 1010 standard into four overvoltage installation categories.

If the DMM can withstand the voltage, the DMM may be damaged but an arc does not start and no arc blast occurs. To protect against transient voltages, protection must be built into the test equipment. For many years, the industry followed standard IEC 348. This standard has been replaced by IEC 1010. A DMM designed to the IEC 1010 standard offers a higher level of protection. A higher CAT number indicates an electrical environment with higher power available, larger short-circuit current available, and higher energy transients. For example, a DMM designed to the CAT III standard is resistant to higher energy transients than a DMM designed to the CAT II standard.

Power distribution systems are divided into categories because a dangerous high-energy transient voltage such as a lightning strike is attenuated (lessened) or dampened as it travels through the impedance (AC resistance) of the system and the system grounds. Within an IEC 1010 standard category, a higher voltage rating denotes a higher transient voltage withstanding rating. For example, a CAT III-1000 V (steady-state) rated DMM has better protection compared to a CAT III-600 V (steady-state) rated DMM. Between categories, a higher voltage rating (steady-state) might not provide higher transient voltage protection. For example, a CAT III-600 V DMM has better transient protection than a CAT II-1000 V DMM. A DMM should be chosen based on the IEC overvoltage installation category first and voltage second.

Overhead Power Lines

People are killed every day from accidental contact with overhead power lines. Overhead power lines are electrical conductors designed to deliver electrical power and located in an aboveground, aerial position. Overhead power lines are suspended from ceramic insulators that are attached to wood utility poles or metal structures. Overhead power lines are generally owned and operated by an electric utility company. Overhead power line conductors 600 V or higher are usually bare (uninsulated), while low-voltage systems such as service drops to buildings consist of insulated conductors.

Electrical power lines should be located far enough overhead or out of reach as to not pose an electrical hazard. Electrical equipment such as transformers and power panels is also isolated by fences, locked in buildings, or buried underground. **See Figure 3-13.** Entrances to electrical rooms and other guarded locations containing exposed energized electrical parts must be marked with a warning sign forbidding unqualified persons to enter.

Utility company electrical workers and linemen are skilled workers who have received extensive and specific training to safely work on and near energized overhead power lines and are equipped with the proper PPE and tools. Workers in other occupations, including residential/commercial/industrial electricians, technicians, engineers, and supervisors, are unqualified (unless trained) to approach overhead power lines closer than an established safe distance. Per NFPA 70E, if the line voltage exceeds 50 kV, the minimum overhead line clearance for all nonqualified individuals is 10′ plus 4″ for every 10 kV over 50 kV.

Overhead power lines should be located far enough overhead as to not pose an electrical hazard.

 SAFETY TIP

According to the NFPA 70E, Standard for Electrical Safety in the Workplace, 2004 Edition, Article 420.10(F)(8), materials shall not be stored in vaults.

Transformer Isolation

WARNING SIGN   FENCE

Figure 3-13. High-voltage equipment must be isolated from personnel and marked with warning signs.

Aerial Lifts. An aerial lift is a movable platform and structure for workers to stand on when working at a height above the floor. Any person or item on a lift must also maintain a safe distance from power lines at all times including during the use and movement of the lift. All scaffolds, persons, and items on lifts must maintain the minimum distance from power lines.

Safety Rules

Safety is a result of consistently following safety rules and practices that protect technicians from the hazards of working with electricity. Electrical safety rules help prevent injuries from electrical energy sources. **See Figure 3-14.**

Electrical Safety Rules

Always comply with the NEC®.

Use approved appliances, components, and equipment.

Keep electrical grounding circuits in good condition. Ground any conductive component or element that does not have to be energized. The grounding connection must be a low-resistance conductor heavy enough to carry the largest fault current that may occur.

Turn OFF, lock out, and tag disconnect switches when working on any electrical circuit or equipment. Test all circuits after they are turned OFF. Insulators may not insulate, grounding circuits may not ground, and switches may not open the circuit.

Use double-insulated power tools or power tools that include a third conductor grounding terminal which provides a path for fault current. Never use a power tool that has the third conductor grounding terminal removed.

Always use protective and safety equipment.

Know what to do in an emergency.

Check conductors, cords, components, and equipment for signs of wear or damage. Replace any equipment that is not safe.

Never throw water on an electrical fire. Turn OFF the power and use a Class C rated fire extinguisher.

Work with another individual when working in a dangerous area or with dangerous equipment.

Learn CPR and first aid.

Do not work when tired or taking medication that causes drowsiness.

Do not work in poorly lighted areas.

Always use nonconductive ladders. Never use a metal ladder when working around electrical equipment.

Ensure there are no atmospheric hazards such as flammable dust or vapor in the area. A live electrical circuit may emit a spark at any time.

Use one hand when working on a live circuit to reduce the chance of an electrical shock passing through the heart and lungs.

Never bypass or disable fuses or circuit breakers.

Extra care must be taken in an electrical fire because burning insulation produces toxic fumes.

LOCKOUT

PPE

DAMAGED CONDUCTOR

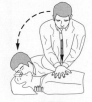

CPR

Figure 3-14. Electrical safety rules are a standard way of working safely all the time.

PERSONAL PROTECTIVE EQUIPMENT (PPE)

Personal protective equipment (PPE) is clothing and/or equipment worn by a technician to reduce the possibility of injury in the work area. The use of PPE is required whenever work may occur on or near energized exposed electrical circuits or when exposure to chemicals if possible. For maximum safety, PPE must be used as specified in NFPA 70E, OSHA Standard Part 1910 *Subpart 1 – Personal Protective Equipment* (1910.132 through 1910.138), and other applicable safety mandates.

All PPE and tools are selected for at least the operating voltage of the equipment or circuits to be worked on or near. Equipment, devices, tools, or test equipment must be suited for the work to be performed. Personal protective equipment includes protective clothing, head protection, eye protection, ear protection, hand protection, foot protection, back protection, knee protection, and rubber insulated matting. **See Figure 3-15.**

DEFINITION

Personal protective equipment (PPE) is clothing and/or equipment worn by a technician to reduce the possibility of injury

Protective Clothing

Protective clothing is clothing that provides protection from contact with sharp objects, hot equipment, and harmful materials. Protective clothing made of durable material such as denim should be snug, yet allow ample movement. Clothing should fit snugly to avoid danger of becoming entangled in moving machinery. Pockets should allow convenient access but should not snag on tools or equipment. Soiled protective clothing should be washed to reduce the flammability hazard.

Arc-resistant clothing must be used when working with live high-voltage electrical circuits. Arc-resistant clothing is made of materials such as Nomex®, Basofil®, or Kevlar® fibers. The arc-resistant fibers can be coated with PVC to offer weather resistance and to increase arc resistance. Arc-resistant clothing must meet the following three requirements:

- Clothing must not ignite and continue to burn.

- Clothing must provide an insulating value to protect the wearer from heat.

- Clothing must provide resistance to the break-open forces generated by the shock wave of an arc.

Head Protection

Head protection requires using a protective helmet. A protective helmet is a hard hat that is used in the workplace to prevent injury from the impact of falling and flying objects, and from electrical shock. **See Figure 3-16.** Protective helmets resist penetration and absorb impact force. Protective helmet shells are made of durable, lightweight materials. A shock-absorbing lining keeps the shell away from the head to provide ventilation. Protective helmets are identified by class of protection against specific hazardous conditions.

Class G, E, and C helmets are used for construction and industrial applications.

Personal Protective Equipment

PROTECTIVE HELMET
SAFETY GLASSES
EAR PLUGS
FIRE-RESISTANT CLOTHING
TINTED FACE SHIELD
LEATHER PROTECTORS
RUBBER INSULATING GLOVES

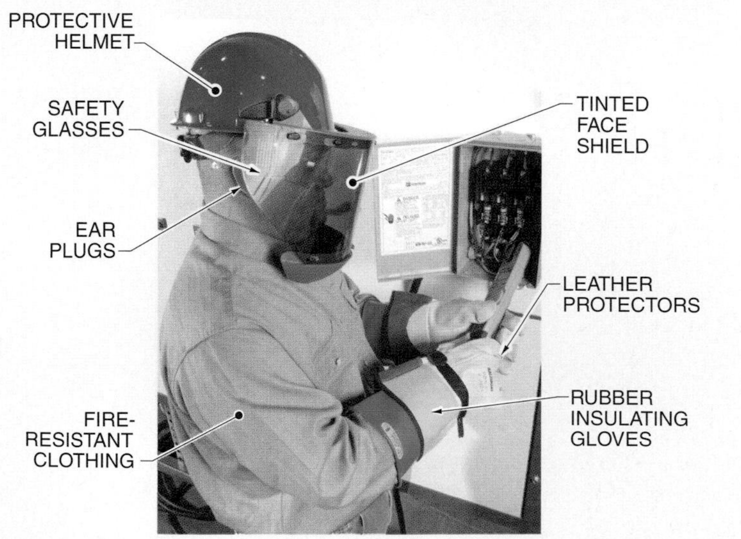

Figure 3-15. *Personal protective equipment is used to reduce the possibility of an injury.*

Class G protective helmets protect against low-voltage shock and burns and impact hazards, and are commonly used in construction and manufacturing facilities. Class E protective helmets protect against high-voltage shock, burns, impact hazards, and penetration by falling or flying objects. Class C protective helmets are manufactured with lighter materials yet provide adequate impact protection.

Head Protection

CROWN STRAPS

LAB SAFETY SUPPLY, INC.
MODEL No. YX27178
ANSI Z89.1-1997
CLASS G, E, C CERTIFIED
TYPE I
MADE IN USA

SHELL

HEADBAND

PROTECTIVE HELMETS

TYPE	IMPACT PROTECTION
I	Impacts from top
II	Impacts from top and side

CLASS	USE
G	General service, limited voltage protection
E	Utility service, high voltage protection
C	Special service, no voltage protection

Figure 3-16. Protective helmets are identified by type and class for protection against specific hazards.

Eye Protection

Eye protection must be worn to prevent eye or face injuries caused by flying particles, contact arcing, and radiant energy. Eye protection must comply with OSHA 29 CFR 1910.133, *Eye and Face Protection*. Eye protection standards are specified in ANSI Z87.1, *Occupational and Educational Eye and Face Protection*. Eye protection includes safety glasses, face shields, and goggles. **See Figure 3-17.**

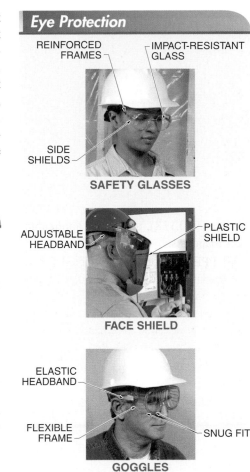

Eye Protection

REINFORCED FRAMES — IMPACT-RESISTANT GLASS

SIDE SHIELDS

SAFETY GLASSES

ADJUSTABLE HEADBAND — PLASTIC SHIELD

FACE SHIELD

ELASTIC HEADBAND

FLEXIBLE FRAME — SNUG FIT

GOGGLES

Figure 3-17. Eye protection must be worn to prevent eye injuries caused by flying particles, contact arcing, or radiant energy.

Safety glasses are an eye protection device with special impact-resistant glass or plastic lenses, reinforced frames, and side shields. Plastic frames are designed to keep the lenses secured in the frame if an impact occurs and minimize the shock hazard when working with electrical equipment. Side shields provide additional protection from flying objects. Tinted-lens safety glasses protect against low-voltage arc hazards.

A face shield is an eye and face protection device that covers the entire face with a plastic shield, and is used for protection from flying objects. Tinted face shields protect against low-voltage arc hazards. Goggles are an eye protection device with a flexible frame that is secured on the face with an elastic headband. Goggles fit snugly against the face to seal the areas around the eyes, and may be used over prescription glasses. Goggles with clear lenses protect against small flying particles or splashing liquids. Tinted goggles are used to protect against low-voltage arc hazards.

Safety glasses, face shields, and goggle lenses must be properly maintained to provide protection and clear visibility. Lens cleaners are available that clean without risk of lens damage. Pitted or scratched lenses reduce vision and may cause lenses to fail on impact.

Hearing Protection

Hearing protection is any device worn to limit the noise entering the ear and includes earplugs and earmuffs. An earplug is an ear protection device made of moldable rubber, foam, or plastic and inserted into the ear canal. An earmuff is an ear protection device worn over the ears. A tight seal around an earmuff is required for proper protection. **See Figure 3-18.**

Power tools and equipment can produce excessive noise levels. Technicians subjected to excessive noise levels may develop hearing loss over a period of time. The severity of hearing loss depends on the intensity and duration of exposure. Noise intensity is expressed in decibels. A decibel (dB) is a unit of measure used to express the relative intensity of sound.

Salisbury

Figure 3-18. *Hearing protection is required whenever workers are exposed to noise equal to or exceeding an 8 hr TWA of 85 dB.*

Ear protection devices are assigned a noise reduction rating (NRR) number based on the noise level reduced. For example, an NRR of 27 means that the noise level is reduced by 27 dB when tested at the factory. To determine approximate noise reduction in the field, 7 dB is subtracted from the NRR. For example, an NRR of 27 provides a noise reduction of approximately 20 dB in the field.

Hand Protection

Hand protection includes gloves worn to prevent injuries to hands from cuts or electrical shock. The appropriate hand protection required is determined by the duration, frequency, and degree of the hazard to the hands. *Rubber insulating gloves* are gloves made of latex rubber and are used to provide maximum insulation from electrical shock. Rubber insulating gloves are stamped with a working voltage range such as 500 V – 26,500 V. *Leather protectors* are gloves worn over rubber insulating gloves to prevent penetration of the rubber insulating gloves and provide added protection against electrical shock. Safety procedures for the use of rubber insulating gloves and leather protectors must be followed at all times. **See Figure 3-19.**

DEFINITION

Rubber insulating gloves are gloves made of latex rubber and are used to provide maximum insulation from electrical shock.

Leather protectors are gloves worn over rubber insulating gloves to prevent penetration of the rubber insulating gloves and provide added protection against electrical shock.

TECH FACT

According to the NFPA 70E, Standard for Electrical Safety in the Workplace, 2004 Edition, Article 420.10(F)(3) and (4), the operating voltage of exposed live parts operating at 50 volts or more of transformer installations shall be indicated by signs or visible markings on the equipment or structures. Oil-insulated transformers installed indoors shall be installed in a vault.

Rubber Insulating Glove Classes

CLASS	MAXIMUM USE VOLTAGE*	COLOR OF LABEL
00	500	Beige
0	1000	Red
1	7500	White
2	17,000	Yellow
3	26,500	Green
4	36,000	Orange

* in V

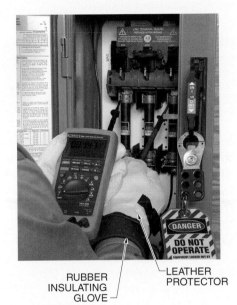

RUBBER INSULATING GLOVE — LEATHER PROTECTOR

Figure 3-19. Hand protection includes gloves worn to prevent injuries to hands caused by cuts or electrical shock.

The primary purpose of rubber insulating gloves and leather protectors is to insulate hands and lower arms from possible contact with live conductors. Rubber insulating gloves offer a high resistance to current flow to help prevent an electrical shock. Leather protectors help protect rubber insulating gloves and add additional insulation.

WARNING: Rubber insulating gloves are designed for specific applications. Leather protectors are required for protecting rubber insulating gloves and should not be used alone. Rubber insulating gloves offer the highest resistance and greatest insulation. Serious injury or death can result from improper use or using outdated and/or the wrong type of gloves for the application.

Leather protectors should be inspected when inspecting rubber insulating gloves. Metal particles or any substance that could physically damage rubber insulating gloves must be removed from a leather protector before it is used.

The entire surface of rubber insulating gloves must be field tested (visual inspection and air test) before each use. In addition, rubber insulating gloves should also be laboratory tested by an approved laboratory every six months. Visual inspection of rubber insulating gloves is performed by stretching a small area (particularly fingertips) and checking for defects such as punctures or pin holes, embedded or foreign material, deep scratches or cracks, cuts or snags, or deterioration caused by oil, heat, grease, insulating compounds, or any other substance that may harm rubber.

Rubber insulating gloves must also be air tested when there is cause to suspect damage. The entire surface of the glove must be inspected by rolling the cuff tightly toward the palm in such a manner that air is trapped inside the glove, or by using a mechanical inflation device. **See Figure 3-20.** When using a mechanical inflation device, care must be taken to avoid overinflation. The glove is examined for punctures and other defects. Puncture detection may be enhanced by holding the glove to the face or ear to listen for the sound of escaping air. Gloves failing the air test should be tagged unsafe and returned to a supervisor.

Proper care of leather protectors is essential for user safety. Leather protectors are checked for cuts, tears, holes, abrasions, defective or worn stitching, oil contamination, and any other condition that might prevent them from adequately protecting rubber insulating gloves. Any substance that could physically damage rubber insulating gloves must be removed before use. Rubber insulating gloves or leather protectors found to be defective shall not be discarded or destroyed in the field, but shall be tagged unsafe and returned to a supervisor.

Pneumatic Glove Tester

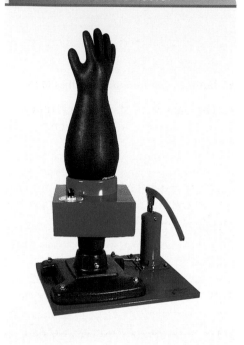

Salisbury

Figure 3-20. *A pneumatic glove tester is used to verify the integrity of electrical gloves.*

Foot Protection

Foot protection is a pair of shoes worn to prevent foot injuries that are typically caused by objects falling less than 4′ and having an average weight of less than 65 lb. Safety shoes with reinforced steel toes protect against injuries caused by compression and impact. **See Figure 3-21.** Insulated rubber-soled shoes are commonly worn during electrical work to prevent electrical shock. Protective footwear must comply with ANSI Z41, *Personal Protection—Protective Footwear.* Thick-soled work shoes may be worn for protection against sharp objects such as nails. Rubber boots may be used when working in damp locations.

Foot Protection

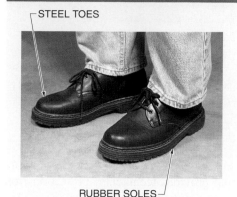

STEEL TOES

RUBBER SOLES

Figure 3-21. *Safety shoes include steel toes and thick rubber soles.*

Back Protection

A back injury is one of the most common injuries resulting in lost time in the workplace. Back injuries are the result of improper lifting procedures. Back injuries are prevented through proper planning and work procedures. Assistance should be sought when moving heavy objects. When lifting objects from the ground, ensure the path is clear of obstacles and free of hazards. When lifting objects, the knees are bent and the object is grasped firmly. The object is lifted by straightening the legs and keeping the back as straight as possible. Keep the load close to the body and keep the load steady.

Long objects such as conduit may not be heavy, but the weight might not be balanced. Long objects should be carried by two or more people whenever possible. When carried on the shoulder by one person, conduit should be transported with the front end pointing downward to minimize the possibility of injury to others when walking around corners or through doorways.

✚ **SAFETY TIP**

Per OSHA 1910.145, signal words must be readable at least 5′ from the sign.

Knee Protection

A kneepad is a rubber, leather, or plastic pad strapped onto the knees for protection. Kneepads are worn by technicians who spend considerable time on their knees or who work in close areas and must kneel for proper access to equipment. Kneepads are secured by buckle straps or Velcro closures.

Rubber Insulating Matting

Rubber insulating matting is a floor covering that provides technicians protection from electrical shock when working on live electrical circuits. **See Figure 3-22.** Dielectric black fluted rubber matting is specifically designed for use in front of open cabinets or high-voltage equipment. Matting is used to protect technicians when voltages are over 50 V. Two types of matting that differ in chemical and physical characteristics are designated as Type I natural rubber and Type II elastomeric compound matting.

LOCKOUT/TAGOUT

Electrical power and other potential sources of energy must be removed when electrical equipment is inspected, serviced, or repaired. To ensure the safety of personnel working with the equipment, all electrical, pneumatic, and hydraulic power is removed and the equipment must be locked out and tagged out. *Lockout* is the process of removing the source of electrical power and installing a lock that prevents the power from being turned on. *Tagout* is the process of placing a danger tag on the source of electrical power, which indicates that the equipment may not be operated until the danger tag is removed. Per OSHA standards, equipment is locked out and tagged out before any installation or preventive maintenance is performed. **See Figure 3-23.**

A danger tag has the same importance and purpose as a lock and is used alone only when a lock does not fit the disconnect device. A danger tag shall be attached at the disconnect device with a tag tie or equivalent and shall have space for the technician's name, craft, and other company-required information. A danger tag must withstand the elements and expected atmosphere for the maximum period of time that exposure is expected. Lockout/tagout is used when:

- Power is not required to be ON to a piece of equipment to perform a task.
- Machine guards or other safety devices are removed or bypassed.
- The possibility exists of being injured or caught in moving machinery.
- Jammed equipment is being cleared.
- The danger exists of being injured if equipment power is turned on.

SAFETY TIP

Personnel should consult OSHA Standard 29 CFR 1910.147 – The Control of Hazardous Energy (Lockout/Tagout) for industry standards on lockout/tagout.

DEFINITION

Rubber insulating matting is a floor covering that provides technicians protection from electrical shock when working on live electrical circuits.

Lockout is the process of removing the source of electrical power and installing a lock that prevents the power from being turned on.

Tagout is the process of placing a danger tag on the source of electrical power, which indicates that the equipment may not be operated until the danger tag is removed.

Rubber Insulating Matting

Figure 3-22. Rubber insulating matting is used to protect workers exposed to live electrical circuits.

Lockout/Tagout

Figure 3-23. *Equipment must be locked out and/or tagged out before installation, preventive maintenance, or servicing is performed.*

Lockout Devices

Lockout devices are lightweight enclosures that allow the lockout of standard control devices. Lockout devices are available in various shapes and sizes that allow for the lockout of ball valves, gate valves, and electrical equipment such as plugs and disconnects.

Lockout devices resist chemicals, cracking, abrasion, and temperature changes. They are available in colors to match ANSI pipe colors. Lockout devices are sized to fit standard size industry control devices. **See Figure 3-24.**

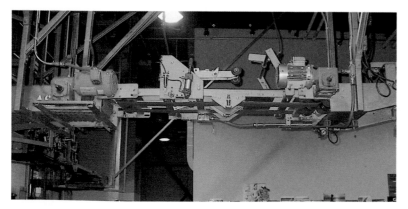

Electric motors used to move conveyors must be locked out before working on the conveyor.

Lockout Devices

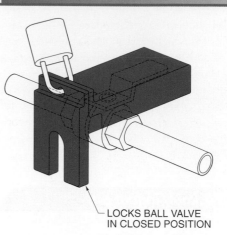

LOCKS BALL VALVE
IN CLOSED POSITION

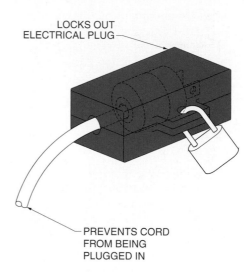

LOCKS OUT
ELECTRICAL PLUG

PREVENTS CORD
FROM BEING
PLUGGED IN

Figure 3-24. *Lockout devices are available in various shapes and sizes that allow for the lockout of standard control devices.*

Locks used to lock out a device may be color-coded and individually keyed. The locks are rust-resistant and are available with various size shackles.

Danger tags provide additional lockout and warning information. Various danger tags are available. Danger tags may include warnings such as "Do Not Start" or "Do Not Operate," or may provide space to enter worker, date, and lockout reason information. Tag ties must be strong enough to prevent accidental removal and must be self-locking and nonreusable.

Lockout/tagout kits are also available. A lockout/tagout kit contains items required to comply with OSHA lockout/tagout standards. Lockout/tagout kits contain reusable danger tags, multiple lockouts, locks, magnetic signs, and information on lockout/tagout procedures. **See Figure 3-25.** A lockout/tagout should be checked to ensure power is removed when returning to work after leaving a job for any reason or when a job cannot be completed in the same day.

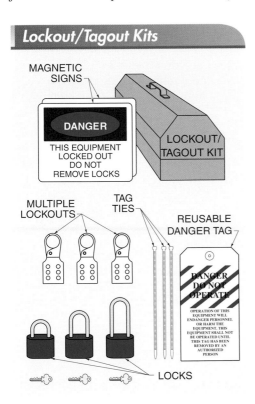

Figure 3-25. Lockout/tagout kits comply with OSHA lockout/tagout standards.

FIRE SAFETY

Fire safety requires established procedures to reduce or eliminate conditions that could cause a fire. Guidelines in assessing hazards of the products of combustion are provided by the NFPA. Prevention is the best strategy to guard against potential fire hazards. Technicians must take responsibility for preventing conditions that could result in a fire. This includes proper use and storage of lubricants, oily rags, and solvents, and immediate cleanup of combustible spills.

Classes of Fires

The five classes of fires are Class A, Class B, Class C, Class D, and Class K. Class A fires include burning wood, paper, textiles, and other ordinary combustible materials containing carbon. Class B fires include burning oil, gas, grease, paint, and other liquids that convert to a gas when heated. Class C fires include burning electrical devices, motors, and transformers. Class D is a specialized class of fires including burning metals such as zirconium, titanium, magnesium, sodium, and potassium. Class K fires include grease in commercial cooking equipment. Fire extinguishers are selected for the class of fire based on the combustible material. **See Figure 3-26.**

TECH FACT

According to the NFPA 70E, Standard for Electrical Safety in the Workplace, 2004 Edition, Article 120.2(E)(3), a lockout device shall include a lock (either keyed or combination). The lockout device shall include a method of identifying the individual who installed the lockout device. Lockout devices shall be suitable for the environment and for the duration of the lockout.

Storage tanks are labeled to identify the potential fire hazard of the material stored within the tank.

Fire Extinguisher Classes

TRASH • WOOD • PAPER

BOXES

Ⓐ **ORDINARY COMBUSTIBLES**

LIQUIDS • GREASE

SOLVENT CEMENT

Ⓑ **FLAMMABLE LIQUIDS**

MOTORS • TRANSFORMERS

ELECTRICAL MOTOR

Ⓒ **ELECTRICAL EQUIPMENT**

ZIRCONIUM • TITANIUM

METAL

Ⓓ **COMBUSTIBLE METALS**

GREASE

DEEP FAT FRYER

Ⓚ **COMMERCIAL COOKING GREASE**

Figure 3-26. *Fire extinguisher classes are based on the combustible material.*

Hazardous Locations

The use of electrical equipment in areas where explosion hazards are present can lead to an explosion and fire. This danger exists in the form of escaped flammable gases such as naphtha, benzene, propane, and others. Coal, grain, and other dust suspended in the air can also cause an explosion. Article 500 of the National Electrical Code® (NEC®) and Article 235 of NFPA 70E cover the requirements of working in hazardous locations. **See Figure 3-27.** Any hazardous location requires the maximum in safety and adherence to local, state, and federal guidelines and laws, as well as in-plant safety rules. Hazardous locations are indicated by Class, Division, and Group.

SAFETY TIP

According to the NFPA 70E, Standard for Electrical Safety in the Workplace, 2004 Edition, Article 420.20 (F)(5), combustible material, buildings, and parts of buildings, fire escapes, and door and window openings shall be safeguarded from fires originating in oil-insulated transformers installed on roofs, or attached to or adjacent to a building.

CONFINED SPACES

A *confined space* is a space large enough and so configured that an employee can physically enter and perform assigned work, that has limited or restricted means for entry and exit, and is not designed for continuous employee occupancy. Confined spaces have a limited means of egress and are subject to the accumulation of toxic or flammable contaminants or an oxygen-deficient atmosphere. **See Figure 3-28.** Confined spaces include, but are not limited to, storage tanks, process vessels, bins, boilers, ventilation or exhaust ducts, sewers, underground utility vaults, tunnels, pipelines, and open top spaces more than 4′ in depth such as pits, tubes, ditches, and vaults.

Hazardous Locations — Article 500

Hazardous Location – A location where there is an increased risk of fire or explosion due to the presence of flammable gases, vapors, liquids, combustible dusts, or easily-ignitable fibers or flyings.

Location – A position or site.

Flammable – Capable of being easily ignited and of burning quickly.

Gas – A fluid (such as air) that has no independent shape or volume but tends to expand indefinitely.

Vapor – A substance in the gaseous state as distinguished from the solid or liquid state.

Liquid – A fluid (such as water) that has no independent shape but has a definite volume. A liquid does not expand indefinitely and is only slightly compressible.

Combustible – Capable of burning.

Ignitable – Capable of being set on fire.

Fiber – A thread or piece of material.

Flyings – Small particles of material.

Dust – Fine particles of matter.

CLASSES	Likelihood that a flammable or combustible concentration is present
I	Sufficient quantities of flammable gases and vapors present in air to cause an explosion or ignite hazardous materials
II	Sufficient quantities of combustible dust are present in air to cause an explosion or ignite hazardous materials
III	Easily ignitable fibers or flyings are present in air, but not in a sufficient quantity to cause an explosion or ignite hazardous materials

DIVISIONS	Location containing hazardous substances
1	Hazardous location in which hazardous substance is normally present in air in sufficient quantities to cause an explosion or ignite hazardous materials
2	Hazardous location in which hazardous substance is not normally present in air in sufficient quantities to cause an explosion or ignite hazardous materials

Class I Division I:

Spray-booth interiors

Areas adjacent to spraying or painting operations using volatile flammable solvents

Open tanks or vats of volatile flammable liquids

Drying or evaporation rooms for flammable vents

Areas where fats and oils extraction equipment using flammable solvents is operated

Cleaning and dyeing plant rooms that use flammable liquids that do not contain adequate ventilation

Refrigeration or freezer interiors that store flammable materials

All other locations where sufficient ignitable quantities of flammable gases or vapors are likely to occur during routine operations

Class II Division I:

Grain and grain products

Pulverized sugar and cocoa

Dried egg and milk powders

Pulverized spices

Starch and pastes

Potato and wood flour

Oil meal from beans and seeds

Dried hay

Any other organic materials that may produce combustible dusts during their use or handling

Class III Division I:

Portions of rayon, cotton, or other textile mills

Manufacturing and processing plants for combustible fibers, cotton gins, and cotton seed mills

Flax processing plants

Clothing manufacturing plants

Woodworking plants

Other establishments involving similar hazardous processes or conditions

HAZARDOUS LOCATIONS		
Class	**Group**	**Material**
I	A	Acetylene
	B	Hydrogen, butadiene, ethylene oxide, propylene oxide
	C	Carbon monoxide, ether, ethylene, hydrogen sulfide, morpholine, cyclopropane
	D	Gasoline, benzene, butane, propane, alcohol, acetone, ammonia, vinyl chloride
II	E	Metal dusts
	F	Carbon black, coke dust, coal
	G	Grain dust, flour, starch, sugar, plastics
III	No groups	Wood chips, cotton, flax, and nylon

Figure 3-27. Article 500 of the NEC® covers hazardous locations.

*A **permit-required confined space** is a confined space that has specific health and safety hazards associated with it.*

*A **non-permit confined space** is a confined space that does not contain or, with respect to atmospheric hazards, have the potential to contain any hazards capable of causing death or serious physical harm.*

Mine Safety Appliances Co.

Figure 3-28. *The air inside a confined space must be tested before entry.*

Confined spaces cause entrapment hazards and life-threatening atmospheres through oxygen deficiency, combustible gases, and/or toxic gases. Oxygen deficiency is caused by the displacement of oxygen by leaking gases or vapors, the combustion or oxidation process, oxygen absorbed by the vessel or product stored, and/or oxygen consumed by bacterial action. Oxygen-deficient air can result in injury or death.

Dangerous gases in a confined space are commonly caused by leaking gases or gases produced in the space such as methane, carbon monoxide, carbon dioxide, and hydrogen sulfide. Air normally contains 21% oxygen. An increase in the oxygen level increases the explosive potential of combustible gases. Finely ground materials including carbon, grain, fibers, metals, and plastics can also cause explosive atmospheres.

WARNING: Confined space procedures vary in each facility. For maximum safety, always refer to specific facility procedures and applicable federal, state, and local regulations.

Confined Space Permits

Confined space permits are required for work in confined spaces based on safety considerations for workers. A *permit-required confined space* is a confined space that has specific health and safety hazards associated with it. OSHA Standard 29 CFR 1910.146 – *Permit-Required Confined Spaces* contains the requirements for practices and procedures to protect workers from the hazards of entry into permit-required confined spaces. A *non-permit confined space* is a confined space that does not contain or, with respect to atmospheric hazards, have the potential to contain any hazards capable of causing death or serious physical harm. These conditions can change with tasks such as welding, painting, or solvent use in the confined space.

SUMMARY

- Codes and standards are developed to protect people and property.

- Electrical work must be done by a qualified person.

- Safety labels are used to alert technicians to workplace hazards.

- Grounding is used to provide a path to earth for currents in order to protect people and property.

- Ground fault circuit interrupters detect an imbalance of current between normal conductor pathways.

- Arc flash and arc blast are electrical hazards in the workplace.

- Overvoltage installation categories are used to specify an area in which a DMM may be used.

- PPE is clothing and/or equipment worn by a technician to reduce the possibility of injury in the work area.

- Lockout/tagout is used to remove all energy from a machine or circuit in order to protect people working on it.

- Specific types of fire extinguishers are used for specific types of fires.

- Article 500 of the NEC® and Article 440 of NFPA 70E cover the requirements for working in hazardous locations.

- Confined space entry is a special hazard that requires detailed procedures to perform safely.

DEFINITIONS . . .

- The *Occupational Safety and Health Administration (OSHA)* is a federal agency that requires all employers to provide a safe environment for their employees.

- The *National Fire Protection Association (NFPA)* is a national organization that provides guidance in assessing the hazards of the products of combustion.

- The *National Electrical Code® (NEC®)* is a standard on practices for the design and installation of electrical products published by the National Fire Protection Association (NFPA).

- The *American National Standards Institute (ANSI)* is a U.S. national organization that helps identify industrial and public needs for standards.

. . . DEFINITIONS . . .

- The *Canadian Standards Association (CSA)* is a Canadian nonprofit membership association for standards development, sale of publications, training, and membership services.

- *Underwriters Laboratories Inc. (UL)* is an independent organization that tests equipment and products to see if they conform to national codes and standards.

- The *International Electrotechnical Commission (IEC)* is an international organization that develops international safety standards for electrical equipment.

- The *National Electrical Manufacturers Association (NEMA)* is a U.S. national organization that assists with information and standards concerning proper selection, ratings, construction, and safety standards for electrical equipment.

- A *qualified person* is a person who is trained in, and has specific knowledge of, the construction and operation, testing, and performance of electrical equipment or a specific task, and is trained to recognize and avoid electrical hazards that might be present with respect to the equipment or specific task.

- A *safety label* is a label that indicates areas or tasks that can pose a hazard to personnel and/or equipment.

- A *danger signal word* is a word used to indicate an imminently hazardous situation which, if not avoided, results in death or serious injury.

- A *warning signal word* is a word used to indicate a potentially hazardous situation which, if not avoided, could result in death or serious injury.

- A *caution signal word* is a word used to indicate a potentially hazardous situation which, if not avoided, may result in minor or moderate injury.

- *Electrical warning signal word* is a word used to indicate a high-voltage location and conditions that could result in death or serious personal injury if proper precautions and procedures are not followed.

- *Explosion warning signal word* is a word used to indicate locations and conditions where exploding parts may cause death or serious personal injury if proper precautions and procedures are not followed.

- An *electrical shock* is a shock that results anytime a body becomes part of an electrical circuit.

- *Grounding* is the connection of portions of the distribution system to earth in order to establish a common electrical reference and a low impedance fault path to facilitate operation of overcurrent protective devices.

- A *separately derived system (SDS)* is a system that supplies electrical power derived (taken from) transformers, storage batteries, solar photovoltaic systems, or generators.

- A *main bonding jumper (MBJ)* is a connection at the service equipment that connects the equipment grounding conductor, the grounding electrode conductor, and the grounded conductor (neutral conductor).

. . . DEFINITIONS

- *An **equipment grounding conductor (EGC)*** is an electrical conductor that provides a low-impedance ground path between electrical equipment and enclosures within the distribution system and takes current back to the source.

- A ***grounding electrode conductor (GEC)*** is a conductor that connects grounded parts of a power distribution system (equipment grounding conductors, grounded conductors, and all metal parts) to the grounding system.

- A ***grounded conductor*** is a conductor that has been intentionally grounded.

- A ***ground fault circuit interrupter (GFCI)*** is a device that protects against electrical shock by detecting an imbalance of current in the normal conductor pathways and opening the circuit.

- ***Arc flash*** is a short circuit through the air.

- An ***arc blast*** is an explosion that occurs when the surrounding air becomes ionized and conductive.

- A ***voltage surge*** is a higher-than-normal voltage that temporarily exists on one or more power lines.

- A ***transient voltage (voltage spike)*** is an unwanted voltage of very short duration in an electrical circuit.

- ***Personal protective equipment (PPE)*** is clothing and/or equipment worn by a technician to reduce the possibility of injury in the work area.

- ***Rubber insulating gloves*** are gloves made of latex rubber and are used to provide maximum isolation from electrical shock.

- ***Leather protectors*** are gloves worn over rubber insulating gloves to prevent penetration of the rubber insulating gloves and provide added protection against electrical shock.

- ***Rubber insulating matting*** is a floor covering that provides technicians protection from electrical shock when working on live electrical currents.

- ***Lockout*** is the process of removing the source of electrical power and installing a lock that prevents the power from being turned on.

- ***Tagout*** is the process of placing a danger tag on the source of electrical power and installing a lock that prevents the power from being turned on.

- A ***confined space*** is a space large enough and so configured that an employee can physically enter and perform assigned work, that has limited or restricted means for entry and exit, and is not designed for continuous employee occupancy.

- A ***permit-required confined space*** is a confined space that has specific health and safety hazards associated with it.

1. What is the purpose of the NEC®?

2. List five requirements for a technician to be a qualified person.

3. Describe the operation of a ground fault circuit interrupter.

4. What is an arc blast?

5. What is a flash hazard analysis?

6. List five types of PPE.

7. What is the purpose of a lockout/tagout procedure?

8. When can a danger tag be used instead of a lock?

9. Which article of the NEC® covers hazardous locations?

10. What is a confined space?

Transformer Connections

Almost all electrical power in use today is generated by 3-phase generators. Transformers are used to change the voltage for transmission and distribution. It is essential that technicians be able to properly connect transformers to ensure safe and reliable operation.

TRANSFORMER CONNECTIONS

Transformers can be connected in various ways. The two common ways of connecting loads and transformers in 3-phase systems are delta and wye connections. Since transformers have connections on both the primary and secondary sides, it is important to understand the various combinations of delta and wye connections. Also, transformer connections must take into account the proper polarity of the transformer.

Voltage Transformation

Although single-phase circuits are not used in the transmission of power nor on the primary side of a distribution circuit, single-phase transformers are used in various applications. The majority of 3-phase transmission circuits in the U.S. are designed with single-phase transformers connected in various ways to obtain single-phase and 3-phase power. Transformers are used for stepping up or stepping down voltage as well as to obtain various phase transformations. The transformer connections that are possible for voltage transformation in power and distribution service include single-phase and 3-phase.

Single-Phase Circuits. The low-voltage winding of a standard distribution transformer connected to a single-phase circuit is normally made with two equal coils. These coils are arranged so that they may be connected in series or parallel. This arrangement permits current to be delivered at two voltages, one twice the other. **See Figure 4-1.**

The secondary winding may have its two coils (each of 120 V) connected in series to provide a load voltage of 240 V. This connection can be used where a motor load is the only load on a circuit. The secondary winding may have its two sections connected in parallel to provide 120 V. This connection is used primarily for lighting loads, although this voltage may also be used for fractional horsepower motors.

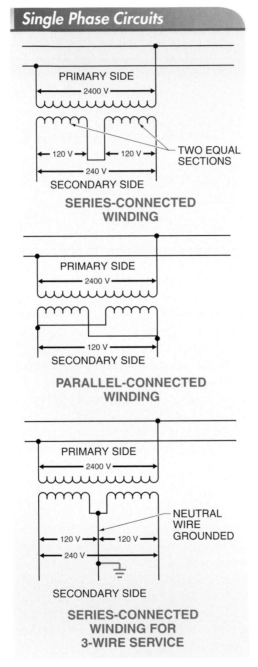

Single Phase Circuits

PRIMARY SIDE
2400 V

120 V — 120 V
TWO EQUAL SECTIONS
240 V
SECONDARY SIDE

SERIES-CONNECTED WINDING

PRIMARY SIDE
2400 V

120 V
SECONDARY SIDE

PARALLEL-CONNECTED WINDING

PRIMARY SIDE
2400 V

NEUTRAL WIRE GROUNDED
120 V — 120 V
240 V
SECONDARY SIDE

SERIES-CONNECTED WINDING FOR 3-WIRE SERVICE

Figure 4-1. The low-voltage winding of a standard distribution transformer connected to a single-phase circuit is normally made with two equal coils. These sections are arranged so that they may be connected in series or parallel.

Three single-phase transformers may be connected together to provide power for both higher voltage motor loads and lower voltage lighting loads. However, it is more common to connect the two coils of the secondary winding of a transformer in series for 3-wire service. The third wire (neutral) of the 3-wire circuit is connected to the secondary winding at the point where the series connection of the coils is made. This permits the use of the 240 V circuit for a motor load and two 120 V circuits for lighting loads.

Today, the majority of dwellings are wired with 3-wire service. Electric ranges, dryers, and central climate-control systems are designed for 3-wire 120 V/240 V operation. In a dwelling, the 3-wire service entrance conductors terminate at the service equipment panel.

Most individual branch circuits carried through dwellings are at 120 V. The circuits which feed large loads, such as air conditioners, electric ranges, or electric water heaters require 240 V, single-phase power. The neutral wire carries only the unbalanced current of the entire circuit. No current flows in the neutral wire if each of the 120 V circuits have equal connected loads. If one circuit is fully loaded and the other circuit carries no load, the entire current of the first circuit flows through the neutral wire. If the second circuit carries a smaller load than the first circuit, the neutral wire carries only the difference in current between the two loads. The 240 V motor load is carried by the two outside wires.

The neutral wire seldom carries as much current as either of the two outside wires. However, the neutral wire should be large enough to safely carry whatever current could flow through the circuit due to a fault such as a short circuit in either one of the two outside wires of a 3-wire circuit. Each of the two outside wires should be fused or provided with circuit breakers to protect both the transformer winding and the connected load against undue overloads, short circuits, or grounds.

Single-Phase Transformer Polarity. Transformer polarity refers to the voltage-vector relationship of the transformer leads as brought outside the transformer case. Because the high- and low-voltage leads are independent of the arrangement of the windings in the magnetic circuit, the polarity of a transformer can be changed by interchanging the position of the two leads of any one winding as brought out of the case.

Regardless of which winding is the primary, the voltage-vector relationship is best understood by considering the induced voltages in the high-voltage and low-voltage windings. Because the induced voltages in both coils are induced by the same flux, the induced voltages must be in the same direction in both windings. Assuming that the induced voltage in the high-voltage winding (H1-H2) is in the same direction as the order of lettering of the leads (H1-H2), the induced voltage in the low-voltage winding is from X1 to X2. **See Figure 4-2.**

ANSI standardization rules state that high-voltage leads brought outside the transformer case are to be marked H1, H2, etc., and low-voltage leads are to be marked X1, X2, etc. The order is such that when H1 and X1 are connected and voltage is applied to the transformer, the voltage between the highest numbered H lead and the highest numbered X lead shall be less than the voltage of the full high-voltage winding. When leads are marked in this manner, the polarity of the transformer is subtractive when H1 and X1 are adjacent to each other and additive when H1 is diagonally located with respect to X1.

Assuming a 1:1 ratio of voltage transformation, if the H1 lead is connected to the X1 lead and voltage is applied to the high-voltage winding, a digital multimeter (DMM) set to measure voltage connected across leads H2 and X2 reads the difference of the two voltages because they are in opposition. In this case, the DMM reads 0 V. For this reason, the polarity of the transformers is subtractive.

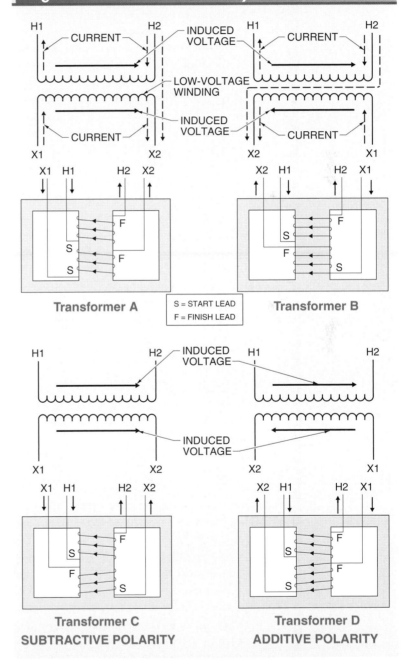

Figure 4-2. Transformer polarity may be subtractive or additive.

If the H1 lead is connected to the X2 lead and voltage is applied to the high-voltage winding, a DMM set to measure voltage connected across leads H2 and X1 reads the sum of the two voltages because they flow in the same direction. The DMM reading is twice the applied voltage to the high-voltage winding. For this reason, the polarity of the transformer is additive.

In transformers that have both windings wound in the same direction with respect to the core, the polarity is changed from subtractive to additive by interchanging the position of the two X leads. In transformers that have both windings wound in opposite directions with respect to the core and the same voltage stresses between similar parts of the windings, the polarity is changed from additive to subtractive by interchanging the position of the two X leads.

Transformers A and C have subtractive polarity, yet the low-voltage windings are wound in opposite directions with respect to each other. Likewise, Transformers B and D have additive polarity, yet the low-voltage windings are wound in opposite directions with respect to each other. The same polarity in each example is obtained by interchanging the numbering of the start and finish leads of the low-voltage windings.

The H1 lead is normally brought out on the right-hand side of the transformer case when facing the high-voltage side. This simplifies the task of connecting transformers in parallel. The polarity of a transformer is not an indication of the direction of the turns of the high- and low-voltage windings around the core nor does it indicate the voltage stresses that exist between the turns of the two windings.

Subtractive polarity has a small advantage over additive polarity in that the voltage stress between external leads is smaller. Under normal operating conditions, with leads insulated from each other, the potential stress between adjacent high- and low-voltage leads is one-half the sum of the high and low voltages for addi-

tive polarity and one-half their difference for subtractive polarity. The advantage of subtractive polarity, which is ordinarily negligible, becomes appreciable for transformers whose primaries and secondaries have very high voltages.

In general, distribution transformers of 200 kVA or less having voltages of 7500 V or less have additive polarity. All other distribution transformers have subtractive polarity. Some power transformers have additive polarity, although the majority have subtractive polarity.

Polarity Test Methods. The methods used to test for polarity and check transformer lead markings include transformer comparison, DC use, and AC use. The method used is the most convenient method.

The polarity of a transformer may be checked by comparison against a transformer of known polarity and correct lead marking, and that has the same ratio as the one to be tested. The polarity is checked by connecting the high-voltage windings of both transformers in parallel. This is done by connecting the H1 and H2 leads. If the leads of the transformer to be tested are not marked, it is assumed that similarly located high-voltage leads on the two transformers are the same. The left-hand low-voltage leads (facing the low-voltage side) of both transformers are connected, leaving the right-hand side leads free. **See Figure 4-3.**

A reduced voltage is applied to the high-voltage windings and the voltage between the two free leads is measured on the low-voltage side. A reading of 0 V indicates that the relative polarity of both transformers is identical and the lead lettering of the low-voltage winding of the transformer under test is the same as that of the known transformer. A reading equal to the sum of the low voltages of both transformers indicates that the polarities of the two units are opposite and the lead lettering of the low-voltage winding of the transformer under test should be reversed to match the known transformer.

Polarity Test Methods

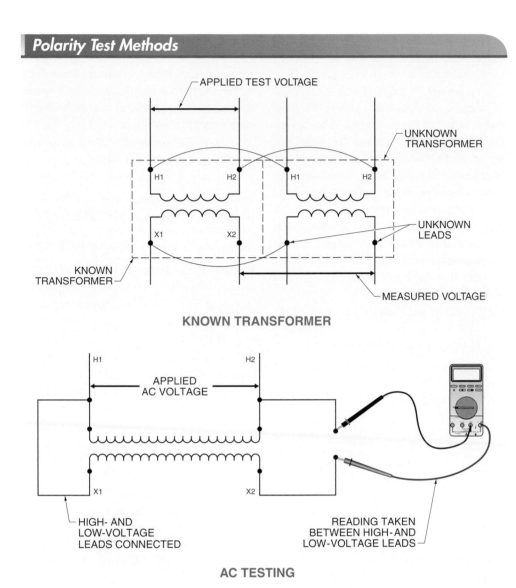

KNOWN TRANSFORMER

AC TESTING

Figure 4-3. *The polarity of a transformer may be determined using a known transformer or with AC voltage.*

The polarity of a transformer may also be determined by the use of DC. With DC passing through the high-voltage winding, a DMM set to measure DC voltage is connected across the outlet terminals of the same winding to obtain a small positive voltage reading. The two DMM leads are transferred directly across the transformer.

For example, the lead from the right-hand, high-voltage terminal is placed on the adjacent low-voltage terminal and the other lead is placed on its adjacent low-voltage terminal. The DC excita-

tion is broken, inducing a voltage in the low-voltage winding which causes a small voltage reading. The polarity is additive if the DMM reading is the same polarity as previously. The polarity is subtractive if the reading is the opposite polarity.

The polarity of a transformer may also be determined by connecting the adjacent left-hand high-voltage and low-voltage leads (H1 and X1) together and applying AC voltage. Any convenient value of AC voltage can be applied to the high-voltage winding and readings taken of the applied voltage and the voltage between the adjacent right-hand high-voltage and low-voltage leads (H2 and X2).

The polarity is subtractive if the voltage between the adjacent right-hand high-voltage and low-voltage leads is less than the applied voltage. This indicates the difference in voltage between the high-voltage and low-voltage windings. The polarity is additive if the voltage between the adjacent right-hand high-voltage and low-voltage leads is greater than the applied voltage. This method is limited to transformers in which the ratio of transformation is 30:1 or less.

Three-Phase Circuits

A 3-phase system originates from three separate windings of an AC generator. The three windings are located within the generator to create three separate voltages which are out of phase with each other by 120 electrical degrees.

A total of six line leads are required if the three circuits are kept separate from the generator to the apparatus using the current. This is impractical, so common wires are used. In forming the common wires, the individual phases of the generator are connected in delta or wye connections.

Delta Connections

A *delta connection* is a 3-phase connection method that has the wire from the ends of each coil connected end-to-end to form a closed loop. A delta connection is developed when the three coils are connected in a 3-coil closed circuit and a line lead is connected at each of the three junction points. A delta connection is often drawn as a symbol diagram with the coils in a triangular shape that resembles the Greek letter delta (Δ). **See Figure 4-4.**

Voltage in Delta Systems. In a delta-connected system, the voltage measured across any two lines is equal to the voltage generated in the coil winding. This is because the voltage is measured directly across the coil winding. For example, if the generated coil voltage is equal to 240 V, the volt-

age between any two lines equals 240 V. Although the three coils are connected to form a closed circuit, no current flows in the circuit at no load because the vector sum of three equal voltages 120° out of phase with each other is always equal to zero. With no voltage impressed across the entire circuit, there can be no current flow.

A *delta connection* is a 3-phase connection method that has the wire from the ends of each coil connected end-to-end to form a closed loop.

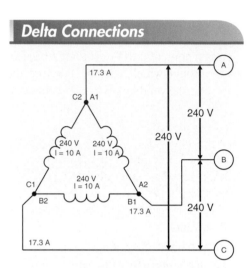

Delta Connections

LINE VOLTAGE = PHASE VOLTAGE

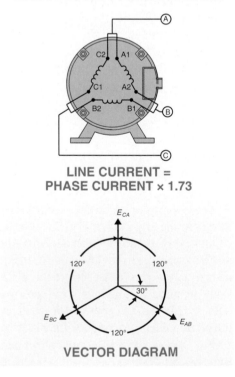

LINE CURRENT = PHASE CURRENT × 1.73

VECTOR DIAGRAM

Figure 4-4. A delta connection has the wires from the ends of each coil connected end-to-end to form a closed loop.

In generator construction, the three windings producing phase A, phase B, and phase C are placed 120° apart in the stator. As the flux field of the rotor rotates around the stator and cuts the windings, power is produced that has three phases 120° out of phase with each other.

Current in Delta Systems. The line current of a 3-phase system must be equal to the current in each of the individual coils. However, the line current, as it reaches the junction point of the two windings, has two paths through which it flows. Therefore, the current in the windings of a delta-connected system is less than the line current. The winding current is actually 57.7% of the line current (equal to the line current divided by 1.73). In a delta-connected system, the line current is equal to the winding current multiplied by 1.73. Therefore, the line current is the vector sum of the two coil currents.

In a balanced system, the phase currents are equal. In a balanced 3-phase, delta-connected system, the line current is equal to 1.73 times the current in one of the coils. For example, if each coil current is equal to 10 A, the line current is equal to 17.3 A (1.73 × 10 = 17.3 A).

Studying the current in a delta-connected system at both 0° and 90° illustrates the current flow in a delta-connected system. **See Figure 4-5.** At 0° with phases A and C at the same potential, there is no current flow between the two points, so these two coils in series have no current. The −5 A on phase A travels through the two coils in series connected to phases A and B. The −5 A on phase C travels through the two coils in series connected to phases C and B.

At the 90° point in time, phase A has −8 A, phase B has 0 A, and phase C has +8 A. With phase B at 0 V, this connects four inductors in parallel with two. The current divides through both circuits depending on the impedance. In one case, current flows through two coils, directly from A to C. In the other case, current flows through four coils, from A to B to C. The current through the two coils in series connected between phases A and C is 5.33 A and through the four coils in series between phases A and C is 2.67 A.

Current in Delta Systems

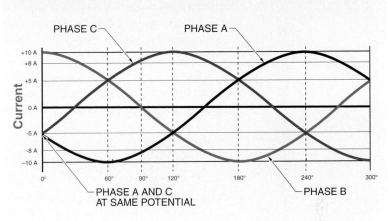

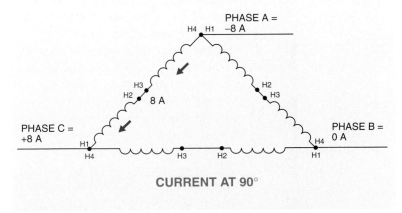

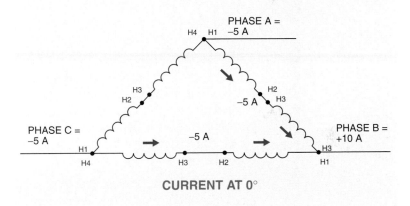

Figure 4-5. *Studying the current in a delta-connected system at both 0° and 90° illustrates the current flow in a delta-connected system.*

A transformer primary is identical to the stator in a polyphase motor, with the only difference being that the secondary does not rotate. The transformer primary sees the same rise and fall of current that the stator sees, but the secondary has much more impedance than the rotor in a motor. When the transformer primary is connected to the line, the self-induction in the primary and the resistance of the winding make up the impedance, which limits the current.

If the secondary is unloaded, then the current limits are at maximum and the current is low. The motor, on the other hand, even if unloaded, offers a short circuit on the secondary when the rotor is at standstill. Two of the three current limits are in play when at locked rotor, so it takes the CEMF of the spinning rotor field, which is a product of mutual induction, to provide the third limit. Many rules apply to both the transformer and the motor when connected in either delta or wye connections.

Wye Connections

A *wye connection* is a 3-phase connection that has one end of each coil connected together and the other end of each coil left open for external connections. A wye (Y) system is also known as a star system. **See Figure 4-6.**

Voltage in Wye Sytems. The three ends can be safely connected at the neutral point because no voltage difference exists between them. As phase A is maximum, phases B and C are opposite to A. If the equal opposing values of B and C are added vectorially, the opposing force of B and C combined is exactly equal to A. For example, if three people are pulling with the same amount of force on ropes tied together at a single point, the resulting forces cancel each other and the resultant force is zero in the center (neutral point).

TECH FACT

The most common type of transformer has the primary delta-connected with the secondary wye-connected.

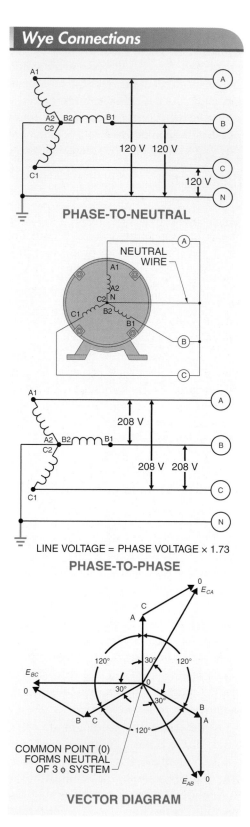

Figure 4-6. *A wye connection has one end of each coil connected together and the other end of each coil left open for external connections.*

DEFINITION

*A **wye connection** is a 3-phase connection that has one end of each coil connected together and the other end of each coil left open for external connections.*

The net effect is a large voltage (pressure) difference between the A1, B1, and C1 coil ends, but no pressure difference between the A2, B2, and C2 coil ends. For example, in a 3-phase, wye-connected lighting circuit, the 3-phase circuit is balanced because the loads are all equal in power consumption. The voltages available from a wye-connected system are phase-to-neutral and phase-to-phase.

In a 3-phase, wye-connected system, the phase-to-neutral voltage is equal to the voltage generated in each coil. For example, if an alternator produces 120 V from A1 to A2, the equivalent 120 V is present from B1 to B2 and C1 to C2. Thus, in a 3-phase, wye-connected system, the output voltage of each coil appears between each phase and the neutral.

In a 3-phase, wye-connected system, the voltage values must be added vectorially because the coils are set 120 electrical degrees apart. In such an arrangement, the phase-to-phase voltage is obtained by multiplying the phase-to-neutral voltage by 1.73. For example, if the phase-to-neutral voltage is 277 V, the phase-to-phase voltage is 480 V (277 × 1.73 = 480).

Similarly, on large wye-connected systems, a phase-to-neutral voltage of 2400 V creates a 4160 V line-to-line voltage, and a phase-to-neutral voltage of 7200 V creates a 12,480 V line-to-line voltage. One of the benefits of wye-connected systems to a utility company is that even though the generators are rated at 2400 V or 7200 V per coil, they can transmit at a higher phase-to-phase voltage with a reduction in losses and can provide better voltage regulation. This is because the higher the transmitted voltage, the less the voltage losses. This is especially important in long rural power lines.

In a wye-connected system, the neutral connection point is grounded and a fourth wire is carried along the system and grounded at every distribution transformer location. This solidly grounded system is regarded as the safest of all distribution systems.

Current in Wye Systems. All large power distribution systems are designed as 3-phase systems with the loads balanced across the phases as closely as possible. The only current that flows in the neutral wire is the unbalanced current. This is normally kept to a minimum because most systems can be kept fairly balanced. The neutral wire is normally connected to a ground such as the earth.

In a 3-phase, wye-connected system, the current in the line is the same as the current in the coil (phase) windings. This is because the current in a series circuit is the same throughout all parts of the circuit. In a balanced circuit, there is no current flow in the neutral wire because the sum of all the currents is zero. **See Figure 4-7.**

Studying the current in a wye-connected system at both 0° and at 90° illustrates the current flow in a wye-connected system. In this system, the generators have three sets of windings in the stators spaced 120° apart and, as the rotating field of the rotor cuts through each set of coils, a potential is induced in each winding that is 120° out of phase with the others.

At 0°, phase A has –5 A and phase C has –5 A. With the phase B potential being at its maximum positive value, the current from the other two phases flows to the positive and the phase B current is a combination of both (10 A). The load on this winding is 3-phase and shows no current on the neutral.

At 90°, phase A has –8 A, phase B has 0 A, and phase C has –8 A. This means that all the line current flows from phase A, through the center tap, and out through the coil connected to phase C. The coil connected to phase B has no current at this time.

This bank of transformers could be three single-phase units with 1200 V coils. With the coils in series and connected in a wye connection, the voltage between phases is 4160 V. All currents below the 0 line represent negative current, and all currents above the line are positive.

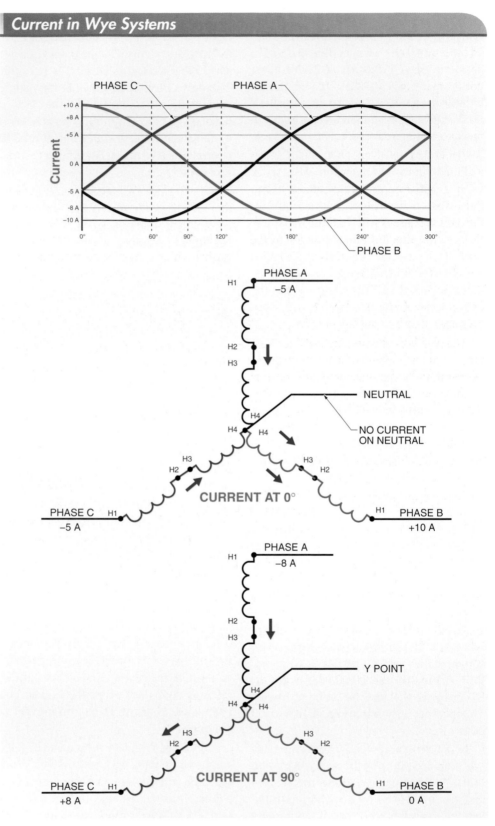

Figure 4-7. *Studying the current in a wye-connected system at both 0° and 90° illustrates the current flow in a wye-connected system.*

A generator at a power plant produces 3-phase power for 3-phase loads in commercial and industrial applications. The phase-to-phase relationship offers three leads, each with a voltage that is 120° out of phase with the others. This phase relationship is the result of placing the coils in the alternator 120 mechanical degrees apart in the stator iron.

A and B are 120° apart as the field rotates and cuts the coils. This means that, as the coils in phase A are seeing the maximum lines of flux and are producing peak voltage, the coils in phase B will see those maximum lines of flux after 120° more rotation by the rotor. This connection is sometimes called common because of the relationship among the three windings. This relationship is one that is symmetrical (balanced), and can be grounded without causing an imbalance.

The windings of three single-phase transformers or one 3-phase transformer are connected in the same manner as are separate windings of a 3-phase generator to make a 3-phase group or a single 3-phase transformer. The primary and secondary windings may be delta- or wye-connected. The winding combinations include delta-to-wye, wye-to-wye, wye-to-delta, delta-to-delta, and open delta connections.

Delta-to-Wye Connections

A 3-phase delta-to-wye connection is often used for distribution where a four-wire secondary distribution circuit is required. The delta-connected primary permits single-phase loads to be connected between the secondary neutral and the 3-phase line leads. The 3-phase wye voltage is 1.73 times the single-phase (line-to-neutral) voltage. **See Figure 4-8.**

With a 3-wire, 2400 V feeder, the delta primary connects with three 2400 V windings and offers a 277 V/480 V, 4-wire secondary for single-phase lighting and 3-phase power for equipment. The single-phase coils provide 277 V to the neutral, and the vector sum of each of these coils provides a 480 V phase-to-phase voltage. The primary system voltage is equal to the

voltage at the coil. The secondary system voltage can be found in two ways. If the voltage across the winding is known, then the line voltage is equal to 1.73 times the winding voltage. If the line voltage is known, the winding voltage is equal to 0.58 times the line voltage.

A dual-wound delta primary may be connected onto the 480 V phase-to-phase bus to provide a 120 V/208 V feeder for a panelboard. The primary delta should not be connected to the fourth wire or neutral terminal. The neutral of the 120 V/208 V secondary can be connected to the neutral busbar which connects the center tap of the secondary to the service ground.

Delta-to-Wye Connections

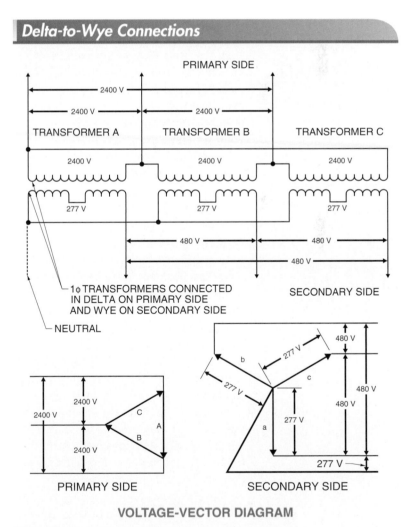

VOLTAGE-VECTOR DIAGRAM

Figure 4-8. A 3-phase delta-to-wye connection is often used for distribution where a four-wire secondary distribution circuit is required.

In some service installations, a 120 V/ 240 V secondary is used for both single-phase and 3-phase loads in a building. If expansion is necessary in a plant, the 3-wire, 240 V bus duct does not provide a neutral for the single-phase loads. A transformer that is readily available in most supply houses has dual-wound 240 V primary coils with dual-wound 120 V secondary coils. This selection gives the user a large number of single-phase circuits along with slots for 2-pole, single-phase as well as some 3-phase power for small equipment.

Wye-to-Wye Connections

In the transmission of power, a generator never produces the high transmission line voltages that are required for transmission over long distances because it is too difficult and too expensive to build such a unit. To transmit the generator voltage over long distances, a transformer is used to step up the output of the generator. If the voltage is increased, the ability to carry a large amount of power on the transmission system is also increased.

A wye-to-wye connection is used when tying together two high-voltage transmission systems of unequal voltage. In this case, it is necessary that the connection used does not cause a shift in phase from the primary to the secondary. For high voltages, the wye-to-wye connection is preferred because the voltage stress of the windings to ground with the neutral grounded is only 57.7% of the voltage stress to ground of similar units connected delta-to-delta. A tertiary winding is provided to suppress the third harmonic voltages which would otherwise appear on the system when a wye-to-wye connection is used for this purpose. A *tertiary winding* is a third winding that is often used in power transformers to provide station power requirements or a tie with synchronous condensers.

Three single-phase transformers may be connected in a wye-to-wye connection to form a 3-phase transformer. **See Figure 4-9.** The 3-phase voltages on the primary and secondary sides are 4160 V

and 480 V, respectively, if the single-phase voltage rating of each transformer is 2400 V and 277 V. The voltage from any line lead to the neutral wire is 2400 V on the primary side and 277 V on the secondary side.

An advantage of wye-to-wye connections is that the voltage of each single-phase unit connected to form a wye-to-wye connection is only 57.7% of the line voltage. This makes the connection suitable for power transmission because the individual transformers are wound for a relatively low voltage, and nearly double the voltage (173%) is obtained on the line.

A wye-to-wye connection also permits the grounding of the neutral on either or both primary and secondary windings. If the primary neutral is carried back to the neutral of the generator, unequal loads may be taken off the secondary windings between the secondary neutral and any of the three line leads. The unbalanced current flows through the neutral, back to the generator. This permits lighting loads to be removed at the neutral, and any of the 3-phase wires and a power load from the 3-phase leads to be removed at the same time.

Disadvantages of a wye-to-wye connection include harmonics that disturb the system, large imbalanced loads that require a fourth wire back to the generator, and difficulty in obtaining a perfectly balanced voltage across the three phases. In many cases, interference with telephone circuits may result unless a tertiary winding is added to provide a path for the circulating currents to flow through.

When using single-phase transformers, this trouble may be avoided by grounding the primary and secondary neutrals. Another concern is that unbalanced loads cannot be carried on the secondary side unless the primary neutral or fourth wire is provided. Finally, it is practically impossible to construct three single-phase units or even a 3-phase unit in which the magnetizing currents of each of the three windings are exactly the same. This makes it impossible to have perfectly balanced voltages in each of the three windings of a wye-to-wye connection.

DEFINITION

*A **tertiary winding** is a third winding that is often used in power transformers to provide station power requirements or a tie with synchronous condensers.*

Wye-to-Delta Connections

A wye-to-delta connection is used for power transmission and distribution. A wye-to-delta connection permits single-phase and 3-phase loads to be drawn simultaneously from the delta-connected secondary at the same voltage. For example, two pump motors, one 3-phase and one single-phase, rated at 240 V each, could be fed from the same service.

With simultaneous single-phase and 3-phase loading, the currents of the two loads are added in their proper phase relations to determine the maximum loads that can be drawn from the connection. In order to prevent overheating, the resultant current in any one of the three transformers due to this loading must not continuously exceed the maximum current rating of the coil.

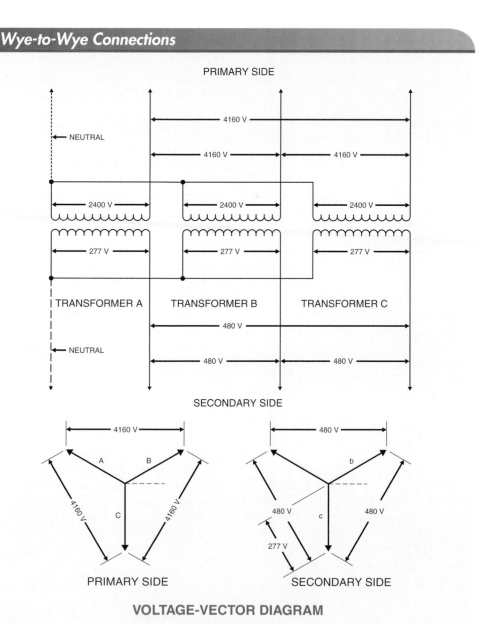

Figure 4-9. *Three single-phase transformers may be connected in a wye-to-wye connection to form a 3-phase transformer.*

A wye-to-delta connection may be made with three single-phase transformers. **See Figure 4-10.** Assuming that only a single-phase load is connected across one of the phases, the maximum load that can be obtained without overloading any part of the circuit is 150% of the maximum rating of the single-phase transformer.

In a wye-to-delta connection of three single-phase transformers, one unit may be disconnected from the circuit and service maintained with the secondary operating in an open delta connection at 57.7% of normal bank capacity. The neutral on the primary side and the neu-tral of the supply source for the primary must be grounded. **See Figure 4-11.** The system is unbalanced and considerable telephone interference may result from such a connection.

TECH FACT

In a wye-connected system, the line voltage is equal to the voltage between a phase conductor and the neutral multiplied by $\sqrt{3}$ (1.73). The voltage between a phase conductor and the neutral is equal to the line voltage divided by $\sqrt{3}$.

Wye-to-Delta Connections

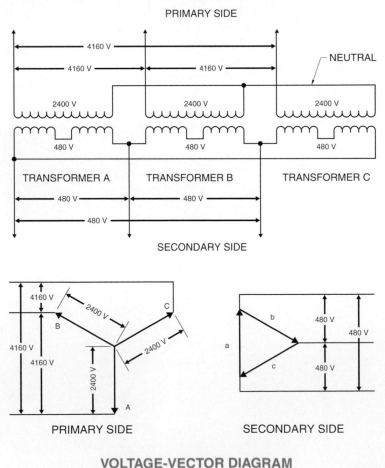

VOLTAGE-VECTOR DIAGRAM

Figure 4-10. *A wye-to-delta connection permits single-phase and 3-phase loads to be drawn simultaneously from the delta-connected secondary at the same voltage.*

Transformer Removed from Service

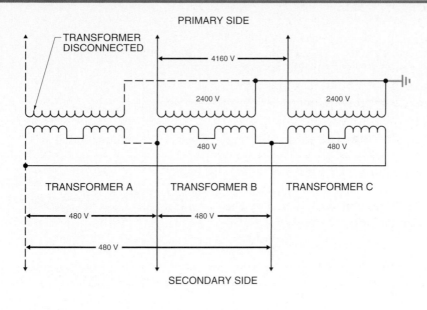

PRIMARY SIDE

TRANSFORMER DISCONNECTED

4160 V

2400 V   2400 V

480 V   480 V

TRANSFORMER A   TRANSFORMER B   TRANSFORMER C

480 V   480 V

480 V

SECONDARY SIDE

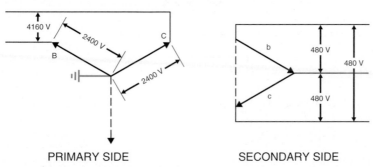

4160 V   2400 V   C

B

2400 V

PRIMARY SIDE   SECONDARY SIDE

b   480 V

480 V

c   480 V

VOLTAGE-VECTOR DIAGRAM

Figure 4-11. In a wye-to-delta connection of three single-phase transformers, one unit may be disconnected from the circuit and service maintained with the secondary operating in open delta connection at 57.7% of normal bank capacity.

Delta-to-Delta Connections

Delta-to-delta connections are used in industry where old equipment is used on new high-voltage 480 V bus duct systems. Often, when equipment arrives at its final destination, the installer discovers single-voltage motors and controls rated for 240 V. This may require a drop in supply voltage to the machine if connections cannot be made to change the controls and motors to the higher voltage. Three single-phase units wired in a delta-to-delta connection can provide power for this circuit. Three single-phase units are often used because they are more common than a 3-phase unit. **See Figure 4-12.**

In the past, delta-to-delta connected transformers were extensively used in electrical distribution systems. With this type of system, it was practical to continue distributing 3-phase power while maintenance was performed on one unit (open delta) of the 3-phase transformer bank. Today, 3-phase transformers are seldom connected in a delta-to-delta connection. However, single-phase transformers connected to form a 3-phase bank are quite common.

Delta-to-Delta Connections

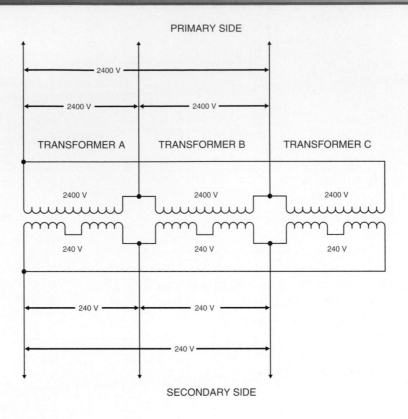

PRIMARY SIDE

TRANSFORMER A   TRANSFORMER B   TRANSFORMER C

2400 V

2400 V   2400 V

2400 V   2400 V   2400 V

240 V   240 V   240 V

240 V   240 V

240 V

SECONDARY SIDE

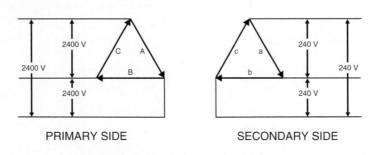

PRIMARY SIDE   SECONDARY SIDE

VOLTAGE-VECTOR DIAGRAM

Figure 4-12. In a delta-to-delta connection, the voltage of each transformer is equal to the 3-phase line voltage and the current in each of the transformers is only 57.7% of the line current.

Delta-to-delta connected transformers must be wound for full-line voltage. A 3-phase, delta-to-delta connection and voltage-vector diagram show that the voltage of each transformer is equal to the 3-phase line voltage. The current in each of the transformers is only 57.7% of the line current. The windings have a greater number of turns than for a wye connection of the same line voltage, while the cross section of the turns is only 57.7% of that for a wye-connected transformer.

One of the disadvantages of a delta-to-delta connection is the absence of an intentional connection to ground on the secondary of the transformer. To obtain a grounded system, one of the corners of the delta secondary is grounded, giving the system a reference to ground. This corner-grounded system is also referred to as a "grounded B phase" system.

The center conductor (phase B) must be used when using a corner-grounded system. According to NEC® Section 200.6, *Use and Identification of Grounded Conductors*, this conductor must be white, grey, or marked white. If the disconnect is used as a service or distribution disconnect, then the center clips cannot be fused. In this case, a shunt must be installed. The shunt usually consists of a piece of copper tube in place of the fuse.

The midpoint of one of the windings may be grounded for 120 V/240 V, single-phase, 3-wire service in a center-grounded delta for lighting, if the secondary is wound in two sections. **See Figure 4-13.** Occasionally, the single-phase unit that is used for the lighting load is made larger than the other two units so that the maximum kilovolt-amps available in the other two units for 3-phase loading may be used.

Delta Grounded Midpoint

PRIMARY SIDE

2400 V
2400 V
2400 V
C   A
B

SECONDARY SIDE

c
b
120 V   120 V
120 V, 1φ CIRCUITS
240 V   240 V
240 V
MIDPOINT GROUNDED

Figure 4-13. *Grounding one of the midpoints of the windings provides 240 V/120 V, 1φ, 3-wire service for lighting applications.*

ABB, Inc.

Three-phase transformers can be wired in delta or wye configurations.

Transformers operating in a 3-phase, delta-to-delta connection have a circulating current flowing in both the primary and secondary windings if their ratios are different. This fails to divide the load properly in each of the windings if their respective impedances and the ratios of reactances to resistances are not equal.

Open Delta Connections

An open delta connection makes 3-phase power available anywhere along the distribution line with the use of only two transformers rather than the usual three.

See Figure 4-14. An open delta connection may be called a V-connection because only two sides of the delta triangle are included. An open delta connection may also be called a V-V connection, since the delta is open on the primary and secondary sides of the transformers.

An open delta connection can be used to supply a small 3-phase load that is expected to increase in the future. This installation requires two single-phase transformers, which keeps the initial investment low but also provides for future load increases.

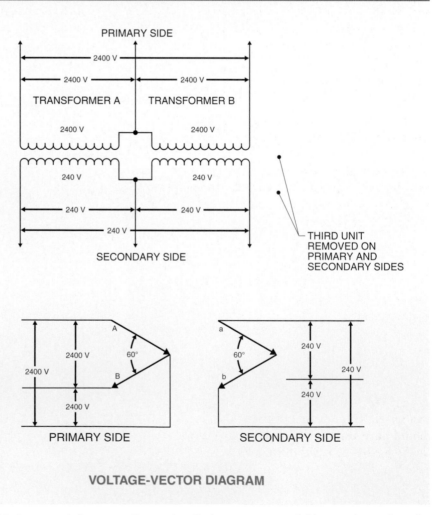

Figure 4-14. An open delta connection makes 3-phase power available anywhere along the distribution line with the use of only two transformers rather than the usual three.

For example, a transformer bank is required for a new industrial building that has an initial 3-phase load of 150 kVA and an anticipated future load of 300 kVA. If two single-phase, 100 kVA transformers are installed and connected in an open delta connection, the present 3-phase capacity of the transformer bank is 57.7% of the normal 3-phase bank capacity or 173 kVA (57.7% × 300 kVA = 173 kVA). As the load increases to the 173 kVA open delta capacity, a third 100 kVA transformer may be installed to increase the capacity of the delta-to-delta connected bank to 300 kVA.

An open delta connection makes it possible to maintain operation in an emergency if one single-phase transformer or one winding of a 3-phase transformer in a delta-to-delta connection becomes defective. For example, a plant that is fed with a delta-to-delta connection has one transformer burned out and the 3-phase power to the plant is lost.

The load on an open delta must be reduced by 42.3% from that of a closed delta, but the plant could continue production with the critical loads until the replacement unit arrives. After the replacement is installed, the plant can go back to full production.

An open delta connection can also be used when the secondary circuits are to supply lighting and power loads. The grounded or neutral conductor of the lighting circuit is taken from a center tap of the 240 V secondary winding. This provides a 120 V/240 V, single-phase, 3-wire lighting circuit.

When an open delta connection supplies a large single-phase load and a small 3-phase load, the two transformers should have different kVA ratings. The single-phase load transformer is larger than the 3-phase load transformer. This application is economical because one small single-phase transformer can be added to a large single-phase transformer to supply a limited 3-phase power load.

Delta Connection with High Leg to Ground

One of the more popular delta-to-delta connections is the 120 V/240 V, 4-wire system with a high leg to ground. These systems are popular and provide 3-phase power for machines and also 120 V for receptacles and lighting. In small industry, including small machine shops and printing shops, the equipment is rated at 240 V, 3-phase.

The 120 V/208 V system does not provide adequate power for these systems. The tolerance on most motors is ±10%, so the lowest acceptable voltage on a 240 VAC motor is 216 VAC and the lowest acceptable voltage on a 230 VAC motor is 207 VAC. The 208 V may drop to 200 V or below at peak load. At these voltage levels, the equipment suffers from overheating.

A delta connection with high leg to ground has a center tap grounded on one of the transformers and provides 240 V, 3-phase power to equipment and 120 V power for lighting and receptacles on two phases to the neutral, which is grounded. **See Figure 4-15.** Phase B and phase C each provide the user 120 V to ground.

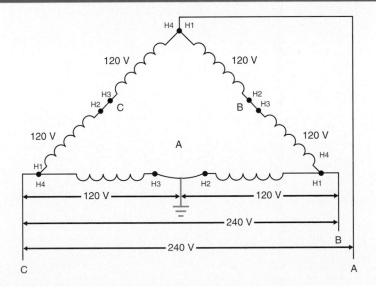

Figure 4-15. A delta connection with a high leg to ground provides 240 V, 3φ power to equipment and 120 V power for lighting and receptacles.

After assembly, the high-voltage leads are marked with the letter H and the low-voltage leads are marked with the letter X.

Three-Phase Transformer Polarity

As a rule, all phases of a 3-phase transformer have the same relative polarity (expressed in terms of single-phase polarity). If the polarity of one phase is subtractive, the polarity of the other two phases is subtractive. If the polarity of one phase is additive, the polarity of the other two phases is additive. The 3-phase polarity is dependent on the interphase connections of the respective outlet leads to the full-phase windings as well as on the polarity of the separate phases.

The three high-voltage leads and the three low-voltage leads which connect to the full-phase windings are marked H1, H2, H3, and X1, X2, and X3, respectively. The markings are applied so that, with the phase sequence of voltage on the high-voltage side in the time order H1, H2, H3, the time order on the low-voltage side is X1, X2, X3. **See Figure 4-16.**

Three-Phase Transformer Polarity

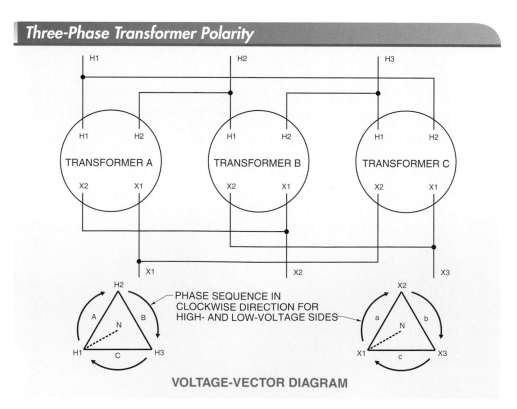

VOLTAGE-VECTOR DIAGRAM

Figure 4-16. *The phase sequence on the secondary of a three-phase transformer is the same as the phase sequence on the primary.*

In the voltage vector diagram of three single-phase transformers connected in a delta-to-delta 3-phase bank, the phase rotation (phase sequence) is in a clockwise direction for both the high-voltage and low-voltage sides. The phase rotation is only relative. The actual phase rotation is dependent on and equal to the phase rotation of the supply voltage.

If a 3-phase motor is first connected across the three leads of the high-voltage side and then transferred directly to similarly numbered leads of the low-voltage side—that is, the motor's leads are transferred from H1 to X1, H2 to X2, and H3 to X3—the motor continues to rotate in the same direction.

The angular-phase displacement between the high- and low-voltage windings as defined by ANSI is the angle between the lines H1-N and X1-N, where N is the neutral point of the voltage-vector diagram. In this circuit, the angle for the connection is 0°. The angular-phase displacement of the 3-phase transformer (or three single-phase transformers) varies

based on the four combinations in which the transformer(s) may be connected. The angular-phase displacement depends on the combination used.

Transformers connected in a delta-to-delta or wye-to-wye connection may have an angular-phase displacement of 0° or 180°. Three single-phase transformers of additive polarity may be connected in a delta-to-delta connection with 180° angular-phase displacement. The angular-phase displacement of 180° is obtained by reversing the delta connection on the low-voltage side.

Wye-to-delta connected transformers, with properly marked leads, always have an angular-phase displacement of 30°. For this reason, all 3-phase to 3-phase connections can be grouped in one of five different groups having an angular-phase displacement of 0°, 30°, or 180°. **See Figure 4-17.** The vector diagram of 3-phase transformers for 3-phase to 3-phase operation is normally included by the manufacturer in the markings on the nameplate or diagram of connections which forms part of the transformer.

Individual transformers can be used to transform the three separate phases.

Transformer Connections . . .

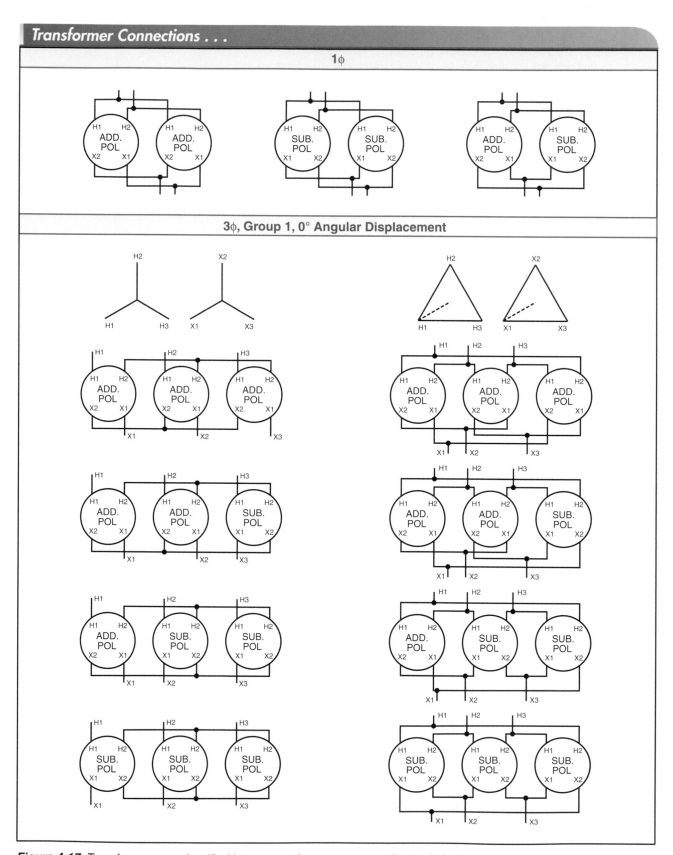

Figure 4-17. Transformers are classified into connection groups according to their phase shift.

. . . *Transformer Connections* . . .

3ϕ, Group 2, 180° Angular Displacement

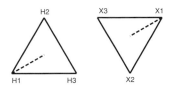 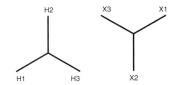

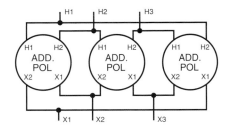 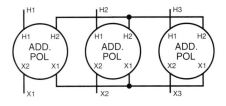

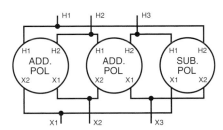 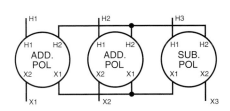

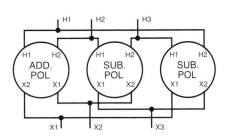 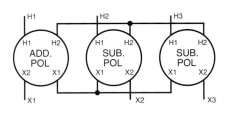

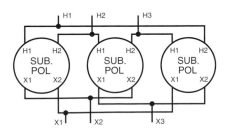 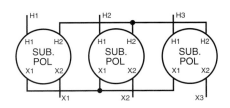

. . . Transformer Connections . . .

3φ, Group 3, 30° Angular Displacement

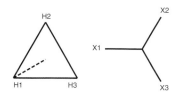
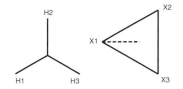
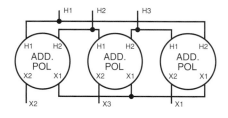
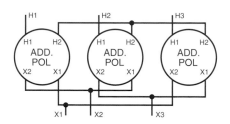
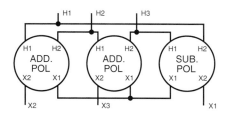
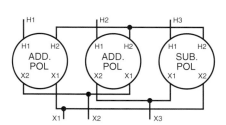
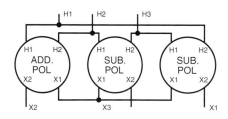
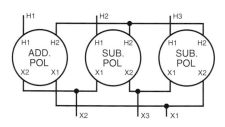
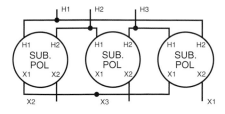
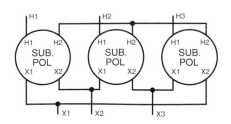

3φ to 6φ, Group 4, 0° Angular Displacement

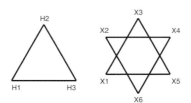

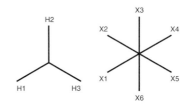

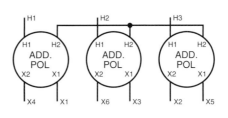

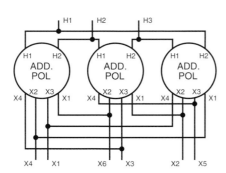

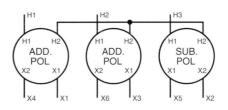

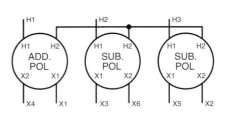

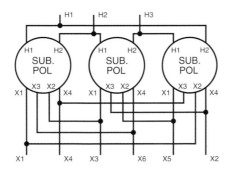

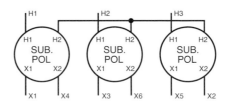

. . . *Transformer Connections*

3φ to 6φ, Group 5, 30° Angular Displacement

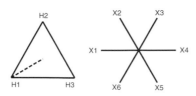

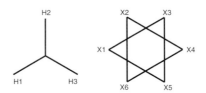

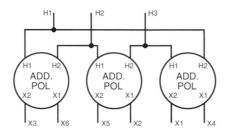

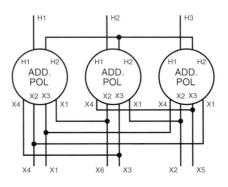

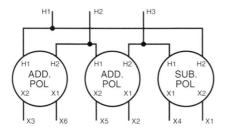

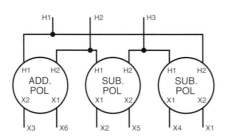

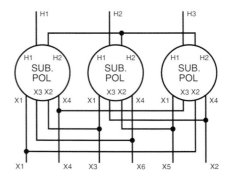

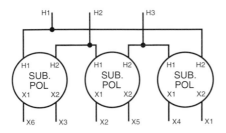

SUMMARY

- The low-voltage winding of a standard distribution transformer connected to a single-phase circuit is normally made with two equal coils that may be connected in series or parallel.

- The polarity of a single-phase transformer can be changed by interchanging the position of the two leads of any one winding.

- A delta connection has three coils connected end-to-end.

- Delta connections offer the ability to remove one unit for repair and keep the system up and running while waiting for a replacement.

- A disadvantage of a delta connection is that the coil is rated for the line voltage and thus makes the high voltage required for transmission difficult.

- Wye connections offer lower voltage-to-ground combined with the higher phase-to-phase voltage for long-distance power distribution.

- The 30° difference in phase displacement makes it impossible to parallel wye and delta transformers on the same feeder.

- The ability to take single-phase loads from line-to-neutral in a wye connection makes it difficult to maintain good balance from phase-to-phase.

DEFINITIONS

- A *delta connection* is a 3-phase connection method that has the wires from the ends of each coil connected end-to-end to form a closed loop.

- A *wye connection* is a 3-phase connection that has one end of each coil connected together and the other end of each coil left open for external connections.

- A *tertiary winding* is a third winding that is often used in power transformers to provide station power requirements or a tie with synchronous condensers.

REVIEW QUESTIONS

1. Explain how to obtain 120 V service from 240 V using a transformer with a dual-wound secondary.

2. Describe how the polarity of an unknown transformer can be determined by applying an AC voltage.

3. Explain how to calculate the amount of current through the coils of a delta-connected system when the line current is known.

4. Explain why the three ends of the coils in a wye-connected system can be safely connected.

5. Explain how to calculate the amount of current through the coils of a wye-connected system when the line current is known.

6. Describe how to obtain a "grounded B phase" system.

CHAPTER

5

Harmonics

Harmonics are a problem that has increased over the last 30 years. Harmonics cause transformers to fail, motors to burn out, circuit breakers to trip (nuisance tripping), and neutral conductors and other parts of a power distribution system to overheat. Severe overheating leads to electrical fires.

Harmonics are caused by nonlinear loads that draw current in pulses, resulting in a distorted waveform. Harmonics are a source of power quality problems that lead to overheating of circuit components. Harmonics mitigating transformers are used to reduce harmonics in power distribution systems.

 TECH FACT

High-frequency harmonics can shorten the operating life or cause failure of electrical equipment.

INTRODUCTION TO HARMONICS

A *harmonic* is voltage or current at a frequency that is an integer (whole number) multiple (2nd, 3rd, 4th, etc.) of the fundamental frequency. For example, when the power supply is 60 Hz AC, the first harmonic (60 Hz) is the fundamental frequency. Other multiples of the fundamental harmonic are the second harmonic (120 Hz), third harmonic (180 Hz), fourth harmonic (240 Hz), etc. When these harmonics are present in a circuit, the resulting waveform consists of the sum of the fundamental and the higher harmonics at every instant. **See Figure 5-1.** The result is a distorted waveform from the contribution of the harmonics.

The basic design of most electrical distribution equipment assumes that the current will be drawn in phase with the voltage in a sinusoidal waveform. Harmonics are produced when the current is drawn in pulses for only a portion of the cycle. When

 **DEFINITION**

*A **harmonic** is voltage or current at a frequency that is an integer (whole number) multiple (2nd, 3rd, 4th, etc.) of the fundamental frequency.*

current is drawn in pulses, the current waveform does not match the shape of the applied voltage waveform. This is called nonlinearity and results in harmonics.

Knowledge of harmonics present on a power line is important for working on any power distribution system. When evaluating power quality, the incoming power, types (linear and nonlinear) and number of loads, and equipment used in the distribution system must all be tested. A power quality meter can be used to measure the amount of voltage and current harmonics on a line. The amount of each harmonic present on the line and related information are indicated by data and the frequency spectrum on the graphic display. **See Figure 5-2.**

Harmonics Classification

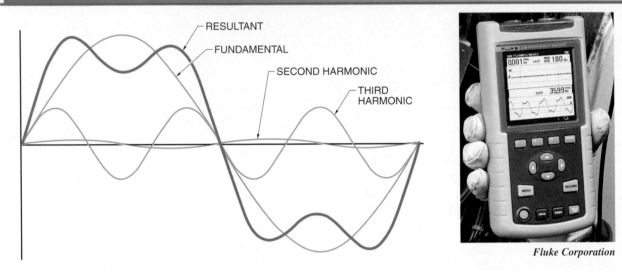

Fluke Corporation

Figure 5-1. *Harmonics are multiples of the fundamental waveform.*

Power Quality Meters

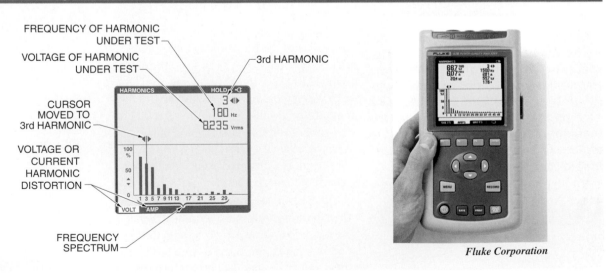

Fluke Corporation

Figure 5-2. *A power quality meter can be used to indicate the presence and magnitude of harmonics.*

Odd- and Even-Numbered Harmonics

Odd-numbered harmonics are odd-number multiples (3rd, 5th, 7th, etc.) of the fundamental. Odd-numbered harmonics add together and increase their effect. Loads that draw odd-numbered harmonics have increased resistance (I^2R) losses and eddy current losses in transformers. If the harmonics are significant, a transformer must be derated to prevent overheating.

Even-numbered harmonics are even-number multiples (2nd, 4th, 6th, etc.) of the fundamental. Even-numbered harmonics are generally fairly small because they tend to cancel each other. If even-numbered harmonics are present, this fact may be used as a troubleshooting tool. Even-numbered harmonics usually indicate that a DC current may be present in the transformer secondary winding. The DC offset is typically caused by half-wave rectification due to a failed rectifier.

A DC offset can cause a transformer to become saturated on alternate half-cycles and draw extremely high currents that can burn out the primary. These problems may be noticeable as a strong vibration and a very loud noise coming from the transformer core. Generally, a DC offset of more than 1% of the rated current can cause problems.

Triplen Harmonics

Triplen harmonics (triplens) are odd multiples of the third harmonic (3rd, 9th, 15th, etc.). Only single-phase loads generate triplen harmonics. Three-phase loads do not generate triplen harmonics. Triplen harmonics can cause problems such as over-loading of neutral conductors, telephone interference, and transformer overheating. Special types of transformers are used to reduce triplen harmonics.

Single-phase electronic loads connected phase-to-neutral, such as 120 V office circuits or 277 V lighting circuits, generate third harmonics with decreasing amounts of the higher odd-numbered harmonics.

Three-phase electronic loads connected phase-to-phase, such as 208 V power supplies or 480 V variable-speed motor drives, do not generate the triplen harmonics, but they do generate significant levels of the other higher-level harmonics. **See Figure 5-3.**

Third Harmonic. Single-phase electronic loads generate third harmonics in addition to smaller amounts of higher odd-numbered harmonics. Of those harmonics, only the triplen harmonics contribute to the problem of high neutral currents. Because of their lower current levels and higher frequencies, the 9th, 15th, and higher triplen harmonics generally distort the neutral current only slightly and often do not have a significant effect on actual rms neutral current.

Since the higher harmonics are relatively smaller, the third harmonic, as a percentage of total rms current, multiplied by three is a fairly good estimate of the percent neutral current that results from three identical non-linear single-phase loads. Thus, the neutral current is at about 100% of the fundamental phase current when the third harmonic is at 33⅓% of the fundamental phase current.

Harmonic Sequence

Harmonic sequence is the phasor rotation with respect to the fundamental (60 Hz) frequency. *Phasor rotation* is the order in which waveforms from each phase (A, B, and C) cross zero. Phasor rotation is simplified by using lines and arrows instead of waveforms to show phase relationships. **See Figure 5-4.** Phase sequence of a harmonic is important because it determines the effect the harmonic has on the operation of loads and components such as conductors within a power distribution system.

Positive Sequence. Positive sequence harmonics (1st, 4th, 7th, etc.) have the same phase sequence as the fundamental harmonic. Positive sequence harmonics cause additional heat in transformers, conductors, circuit breakers, and panels in a power distribution system. A positive sequence harmonic rotates in the same direction as the fundamental in an induction motor.

DEFINITION

Harmonic sequence is the phasor rotation with respect to the fundamental (60 Hz) frequency.

Phasor rotation is the order in which waveforms from each phase (A, B, and C) cross zero.

Source of Triplens

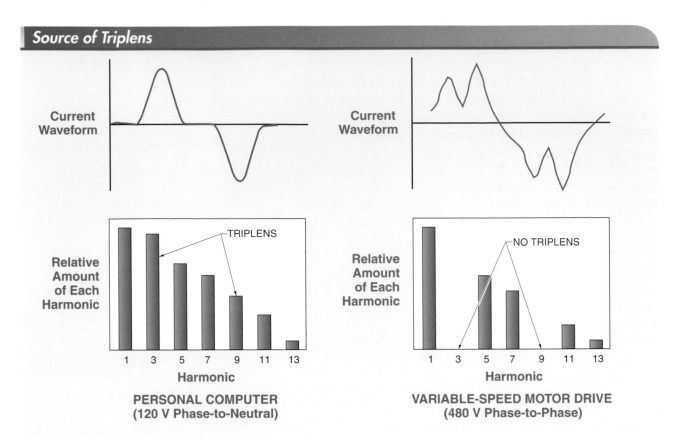

Figure 5-3. *Triplen harmonics are generated by circuits wired phase-to-neutral.*

Harmonic Sequence

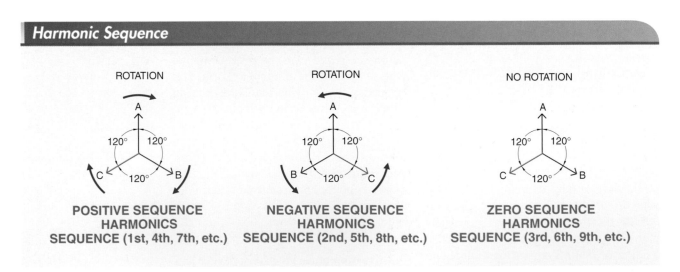

Figure 5-4. *Harmonic sequence is the phasor rotation with respect to the fundamental.*

Negative Sequence. Negative sequence harmonics (2nd, 5th, 8th, etc.) have the opposite phase sequence to the fundamental harmonic. Like positive sequence harmonics, negative sequence harmonics cause additional heat in power distribution system components such as transformers, conductors, circuit breakers, and panels. A negative sequence harmonic rotates in the opposite direction from the fundamental in an induction motor. The reverse rotation is not enough to cause the motor to reverse direction, but it does reduce the forward torque of the motor. The reduced torque causes a higher motor current to be drawn and results in excessive heating.

Zero Sequence. Zero sequence harmonics (3rd, 6th, 9th, etc.) do not produce a rotating field in either direction. However, zero sequence harmonics do cause component and system heating. Zero sequence harmonics do not cancel, but can add together in the neutral conductor of 3-phase, 4-wire systems. **See Figure 5-5.** Single-phase devices with rectifier power supplies, such as fluorescent lighting with electronic ballasts, computers, copiers, and other similar electronic devices, are significant contributors to current on neutral wires.

Total Harmonic Distortion

Total harmonic distortion (THD) is the amount of harmonics on a line compared to the fundamental frequency of 60 Hz. The THD considers all of the harmonic frequencies on a line. The THD of a pure sine waveform with no higher harmonics, such as the ideal voltage supply, is 0%. A value of THD greater than zero means the sine waveform has become distorted. THD is often given as a percentage, such as 5% or 50%. THD can be measured for current and voltage.

Current THD is caused by nonlinear loads that draw pulses of current. Voltage THD is caused by the harmonic currents on the line. The current flowing through a transformer causes a voltage drop across the coil. When current flows in pulses, the voltage will also be in pulses. High voltage distortion is a problem because voltage distortion becomes a carrier of harmonics to linear loads such as motors. Voltage harmonics cause problems (extra heat) in the power distribution system and to the loads connected to the system.

Measuring THD. When troubleshooting a circuit for harmonics, the voltage THD and the current THD should be measured. For best results, the voltage THD should not exceed 5% and the current THD should not exceed 20% of the fundamental frequency. For an accurate measurement of the THD in a system, the THD should be measured at the transformer instead of at the harmonic-generating loads. **See Figure 5-6.** Measuring THD at the loads provides the highest THD reading because THD cancellation has not occurred along the system.

When THD current is measured during full load, the THD is approximately equal to the total demand distortion (TDD). *Total demand distortion (TDD)* is the ratio of the current harmonics to the maximum load current. A THD measurement is taken when testing or troubleshooting a system. The TDD is different from the THD because TDD is referenced to the maximum current measurement taken over time. The THD is a measurement of current on a power line only at the specific time of the measurement. The purpose of the TDD measurement is to account for situations where the THD is relatively high, but the total load is fairly low. In this type of situation, the TDD is relatively low and overheating is minimized.

DEFINITION

Total harmonic distortion (THD) is the amount of harmonics on a line compared to the fundamental frequency of 60 Hz.

Total demand distortion (TDD) is the ratio of the current harmonics to the maximum load current.

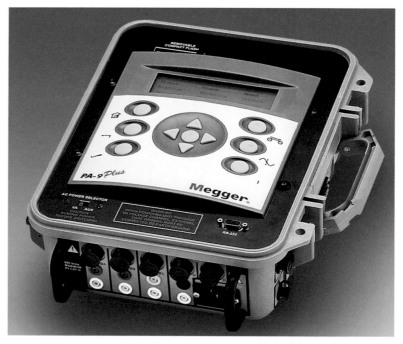

Megger Group Limited
Power quality analyzers are used to measure total harmonic distortion.

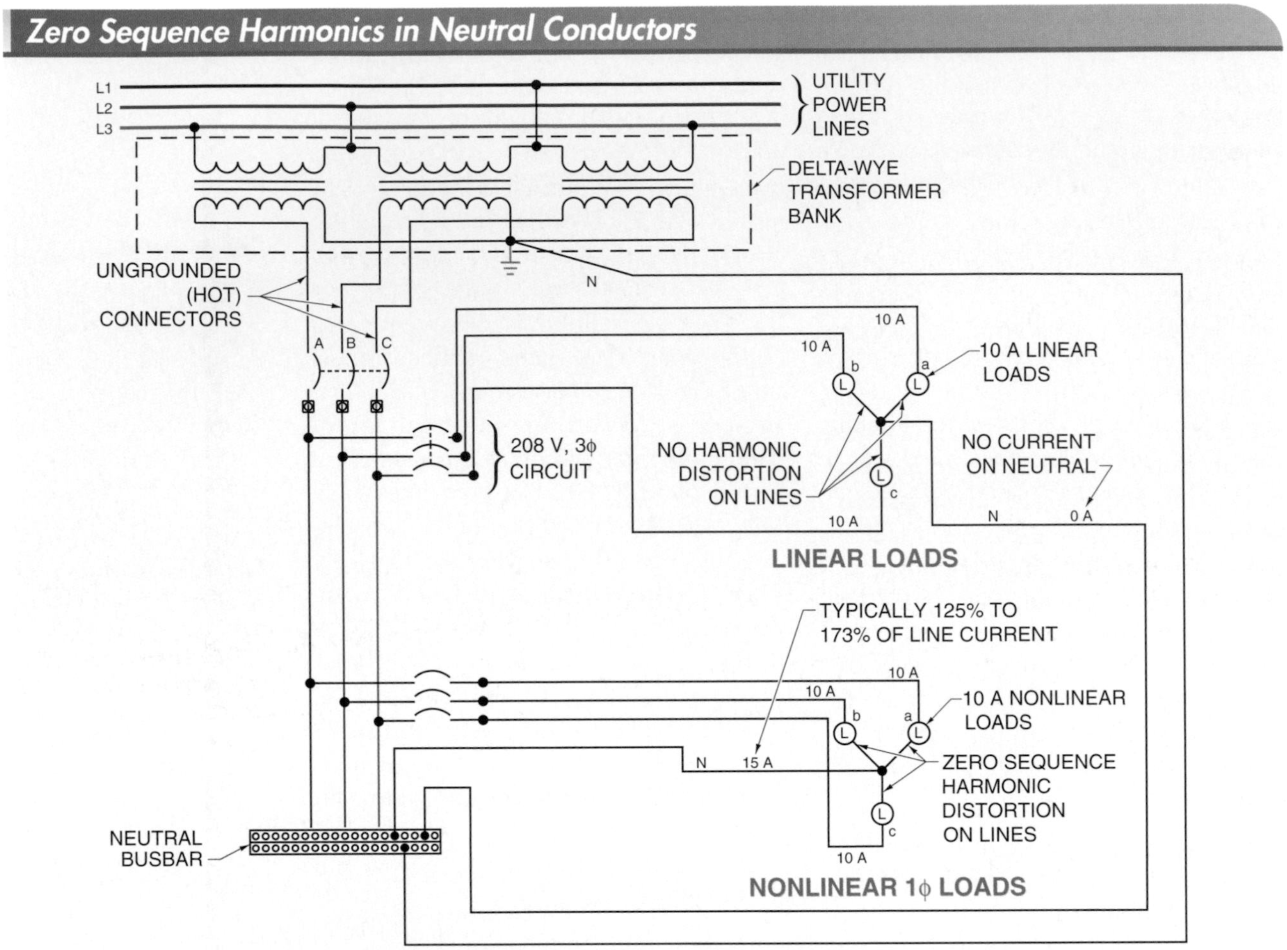

Zero Sequence Harmonics in Neutral Conductors

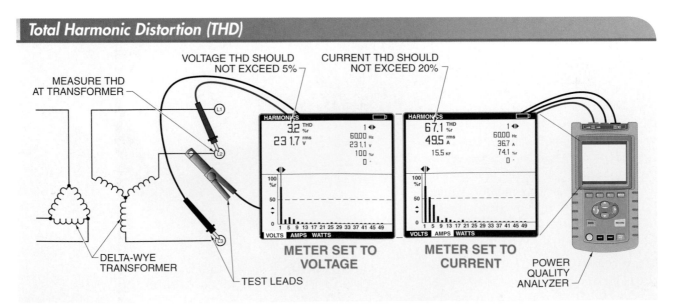

Figure 5-5. *Zero sequence harmonics add together in the neutral conductor.*

Total Harmonic Distortion (THD)

Figure 5-6. *Total harmonic distortion (THD) should be measured at the transformer, not at the load.*

Power Factor. *Power factor* is the ratio of true power to apparent power in a circuit or distribution system. Any AC circuit consists of real, reactive, harmonic, and apparent (total) power. *True power* is the power, in W or kW, used by motors, lights, and other devices to produce useful work. *Reactive power* is the power, in VAR or kVAR, stored and released by inductors and capacitors. Reactive power shows up as a phase displacement between the current and voltage waveforms. *Harmonic power* is power, in VA or kVA, lost to harmonic distortion. *Apparent power* is the power, in VA or kVA, that is the sum of true power, reactive power, and harmonic power.

Displacement power factor is the ratio of true power to apparent power due to the phase displacement between the current and voltage. **See Figure 5-7.** Capacitors can usually be added to a circuit or distribution system to correct displacement power factor. Displacement power factor is calculated as follows:

$$PF = \cos(\theta)$$
where
PF = displacement power factor
θ = phase displacement (in degrees)
Note: The term *DPF* or PF_D is sometimes used instead of *PF* to describe displacement power factor.

The presence of harmonics complicates the discussion of power factor. *Distortion power factor* is the ratio of true power to apparent power due to THD. Capacitors cannot be added to a circuit or distribution system to compensate for distortion power factor. The impedance of capacitors decreases with increasing frequency. Therefore, a capacitor can become a sink for high-frequency harmonics. Special types of transformers or tuned harmonic filters consisting of capacitors and inductors are used to correct distortion power factor. The distortion power factor is calculated as follows:

$$PF_{THD} = \sqrt{\frac{1}{1+(THD)^2}}$$

where
PF_{THD} = distortion power factor
THD = total harmonic distortion

The total power factor is the product of the displacement power factor and the distortion power factor and is calculated as follows:

$$PF_{Tot} = PF \times PF_{THD}$$
where
PF_{Tot} = total power factor
PF = displacement power factor
PF_{THD} = distortion power factor

For example, what is the total power factor when the displacement between voltage and current is 25° and the THD is 49% (0.49)? The displacement power factor is calculated as follows:

$$PF = \cos(\theta)$$
$$PF = \cos(25°)$$
$$PF = \textbf{0.906}$$

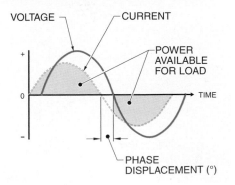

Figure 5-7. *The displacement power factor accounts for the amount of power that is actually available for a load.*

 TECH FACT

Sags and swells may cause reset and loss of data in computer systems.

 **DEFINITION**

Power factor is the ratio of true power to apparent power in a circuit or distribution system.

True power is the power, in W or kW, used by motors, lights, and other devices to produce useful work.

Reactive power is the power, in VAR or kVAR, stored and released by inductors and capacitors.

Harmonic power is power, in VA or kVA, lost to harmonic distortion.

Apparent power is the power, in VA or kVA, that is the sum of true power, reactive power, and harmonic power.

Displacement power factor is the ratio of true power to apparent power due to the phase displacement between the current and voltage.

Distortion power factor is the ratio of true power to apparent power due to THD.

The distortion power factor is calculated as follows:

$$PF_{THD} = \sqrt{\frac{1}{1+(THD)^2}}$$

$$PF_{THD} = \sqrt{\frac{1}{1+(0.49)^2}}$$

$$PF_{THD} = \sqrt{\frac{1}{1+(0.2401)}}$$

$$PF_{THD} = \sqrt{\frac{1}{1.2401}}$$

$$PF_{THD} = \sqrt{0.8064}$$

$$PF_{THD} = \mathbf{0.898}$$

The total power factor is calculated as follows:

$$PF_{Tot} = PF \times PF_{THD}$$
$$PF_{Tot} = 0.906 \times 0.898$$
$$PF_{Tot} = \mathbf{0.814}$$

It is important to know the total power factor because it relates to apparent power. Apparent power is used to size the elements of a power distribution system. **See Figure 5-8.**

Current Crest Factor

The *current crest factor* is the peak value of a waveform divided by the rms value of the waveform. The purpose of a current crest factor is to give an idea of how much distortion is occurring in a waveform. The current crest factor is calculated as follows:

$$CCF = \frac{I_{peak}}{I_{rms}}$$

where
CCF = current crest factor
I_{peak} = peak value (in A)
I_{rms} = root mean square value (in A)

DEFINITION

*The **current crest factor** is the peak value of a waveform divided by the rms value of the waveform.*

For example, what is the current crest value of a perfect sine waveform? In a perfect sine waveform with a peak value of 1, the rms value is 0.707.

$$CCF = \frac{I_{peak}}{I_{rms}}$$

$$CCF = \frac{1}{0.707}$$

$$CCF = \mathbf{1.414}$$

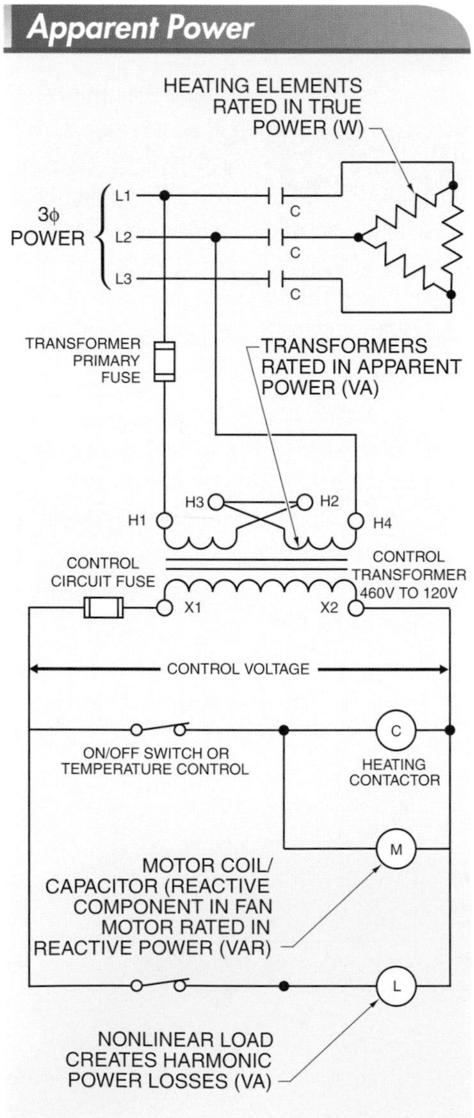

Apparent Power

Figure 5-8. Transformers and other power distribution equipment are sized based on the apparent power in a circuit.

A high current crest factor can cause overheating of circuits and devices. A typical distorted current waveform on a 120 V circuit supplying digital devices like computers may have a current crest factor of about 2 to 6. **See Figure 5-9.** In general, a circuit with a higher current crest factor has more energy contained in the higher harmonics. A power source must be able to supply the maximum power required by the circuit at the required voltage and current. A typical backup power system, such as a computer uninterruptible power source, has the capability of supplying a current crest factor of 3 at full load, but can supply higher crest factors at lower loads.

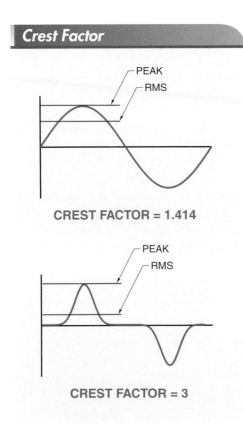

Crest Factor

PEAK
RMS

CREST FACTOR = 1.414

PEAK
RMS

CREST FACTOR = 3

Figure 5-9. The current crest factor is the ratio of the peak current to the rms current.

Source Impedance. Source impedance has an effect on the crest factor created by a nonlinear load. Once the voltage rises to a predetermined point, the power supply starts charging a smoothing capacitor.

With low source impedance, the current drawn by the capacitor is high and the duration of the charging cycle is short. Higher impedance does not allow as much current to be drawn and extends the time it takes to charge the capacitor. The extended charge time has the effect of reducing the crest factor. The source impedance can be increased by adding line reactors or drive isolation transformers.

SOURCES OF HARMONICS

Harmonic distortion is created by electronic circuits that draw current in short pulses, such as variable-speed motor drives, personal computers, printers, electronic ballasts used in lighting applications, and many types of medical test equipment. Harmonics are especially a problem wherever there are a large number of nonlinear single-phase loads. Equipment efficiency is improved when electronic circuits draw current in short pulses, but this causes harmonic distortion on the power lines.

Nonlinear Loads

There are two very common types of non-linear loads that cause harmonics. The first type of nonlinear loads includes computers, copiers, and other similar electronic 120 V devices that use switched-mode power supplies. These loads are major contributors to neutral current in 208Y/120 V systems. In addition, variable-speed motor drives use similar power supplies and operate at many voltages. The second type of single-phase nonlinear loads includes 277 V electronic lighting ballasts, which predominate in 480Y/277 V distribution systems.

Switched-Mode Power Supplies. A *switched-mode power supply (SMPS)* is a power supply for electronic devices that includes an internal control circuit that quickly switches the load current ON and OFF in order to deliver a stable output voltage. A switched-mode power supply may also be known as a switching power supply. Switched-mode power supplies are

DEFINITION

*A **switched-mode power supply (SMPS)** is a power supply for electronic devices that includes an internal control circuit that quickly switches the load current ON and OFF in order to deliver a stable output voltage.*

commonly used in electronic equipment because they are inherently more efficient than older, linear power supply designs. However, an SMPS is much more complex and harder to design correctly.

The first task of the SMPS is to rectify the AC input. A typical single-phase SMPS is designed to use a full-wave bridge rectifier. The rectified DC voltage charges a smoothing capacitor that is used to even out the voltage. **See Figure 5-10.** The diodes are forward biased only when the voltage exceeds a certain minimum value. Therefore, current only flows during the part of the cycle where the voltage exceeds the minimum value. The capacitor also draws current only at the voltage peak because it does not fully discharge between the ripples of the rectified DC power. During the rest of the sine wave, the capacitor draws no current. This results in a circuit that draws current only during the peak of each half cycle of the AC sine wave.

Harmonics can also cause "flat-topping" of the voltage waveform, lowering circuit peak voltage. Flat-topping occurs when many loads draw current in pulses at the maximum voltage level and overload the power source. When flat-topping occurs, the capacitor is not fully charged because of the lowered voltage. In severe cases

of flat-topping or when flat-topping and voltage sags occur, computers or other electronic equipment can continually reset because of insufficient peak voltage to keep the capacitor charged.

Because the current waveform is significantly distorted, an SMPS circuit has a low power factor. The simplest types of SMPS have a power factor of about 0.6. A more sophisticated SMPS has circuits that force the input current to follow the sine waveform of the voltage input. For DC inputs, the rectification step is not needed.

The next step in SMPS operation is the inverter stage. **See Figure 5-11.** The inverter changes the DC to AC by switching it ON and OFF at a high frequency. The frequency is usually selected to be above the audible range to minimize noise. The inverted AC can be isolated from the source by being sent through a relatively small high-frequency step-down transformer. If DC output is required, the transformer output is then rectified and filtered again. In other power supply designs, the step-down transformer may be ahead of the rectifier section. In either case, these loads are characterized as nonlinear because the input current is significantly distorted as compared to the ideal current waveforms.

Switched-Mode Power Supplies

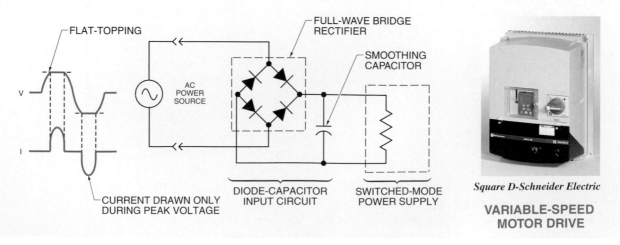

Figure 5-10. Switched-mode power supplies cause harmonics when the current is drawn only during peak voltage.

SMPS Inverters

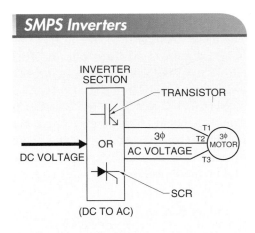

Figure 5-11. The SMPS inverter switches the DC current at the frequency required for the application.

Lighting Ballasts. Lighting can account for up to 40% of energy use in many commercial buildings. This makes lighting a common target for energy efficiency initiatives. Energy-efficient fluorescent or high-intensity discharge (HID) lighting is frequently used to replace incandescent lighting.

Fluorescent and HID lamps require a ballast. A *ballast* is a controller responsible for providing the initial startup voltage and maintaining a constant current through a lamp. Older magnetic ballasts use power at the line frequency of 60 Hz. Modern electronic ballasts use diodes to convert the 60 Hz power supply to DC. The DC is then converted to high-frequency AC for use in the lamp. The frequency in electronic ballasts can be up to about 60 kHz. Ballasts can create harmonics in the power lines. In addition, ballasts can create a phase shift between current and voltage, resulting in a low power factor.

The electronic ballast industry has universally adopted standards that establish maximum allowable current THD. The standards are outlined in ANSI Standard C82.11. This ANSI standard put a limit of 32% current THD on lighting ballasts.

See Figure 5-12. The ANSI specification is quite comprehensive in that it puts limits on specific low-order harmonics (2nd and 3rd), high-order harmonics (greater than 11th), triplen harmonics, and THD. Ballast manufacturers have implemented passive filtering to reduce harmonics, and many modern ballasts operate at 20% THD or less.

Ballast Current THD Limits

HARMONIC	MAXIMUM CURRENT
Fundamental	100%
2nd	5%
3rd	30%
>11th	7%
Triplens	30%
THD	32%

Advance Transformer Co.

Figure 5-12. ANSI specifications put limits on the allowable THD in lighting ballasts.

NEUTRAL CURRENTS

Three-phase nonlinear loads connected phase-to-phase, such as electric power converters, variable-speed motor drives, silicon controlled rectifier (SCR) controllers, and other similar devices, create harmonics but do not contribute to neutral currents. Neutral currents are not a concern for these circuits since there is no neutral wire.

Neutral currents are a concern for circuits connected phase-to-neutral, such as 120 V commercial and residential circuits. Single-phase devices with rectifier power supplies, such as fluorescent lighting with electronic ballasts, computers, copiers, and other similar electronic devices, are significant sources of harmonics. When these types of nonlinear single-phase loads are connected line-to-neutral in a 3-phase wye-connected power system, the neutral conductors in the 3-phase feeders can carry significant

> **DEFINITION**
>
> *A **ballast** is a controller responsible for providing the initial startup voltage and maintaining a constant current through a lamp.*

levels of current, even when the loads are balanced on the three phases.

The amount of neutral current is related to the amount of THD. This is true whether the loads are balanced or not. The actual neutral current waveforms have third harmonic components, but pure third harmonic sine waveforms are not flowing in the lines. Building electrical systems are typically designed so that the single-phase loads in a facility draw from different phases of the three-phase supply. This allows the load currents to be balanced on the three phases. Because of the switched-mode power supplies, the current is drawn in pulses. The pulses add up and return on the neutral as pulses, not a sine waveform.

TECH FACT

The THD at a typical electrical wall outlet in the United States is about 3%.

Balanced Load

The ideal situation is for the loads to be balanced across all three phases. This allows the same current to flow through each phase conductor. The loads can be considered balanced even if the loads are slightly unbalanced because the effect of a small imbalance is negligible. The amount of neutral current varies depending on the load on the phases.

Low Load. At low load, such as when supplying individual nonlinear loads, the current pulses typically are narrow and do not overlap on the neutral. **See Figure 5-13.** This means that no more than one phase of the three-phase system carries current at any instant of time. The currents no longer return on the phase lines as expected in a three-phase system and the only return path for current is the neutral conductor. As a result, the number of current pulses accumulated on the neutral is three times that in the individual phases.

Low-Load Neutral Current

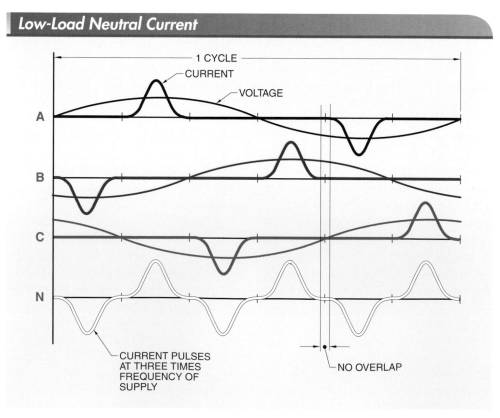

Figure 5-13. The maximum low-load neutral current is 173% of line current.

The rms current increase, from one to three current pulses in a common time interval, is 173%. This means that the maximum current on the neutral when supplying single-phase loads is 173% of the phase current. This does not mean that the neutral is carrying 173% of its capacity.

High Load. Every switched-mode power source is slightly different from every other power source, even when they are the same model. This is due to slight differences in the diodes, resistors, and other components in the power supply. When multiple loads are connected to the same phase, the minor differences in the power sources cause the timing and duration of the current pulses to vary slightly. As the pulses are combined on the neutral wire, the minor variations in the pulses add up to wider pulses than from the single individual loads. For example, an increasing number of computers on a circuit results in wider peaks. **See Figure 5-14.**

 TECH FACT

Power quality meters for 3-phase power circuits should be used when checking 3-phase transformer connections.

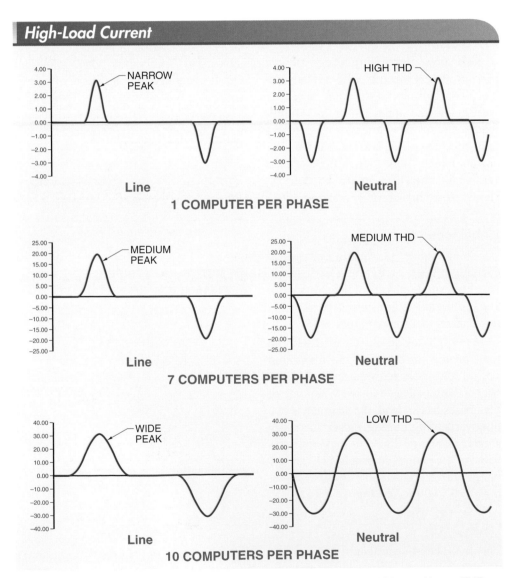

Figure 5-14. An increasing number of loads results in wider pulse widths and lower THD.

In many typical systems, the pulses start overlapping when about 7 unfiltered loads are placed on each phase. When the pulses overlap, more than one phase is conducting at a time and some current is being returned on the phase lines. As the pulse widths increase, the waveform becomes more and more like a pure sine waveform. The neutral current maximum is 173% of phase current, and it drops off for highly loaded circuits. This also means that the THD decreases and the neutral current decreases as a percent of phase current. The THD in typical computer load currents ranges from about 40% THD on systems that have many computers to about 150% THD on systems that have few computers on a circuit.

This shows that lightly loaded circuits can have a high neutral current relative to the phase current but the total current is well below the ampacity limits. Highly loaded circuits have lower neutral current relative to the phase current, but the total current may be high enough to be of concern. In this case, the neutral conductor size should be evaluated for its total capacity.

Unbalanced Load

For linear loads, the maximum neutral current is 100%, regardless of balance. However, single-phase nonlinear loads can create elevated neutral currents in severely unbalanced loads. **See Figure 5-15.**

If two phases are at full load with no load on the third phase, the maximum neutral current can be evaluated in the same way as with balanced loads. When one of the phases has no load, the neutral has two nonoverlapping current pulses for every single pulse on the line. The maximum rms value of the neutral current is 141% of the phase current. This value occurs only at high THD values. **See Figure 5-16.**

Neutral Sizing

The presence of neutral current is a major problem because neutral wires were traditionally designed to be the same ampacity

as phase conductors. In addition, the National Electrical Code® prohibits neutral conductor overcurrent protection. An exception to this is when the overcurrent device opens all conductors of the circuit including the neutral. Therefore, proper sizing of neutral conductors is a concern when supplying large numbers of single-phase nonlinear loads. However, it is very important to distinguish between neutrals that carry a high percentage of phase current and neutrals that are overloaded.

For linear loads, the amount of neutral current is never more than the phase current, regardless of the balance. The neutral can be sized to be the same as the phase conductors. For the typical 20% THD from lighting ballasts, the maximum neutral current for unbalanced loads is about 105% of line current. It is unlikely that this level of current would ever be seen in practice. Therefore, it is not reasonable to design fluorescent lighting circuits with oversized neutrals.

TECH FACT

In a circuit that includes only linear loads, voltage and current are both sine waveforms, even when out of phase.

Unbalanced Loads

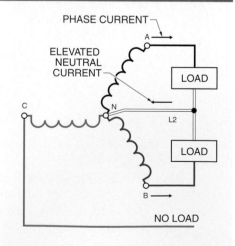

Figure 5-15. Unbalanced single-phase nonlinear loads can cause elevated neutral current.

Maximum Neutral Current

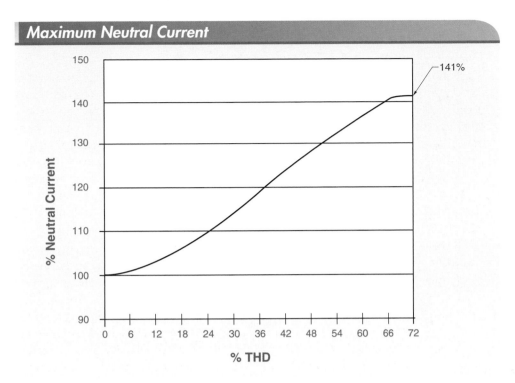

Figure 5-16. *The maximum neutral current for unbalanced single-phase nonlinear loads increases with increases in THD.*

However, the THD for computer loads can be over 40% on highly loaded circuits. For nonlinear loads with a separate neutral for each circuit, this can result in neutral currents that are from 125% to over 173% of phase current. Therefore, oversized neutrals should be investigated for 208Y/120 V systems. In some cases, several circuits are wired with a common neutral. In this case, the neutral current may be more than double the phase current. **See Figure 5-17.**

REDUCING HARMONICS PROBLEMS

There are several ways to reduce the problems of harmonics in a circuit or power distribution system. A K-rated transformer is designed to withstand the overheating problems created by harmonics. A harmonics mitigating transformer is designed to reduce problems by reducing or canceling harmonics. In addition, harmonic filters are occasionally used to reduce harmonics.

K-Rated Transformers

ANSI Standard C57.110-1986 defined a K-factor to evaluate how much harmonic current a circuit draws and to determine the heating effect of that harmonic current. Based on the circuit K-factor, transformers are manufactured with a K-rating. It is important to note that K-rated transformers do not reduce harmonics. The K-rating indicates the relative ability of a transformer to withstand the harmful effects of harmonics. K-rated transformers increase the size of the core, increase the size of the neutral conductor, and use special winding techniques to reduce eddy current and skin effect losses.

 TECH FACT

The occurrence of sags and swells may indicate a weak power distribution system. In such a system, voltage will change dramatically when a large motor or welding machine is switched on or off.

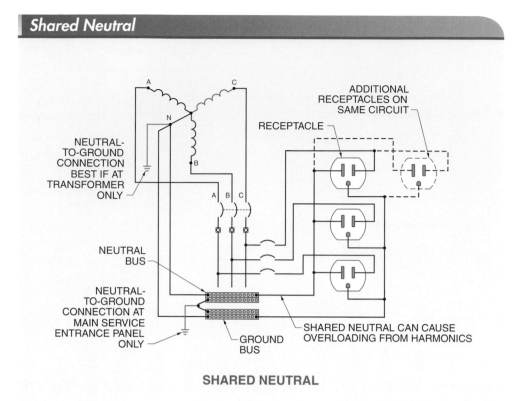

Shared Neutral

NEUTRAL-TO-GROUND CONNECTION BEST IF AT TRANSFORMER ONLY

ADDITIONAL RECEPTACLES ON SAME CIRCUIT

RECEPTACLE

NEUTRAL BUS

NEUTRAL-TO-GROUND CONNECTION AT MAIN SERVICE ENTRANCE PANEL ONLY

GROUND BUS

SHARED NEUTRAL CAN CAUSE OVERLOADING FROM HARMONICS

SHARED NEUTRAL

Figure 5-17. A shared neutral can carry very high harmonic current.

Measuring K-Factor. In any system containing harmonics, the K-factor can be measured with a power quality analyzer. **See Figure 5-18.** A K-factor of 1 indicates a linear load. A higher K-factor indicates increased heating from harmonics. For example, a circuit with a K-factor of 2 has twice the heating effect of a circuit with a K-factor of 1.

Standard K-ratings are 1, 4, 9, 13, and 20. In addition, K-ratings of 30, 40, and 50 exist, but loads with K-factors greater than 20 are relatively rare. Computer rooms typically have K-factors of 4 to 9. Areas with many single-phase computers have K-factors of 13 to 17. The K-rating of the transformer should exceed the K-factor of the circuit, or the transformer should be derated.

Circuit Load. Guidelines have been developed that recommend a K-factor based on the predominant type of load on a circuit. **See Figure 5-19.** When specifying a

transformer based on the K-factor, more is not better. A transformer with a K-factor greater than needed has its own set of problems. Typical problems include increased inrush current, higher eddy current core losses, and a larger footprint.

The K-factor of a circuit generally decreases with more loads on the circuit, just as the THD decreases with more identical loads on a circuit. Typically, by the time 10 or 20 devices are on-line simultaneously, the combined K-factor at the bus of the distribution panel is reduced by a factor of 3 or more.

TECH FACT

Doubling the cross-sectional area of a conductor reduces resistance by half. K-rated transformers use heavier gauge wires for the primary and secondary coils to reduce the resistance heating.

K-Factor

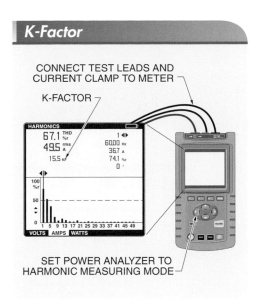

CONNECT TEST LEADS AND
CURRENT CLAMP TO METER

K-FACTOR

SET POWER ANALYZER TO
HARMONIC MEASURING MODE

Figure 5-18. K-factor can be measured with a power analyzer.

Harmonics Mitigating Transformers

A *harmonics mitigating transformer (HMT)* is a transformer designed to reduce the harmonics in a power distribution system. Some styles of HMTs are referred to as phase-shifting transformers. HMTs generally work on the principle of combining the waveforms in ways where the positive part of one waveform adds to the negative part of another waveform. This addition results in complete cancellation when the phase loads are perfectly balanced and a much smaller harmonic even when imperfectly balanced. The end result is that harmonics are prevented from propagating through the power distribution system but they remain in the secondary windings and the circuit.

An HMT works best when located close to the load. This generally means that HMTs are located at scattered locations throughout a facility. Therefore, HMTs are usually available with kVA ratings of about 100 kVA or less.

Delta-Wye Wiring. A common transformer wiring arrangement has the primary wound in a delta configuration with the secondary wound in a wye configuration. **See Figure 5-20.** Delta-wye transformers have higher impedance to the flow of harmonic currents than to the fundamental current. This reduces the current harmonics, but higher impedance in the transformer causes a relatively higher voltage drop from the harmonic currents. The high voltage drop contributes to voltage THD and flat-topping. This voltage THD can be distributed upstream throughout the power distribution system. At full load, a delta-wye transformer can produce voltage THD above that recommended by the IEEE Standard 519-1992, *IEEE Recommended Practices and Requirements for Harmonic Control in Electrical Power Systems*.

DEFINITION

A *harmonics mitigating transformer (HMT)* is a transformer designed to reduce the harmonics in a power distribution system.

K-Factor for Common Loads

Predominant Load	K-Factor
Incandescent lighting Resistance heating Motors without solid state drives Control transformers/electromagnetic control devices	K-1
UPS without input filtering Welders Solid state controls other than variable-speed drives	K-4
UPS with input filtering Multiwire receptacle in classrooms and general care areas of health care facilities	K-13
Mainframe computer loads Variable-speed motor drives Multiwire receptacle circuits in critical care areas of health care facilities	K-20
Other loads producing large amounts of harmonics	K-30, K-40, K-50

Figure 5-19. The K-factor of common loads can be estimated.

Delta-Wye Transformers

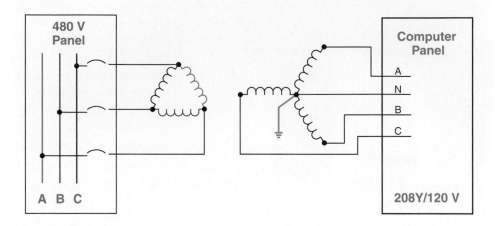

480 V Panel

A B C

Computer Panel

A
N
B
C

208Y/120 V

Figure 5-20. Delta-wye transformers are used to prevent triplen harmonics from propagating back into the power system.

The transformer impedance causes some attenuation of the harmonics and limits the crest factor allowed downstream. While the triplen harmonics are attenuated, the other higher order harmonics are not affected. However, if an HMT is being used to provide additional source impedance and/or phase shift from other nonlinear loads elsewhere in the facility, some benefit may be achieved.

Zigzag Windings. Another common HMT design is to wind the secondary of each phase in a zigzag configuration to eliminate the triplen harmonics. The zigzag is accomplished by winding half of the secondary turns of one phase of the transformer on one leg of the 3-phase transformer, with the other half of the secondary turns on an adjacent phase. **See Figure 5-21.** With all of the triplen harmonics in phase with each other, the triplen harmonic currents produce ampere-turn fluxes that cancel each other such that no currents are induced in the primary winding.

Phase-Shifting. Three-phase loads do not generate triplen harmonics. Therefore, harmonic problems in situations where

3-phase loads dominate are primarily from currents flowing at the 5th, 7th, 17th, 19th, or higher harmonics. An HMT can use dual secondary windings or pairs of transformers to reduce these harmonics. A bank of two or more transformers with a 30° phase shift between them can be used to treat these harmonics. This degree of phase shift is chosen to ensure that the harmonic components of one secondary are out of phase with those of another.

In the case where a transformer is supplying both single-phase and 3-phase loads, a combination approach is needed. Pairs of delta zigzag transformers with a 30° phase shift are often used as part of a separate transformer bank. The 30° phase shift between the transformers reduces the 5th, 7th, 17th, and 19th harmonics. The secondary zigzag windings greatly reduce the triplen harmonics.

TECH FACT

Voltage sags during startup can be recorded using a power quality meter.

Zigzag Winding

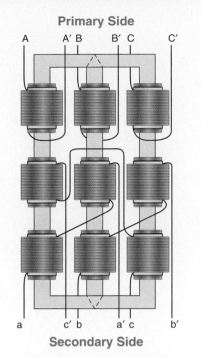

Primary Side

Secondary Side

TRANSFORMER WIRING

NOTE: INSULATION REMOVED FOR CLARITY

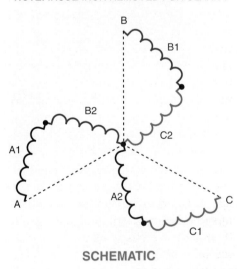

SCHEMATIC

Figure 5-21. A zigzag winding is used to cancel triplen harmonics.

For best results, it is necessary to balance the single-phase, line-to-neutral, nonlinear load between two panels that are being fed by two different HMTs. One HMT should be a delta-zigzag, which has a 0° phase shift. The second HMT should be either a delta-wye or a wye-zigzag, each of which has a 30° phase shift. Using the two transformers will help attenuate the 5th, 7th, 17th, and 19th harmonics. Also, the harmonic attenuation will be more effective when the loads are balanced.

For example, if a main power panel feeding single-phase nonlinear loads requires 200 A, it is better to use two separate panels of 100 A each. **See Figure 5-22.** Two transformers are used to feed the separate panels. One transformer is wired in a delta-zigzag configuration and the other transformer is wired in a delta-wye or a wye-zigzag configuration. The two transformers are 30° out of phase with each other. The computer loads draw current in pulses and the harmonics move back through the transformers to the main power panel. The harmonics add together so that the overall system draws current in a waveform with very low THD.

HMT Impedance. The two HMTs should have the same impedance values, be located close to the source bus, and have the same load harmonic profiles. With a zigzag secondary, the impedance is less than the transformer nameplate impedance rating. In a delta-wye or delta-delta transformer, the single-phase impedance is the same as the positive and negative sequence impedance. This is the impedance on the nameplate.

With a delta-zigzag or a wye-zigzag transformer, the phase to neutral impedance is approximately 75% to 85% of the positive and negative sequence impedance. This results in a higher fault current in the event of a single-phase fault to neutral or ground. **See Figure 5-23.** This may require an overcurrent protection device with a higher rating. The impedance value given on the nameplate of the transformer is the positive/negative sequence impedance. Therefore, it is best to assume that any fault current is about 133% of a calculated fault current. This is very important when conducting a coordination study for arc flash protection.

A **harmonic filter** is a
device used to reduce
harmonic frequencies
and THD.

DEFINITION

A **harmonic filter** is a
device used to reduce
harmonic frequencies
and THD.

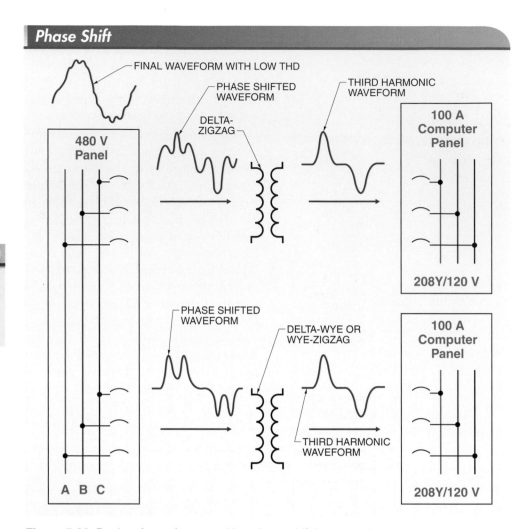

Figure 5-22. *Banks of transformers with a phase shift between them are used to cancel out harmonics.*

HMT Impedance

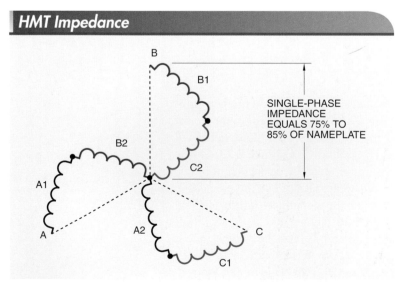

SINGLE-PHASE
IMPEDANCE
EQUALS 75% TO
85% OF NAMEPLATE

Figure 5-23. *The single-phase impedance of a zigzag transformer is about 75% to 85% of the nameplate impedance.*

Harmonic Filters

A *harmonic filter* is a device used to reduce harmonic frequencies and THD. A single-phase harmonic filter is used to reduce the harmonics from nonlinear single-phase loads by minimizing the third and other triplen harmonics. Three-phase harmonic filters, also called trap filters, are used to reduce harmonics produced by single-phase nonlinear loads connected to a 3-phase system, or 3-phase loads such as AC variable-speed motor drives connected to the system. The primary function of a 3-phase harmonic filter is to reduce the fifth and seventh harmonic currents generated by six-pulse (six-diode) converters that change AC to DC. The filter is usually tuned to just below the fifth harmonic and offers a low-impedance path that traps the fifth and most of the seventh harmonic. Harmonic filters should be installed as close as possible to the nonlinear load. With 3-phase drives, they are typically installed at the service equipment.

Harmonic filters may include different types of circuits or components designed to reduce harmonic currents, such as combinations of capacitors, inductors, and other components.

Harmonic filters are typically classified as passive harmonic filters and active harmonic filters.

Passive Harmonic Filters. A passive harmonic filter uses capacitors and inductors that are tuned to remove particular harmonic frequencies. **See Figure 5-24.** The passive harmonic filter works like a band-pass or low-pass filter in an electronic circuit. It allows low frequencies (60 Hz) to pass through unchanged while removing higher frequencies at 180 Hz and above. Passive harmonic filters can be difficult to use because they often cause other problems like ringing, unwanted resonances, and overcompensation. Single-phase harmonics sources like SMPSs generally do not generate very much phase shift between current and voltage. Therefore, a passive filter can easily cause a circuit to switch from lagging to leading. In addition, passive harmonic filters tend to be fairly large and can be somewhat expensive.

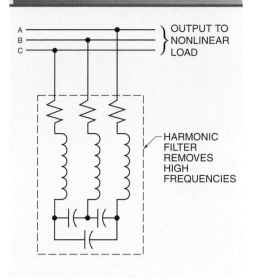

Passive Harmonic Filters

OUTPUT TO NONLINEAR LOAD

HARMONIC FILTER REMOVES HIGH FREQUENCIES

Figure 5-24. A passive harmonic filter uses a set of resistors, capacitors, and inductors tuned to remove harmonic frequencies.

Active Harmonic Filters. An active harmonic filter uses electronics to provide a variable impedance to remove harmonics from the circuit or to generate an adaptive current waveform that is 180° out of phase with the harmonics. **See Figure 5-25.** Active filters have typically been very expensive and not widely available. However, advances in electronics are making these types of devices more available and cost effective.

FOURIER ANALYSIS

When harmonics are present, distorted voltage and current waveforms are present on the lines. The distorted waveforms must be analyzed to determine the type and quantity of harmonics that are present. When harmonics are present, the higher order harmonics add together with the fundamental to produce the resultant waveform. The resultant waveform is the waveform measured by a power quality analyzer and includes all harmonics. Fourier analysis is used to analyze the waveforms. There are many types of Fourier analysis, but the simplest concept is the Fourier transform.

TECH FACT

When sizing 3-phase and neutral conductors for a nonlinear load application, the NEC® requires that the neutral be considered a current-carrying conductor. This results in 4 current-carrying conductors in a conduit or raceway. A correction factor or ampacity adjustment must be applied. It is important to note that only the balanced portion of harmonics can be cancelled in an HMT. When there are unbalanced loads, the difference between the harmonics remains on the neutral line.

Active Harmonic Filters

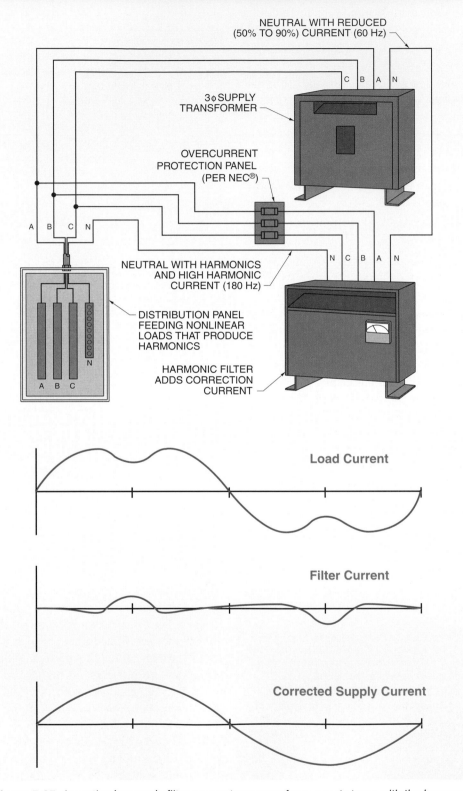

Figure 5-25. *An active harmonic filter generates a waveform out of phase with the harmonics to remove the desired harmonic frequencies.*

Fourier Transform

The resultant waveform must be analyzed with a Fourier transform to learn about the harmonics present. The *Fourier transform* is the mathematical method of converting a time-based waveform, like a sine waveform, into frequency-based information. This is equivalent to breaking down (decomposing) a periodic waveform into a series of sine waveforms that can be added together to reproduce the original waveform. **See Figure 5-26.** The Fourier transform is used to represent the frequency and magnitude of the harmonics present on the lines. The output of a power quality analyzer gives the relative magnitude of each harmonic.

Combining Waveforms

The Fourier transform is a way of decomposing a periodic waveform into its basic parts. Before the waveforms can be decomposed, the waves must have combined on the lines. This can happen at any point where two or more wires come together in a junction. This typically happens at the connection points of the windings within a transformer or at a common bus feeding two or more transformers.

When sine waveforms combine, they add together and the resultant waveform has a value equal to the sum of the individual values. If one of the waveforms is positive and the other is negative and both are at the same numerical value, they are said to cancel. If two current waveforms are exactly out of phase with each other with one equal to +10 A and the other equal to −10 A, the resulting current at that instant has a value of 0.

TECH FACT

Jean Baptiste Joseph Fourier developed the idea behind Fourier analysis. Fourier analysis takes a time-varying signal, such as a source with harmonics, and transforms it into the frequency components that make up the signal.

HARMONICS

Harmonics can show up in many unexpected ways. Sometimes it only takes the addition of one or a few new loads to cause a system to change. For example, a large school used standard AC drives for an existing air handling system for many years. Over the years, many other loads, like computers, were added to the system. Finally, variable-speed drives were added to the air handling system to counter a problem with negative pressure inside the building. After completing the installation, the disconnects were enabled and the blowers were started.

Within a very short time, the installer received complaints about the clocks running fast and the bells going off at unexpected times. The contractor bypassed the variable-speed drives and the problems ceased. A review of the school wiring system revealed that all the clocks and bells were controlled by signals through the AC system in the school. This is what kept all the clocks at the same time and all the bells going off at the same instant. The harmonic signature on the AC had tricked the clocks into running fast. Harmonics mitigating transformers were ordered and installed and the problem was fixed.

DEFINITION

*The **Fourier transform** is the mathematical method of converting a time-based waveform, like a sine waveform, into frequency-based information.*

ASI Robicon

Large arc furnaces are a significant source of power line harmonics.

Fourier Analysis

Fluke Corporation

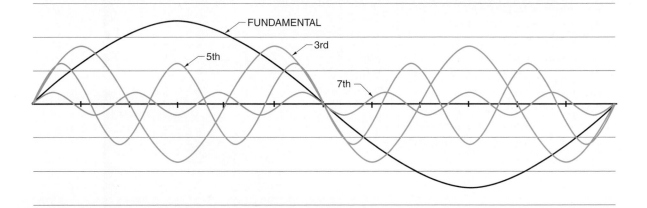

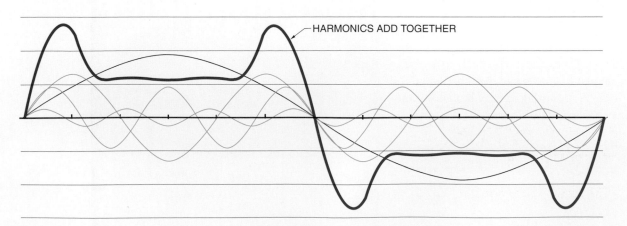

NOTE: Relative height of current waveforms equals relative height of Fourier transform peaks.

Figure 5-26. *Fourier analysis is used to decompose a distorted wave into its component harmonics.*

If two current waveforms are out of phase with each other and they do not exactly cancel, the waveforms add. When an SMPS draws current in pulses, the pulses are separated from each other. If one of these waveforms is shifted 60° relative to the other and the waveforms are added, the resultant has two peaks for every peak of the original. **See Figure 5-27.** This is the type of waveform shift and recombination that happens in transformers with a standard delta-wye winding or a wye-zigzag winding. The combined waveform is found on the line side of either a standard delta-wye or wye-zigzag transformer that feeds single-phase, line-to-neutral, nonlinear loads. This results in the cancellation of the triplen harmonics.

a result of canceling the 5th, 7th, 17th, and 19th harmonics in addition to the triplen harmonics that were canceled by the initial phase shift. This additional phase shift can be accomplished with delta-zigzag transformers in parallel with wye-zigzag or delta-wye transformers. The combined waveform will be seen at the power busbar upstream of the transformer pair.

If the load on a transformer changes, the waveforms get out of balance and do not cancel. Additional combinations of phase shifts could be designed to eliminate more harmonics, but the benefits would be very small. The transformer bank would have to be modified every time the load changes. This can happen whenever a computer is turned on or off or the load on a variable-speed motor drive changes. It is not practical for this type of transformer bank to be modified with every load change.

Triplen Harmonic Cancellation

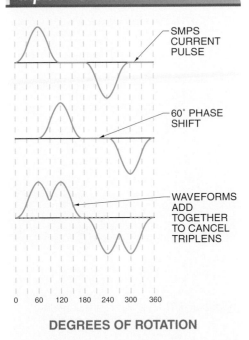

Figure 5-27. Waveforms add together when combined at a wiring junction. A 60° phase shift cancels triplen harmonics.

The combination waveform created by a delta-wye or a wye-zigzag transformer can be combined with two other waveforms with a phase shift to create a new waveform. **See Figure 5-28.** This waveform is

Higher Order Harmonic Cancellation

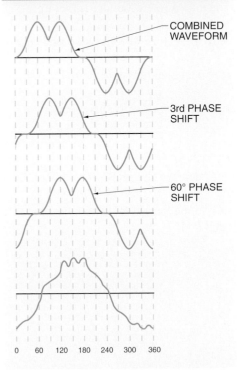

Figure 5-28. Further waveform shifts are used to cancel higher order harmonics.

SUMMARY

- A harmonic is voltage or current at a frequency that is an integer (whole number) multiple (2nd, 3rd, 4th, etc.) of the fundamental frequency.

- Harmonics are caused by nonlinear electrical devices.

- Nonlinearity is the condition where the current waveform does not match the voltage waveform that caused it.

- Even-numbered harmonics cancel each other. Odd-numbered harmonics cause problems with excessive current flow that results in overheating.

- Triplen harmonics are generated by single-phase loads only.

- Three-phase loads generate higher order harmonics.

- Total harmonic distortion (THD) is the amount of harmonics on a line compared to the fundamental frequency of 60 Hz.

- Power factor is the ratio of real power to apparent power in a circuit or distribution system.

- A switched-mode power supply (SMPS) uses diodes and capacitors to rectify AC power supplies. An SMPS generates significant amounts of harmonics.

- Lighting ballasts are manufactured to standards that limit THD.

- Single-phase loads cause high currents to flow on neutral wires.

- K-rated transformers are designed to withstand the harmful effects of harmonics.

- Delta-wye transformers are often used to attenuate triplen harmonics.

- An HMT with zigzag wiring in the secondary wiring can be used to cancel triplen harmonics.

- A phase-shifting transformer can be used to cancel higher order harmonics.

- The Fourier transform is a mathematical technique used to decompose a periodic waveform into the harmonics that generated it.

DEFINITIONS

- A *harmonic* is voltage or current at a frequency that is an integer (whole number) multiple (2nd, 3rd, 4th, etc.) of the fundamental frequency.

- *Harmonic sequence* is the phasor rotation with respect to the fundamental (60 Hz) frequency.

- *Phasor rotation* is the order in which waveforms from each phase (A, B, and C) cross zero.

- *Total harmonic distortion (THD)* is the amount of harmonics on a line compared to the fundamental frequency of 60 Hz.

- *Total demand distortion (TDD)* is the ratio of the current harmonics to the maximum load current.

- *Power factor* is the ratio of true power to apparent power in a circuit or distribution system.

- *True power* is the power, in W or kW, used by motors, lights, and other devices to produce useful work.

- *Reactive power* is the power, in VAR or kVAR, stored and released by inductors and capacitors.

- *Harmonic power* is power, in VA or kVA, lost to distortion.

- *Apparent power* is the power, in VA or kVA, that is the sum of true power, reactive power, and harmonic power.

- *Displacement power factor* is the ratio of true power to apparent power due to the phase displacement between the current and voltage.

- *Distortion power factor* is the ratio of true power to apparent power due to THD.

- The *current crest factor* is the peak value of a waveform divided by the rms value of the waveform.

- A *switched-mode power supply (SMPS)* is a power supply for electronic devices that includes an internal control circuit that quickly switches the load current ON and OFF in order to deliver a stable output voltage.

- A *ballast* is a controller responsible for providing the initial startup voltage and maintaining a constant current through a lamp.

- A *harmonics mitigating transfomer (HMT)* is a transformer designed to reduce the harmonics in a power distribution system.

- A *harmonic filter* is a device used to reduce harmonic frequencies and THD.

- The *Fourier transform* is the mathematical method of converting a time-based waveform, like a sine waveform, into frequency-based information.

REVIEW QUESTIONS

1. Describe how a linear load and a nonlinear load draw current in different ways.

2. What is a harmonic?

3. List three common nonlinear devices.

4. Explain how nonlinear single-phase loads cause high currents on the neutral wires of a 3-phase system.

5. Describe the THD limits placed on electronic ballasts.

6. List the main components associated with a switched-mode power supply (SMPS).

7. Describe displacement power factor, distortion power factor, and total power factor.

8. Explain how a neutral wire can carry more than double the phase current when the maximum current on a circuit from the third harmonic is 173% of phase current.

9. Describe the meaning of the K-factor of a circuit.

10. Describe how the Fourier transform is used to analyze harmonics.

Power Generation and Distribution

In the production of electrical power, generators use a variety of power sources to turn a turbine or rotor. A transformer is used to step up the voltage from the generator to enable the lines to carry large amounts of power over the distance required to reach the customer. Other transformers are used at substations and at the point of use to step down the voltage. The power grid that provides electricity for users consists of many generators connected in parallel. All the generators must be synchronized before being connected to the power grid.

POWER GENERATION

All electrical and electronic circuits require a power source. The type and level of power required depends on the size and power requirements of the load and circuit. Very small circuits, such as an electric wristwatch, require very little power. In an electric watch, one low-voltage battery (1.5 V) provides enough power to operate the electrical system for a year or more. A large circuit, such as an industrial plant, may require thousands of kilowatts of power. An industrial plant receives power through a transformer from a high-voltage source provided by the local electric utility or from a local power generation facility.

Generators

Generators, or alternators, are used to produce power for common loads used in residential, commercial, and industrial applications. A *generator* is a machine that converts mechanical energy into electrical energy by means of electromagnetic induction. The output of a generator may be connected directly to the load, connected to transformers, or connected to a rectifier circuit to produce DC power. In small generator applications such as portable generators used by emergency crews, the output of the generator is normally connected directly to the load.

A generator produces electricity when a rotating coil (rotor) cuts through magnetic flux. **See Figure 6-1.** The magnetic flux is produced by the magnetic field that is pres-

DEFINITION

*A **generator** is a machine that converts mechanical energy into electrical energy by means of electromagnetic induction.*

ent between the north and south poles of a permanent magnet, or by the electromagnetic field of an electromagnet. As the rotor moves through the magnetic field, electric current flows through the wire coils of the rotor. The voltage produced by a generator depends on the strength of the magnetic field, the number of turns, and the speed of rotation of the rotor. A stronger magnetic field, more turns, and faster rotation produce a higher voltage.

Voltage. Utility generators often produce voltages at about 10 kV to 20 kV, but this voltage is not used to transmit large amounts of power over a long distance because of resistance losses in the transmission lines. All conductors have some resistance to the flow of current. This results in resistance heating of the conductor and a voltage drop. The resistance heating increases with increases in the amount of current. Therefore, the most efficient way to send electric power over transmission lines is at the lowest current possible. Transformers are used to step up the voltage at the generator. Many transmission lines carry about 345 kV to 750 kV. Transformers are used again to step down the voltage at the point of use.

Industrial motor drives use 3-phase power to supply electric motors.

Frequency. In North America, power is generated at a frequency of 60 Hz. There are a few isolated systems that use 25 Hz, but they are not part of the power grid. In Europe, the Middle East, and Africa, power is generated at a frequency of 50 Hz. In Asia and Central and South America, some countries use 50 Hz power and others use 60 Hz power.

Three-Phase Power

All generators operate on the same principle of induction as a transformer. The three requirements for induction are a conductor, a magnetic field, and relative motion between the conductor and the magnetic field. Single-phase power uses one coil rotating through the magnetic field. Each rotation of the armature in a single-coil (1ϕ) AC generator produces one complete alternating current cycle. The power in a single-phase circuit drops to zero twice every cycle as the voltage rises and falls.

Utilities normally generate 3-phase (3ϕ) AC power by rotating three coils through the magnetic field. Each rotation of the armature produces three complete alternating current cycles that are 120° out of phase. **See Figure 6-2.** One advantage of 3-phase AC power is that more power can be delivered over existing systems without replacing the conductors. The power in a 3-phase circuit varies during a cycle, but never falls to zero. This results in better performance in 3-phase motors and other loads. Transformer designs are readily available for 3-phase power as well as single-phase power.

SAFETY TIP

Never assume that phase A (L1), phase B (L2), and phase C (L3) are actually the same throughout a structural distribution system; use a phase sequence test instrument to identify which lines are powered and which power line is phase A, phase B, and phase C.

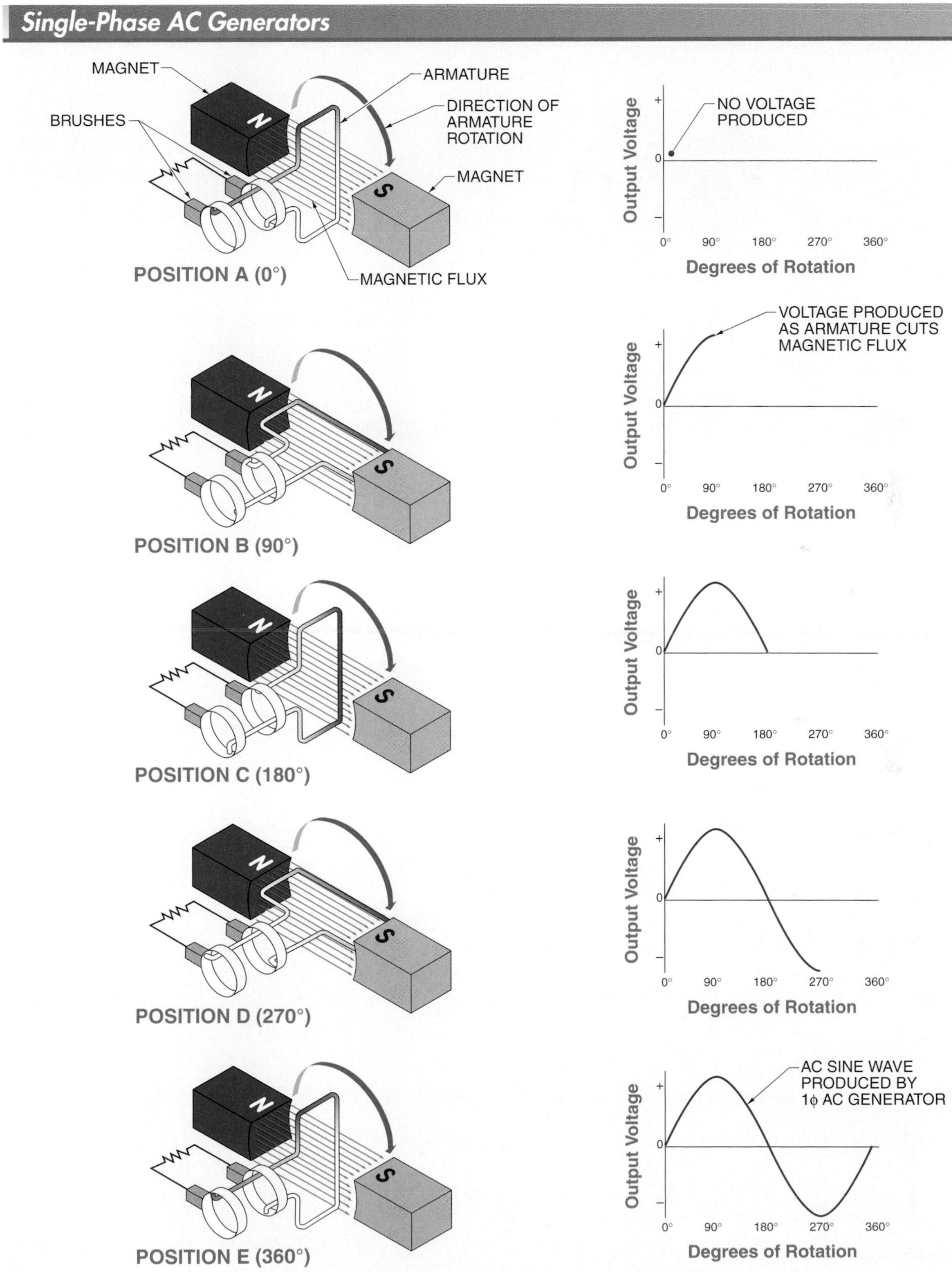

Single-Phase AC Generators

Figure 6-1. *A generator produces electricity when magnetic lines of force are cut by a rotating wire coil.*

Three-Phase AC Generators

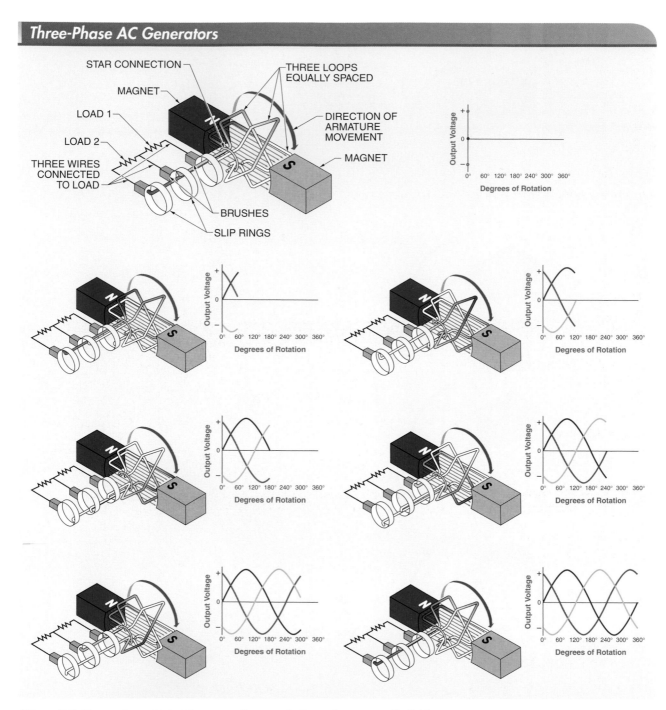

Figure 6-2. Three-phase generators pass three coils through a magnetic field.

DEFINITION

*The **phase voltage** is the voltage measured across the windings.*

Winding and Line Relationships. The *winding voltage,* or *phase voltage,* is the voltage measured across the windings. The *line voltage* is the voltage measured line-to-line. The three winding voltages are often represented by three phasors of equal length that are 120° apart. The three line voltages are often represented by three vectors connecting the tips of the three winding voltages. For a delta connection, the winding voltages are equal to the line voltages. For a wye connection, the line voltages are greater than the winding voltages. **See Figure 6-3.** The actual relationship for wye connections is as follows:

$$E_{Line} = E_{Winding} \times \sqrt{3}$$
$$E_{Winding} = E_{Line} \div \sqrt{3}$$

where

E_{Line} = line voltage (in V)

$\sqrt{3}$ = constant (usually rounded to 1.73)

$E_{Winding}$ = winding voltage (in V)

The factor $\sqrt{3}$ comes from the relative magnitudes of the winding and line voltages. This is approximately equal to 1.73205 and is often rounded to 1.73. For example, if the winding voltage is 120 V, what is the line voltage? The line voltage is calculated as follows:

$$E_{Line} = E_{Winding} \times \sqrt{3}$$
$$E_{Line} = 120 \times \sqrt{3}$$
$$E_{Line} = 120 \times 1.73$$
$$E_{Line} = \mathbf{208\ V}$$

The *line current* is the current flow through the lines to a load. The *winding current,* or *phase current* is the current flow through the individual windings. For a wye connection, the current flowing in each line is equal to the current flowing in its respective winding. For a delta connection, the line currents are greater than the winding currents. For a delta connection, the actual relationships are as follows:

$$I_{Line} = I_{Winding} \times \sqrt{3}$$
$$I_{Winding} = I_{Line} \div \sqrt{3}$$

where

I_{Line} = line current (in A)

$\sqrt{3}$ = constant (usually rounded to 1.73)

$I_{Winding}$ = winding current (in A)

For example, what is the line current in a delta circuit when the winding current is 10 A? The line current is calculated as follows:

$$I_{Line} = I_{Winding} \times \sqrt{3}$$
$$I_{Line} = 10 \times \sqrt{3}$$
$$I_{Line} = 10 \times 1.73$$
$$I_{Line} = \mathbf{17.3\ A}$$

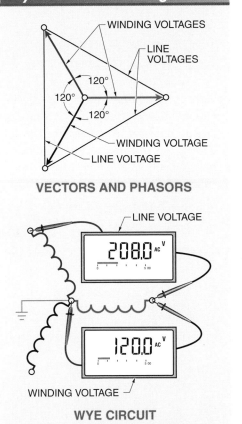

Wye Connection Voltages

VECTORS AND PHASORS

WYE CIRCUIT

Figure 6-3. *Three-phase voltages can be represented as vectors and phasors, or circuit diagrams*

DEFINITION

*The **line voltage** is the voltage measured line-to-line.*

*The **line current** is the current flow through the lines to a load.*

*The **phase current** is the current flow through the individual windinghs.*

TECH FACT

Most motor control circuits are powered from step-down transformers to reduce the voltage to the control circuit. A step-down transformer reduces the voltage to the control circuit to a level of 480 V, 240 V, or 120 V as needed.

Power Calculations. The power in 3-phase wye-connected circuits can be calculated in two different ways, depending on whether the line voltage or the winding voltage is known. The two methods used are as follows:

$$P = \sqrt{3} \times E_{Line} \times I_{Line} \times PF$$
or
$$P = 3 \times E_{Phase} \times I_{Phase} \times PF$$
where
P = power (in W)
E_{Line} = line voltage (in V)
I_{Line} = line current (in A)
PF = power factor
E_{Phase} = phase voltage (in V)
I_{Phase} = phase current (in A)

For example, what is the power used in a circuit that draws 10 A at 480 V (line voltage) with a power factor of 0.9? The line current is equal to the winding current in a wye connection. The power is calculated as follows:

$$P = \sqrt{3} \times E_{Line} \times I_{Line} \times PF$$
$$P = \sqrt{3} \times 480 \times 10 \times 0.9$$
$$P = \textbf{7480 W}$$
or
$$P = \textbf{7.48 kW}$$

Distribution substations contain large utility transformers.

Power Sources. Mechanical force, such as steam, water, or an engine, must be used to rotate the coils in the generators. Large utility generators are usually powered by steam turbines. Natural gas, nuclear energy, coal, and oil provide power for these turbines. Alternatively, hydropower at large dams is used to drive mechanical turbines. **See Figure 6-4.**

Engines fueled by natural gas, propane, diesel fuel, or gasoline often power standby power units that provide emergency power to critical facilities. AC generators come in many sizes, depending on the power requirements that must be met. Small generators can deliver hundreds to thousands of watts of power. Large utility generators can deliver millions to billions of watts of power.

Landfill Generators. A method of generating electricity that is growing in popularity is the use of landfill generators. These units run on methane gas created by the decomposing trash in landfills. The gas is recovered and used to power the units. The gas is piped to a compressor and then fed to a series of generators with the prime movers being large gas-fired engines. The most efficient systems also capture the waste heat as hot water or steam for use in nearby industry.

POWER DISTRIBUTION

The grid network that feeds North America is a large system of transmission lines fed from many power plants. All the phases on the entire grid are at the same polarity at the same time. The high potentials that are carried on these lines require large towers to carry the lines and keep them isolated from the public. **See Figure 6-5.** Large insulators are used to isolate the high-voltage lines from the towers.

Transmission

Power is sent across the grid network to the place where it is needed, usually close to the larger population areas. At that point, the very high-voltage power is sent to substations called distribution subs. Transformers are used to step down to a lower voltage, usually 69 kV. From the distribution sub, the power is then sent to a number of substations around the area with units called substation transformers. Again, transformers are used to step down to a lower voltage, sometimes 13.8 kV, for feeders supplying the city. **See Figure 6-6.** The distribution system always includes some type of transformer to connect to the lines and make the power available at the needed voltage.

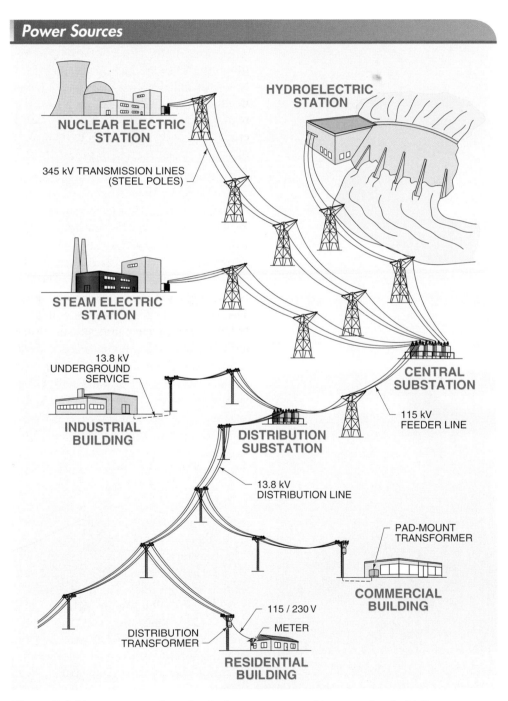

Power Sources

HYDROELECTRIC STATION

NUCLEAR ELECTRIC STATION

345 kV TRANSMISSION LINES (STEEL POLES)

STEAM ELECTRIC STATION

13.8 kV UNDERGROUND SERVICE

INDUSTRIAL BUILDING

DISTRIBUTION SUBSTATION

CENTRAL SUBSTATION

115 kV FEEDER LINE

13.8 kV DISTRIBUTION LINE

PAD-MOUNT TRANSFORMER

COMMERCIAL BUILDING

115 / 230 V

METER

DISTRIBUTION TRANSFORMER

RESIDENTIAL BUILDING

Figure 6-4. Many sources of mechanical energy are used to generate electricity.

Transmission Towers

INSULATOR   TOWER

Figure 6-5. Large towers are used for transmission lines.

Substations

Figure 6-6. Substations are used to reduce the voltage on transmission lines to the lower voltages used for distribution lines.

System Shutdown

The electrical distribution grid consists of many interconnected systems that work together to deliver power wherever it is needed. All electrical distribution systems include shutoff devices to protect the equipment in the case of a malfunction or overload. These shutoff devices operate in a similar way to overcurrent protection devices used in industrial and commercial operations. In the case of an overload, the shutoff devices are intended to shut down a part of the system for repairs while allowing the rest of the system to operate normally.

On a small scale, a power generator can be shut down for many reasons. A generator is typically shut down when some type of overload device detects a problem or when the demand drops. On a large scale, a power generator can be shut down when there are problems with the electrical grid.

On August 14, 2003, there was a massive power blackout on the East coast of the U.S. and Canada. This time of year is always a time of peak demand. When power lines are operating at their maximum power load, they get hot, and can sag slightly and reduce the clearances between the lines and nearby vegetation. The electric current in a high-voltage power line may arc to a tall-growing tree if clearances are not maintained. **See Figure 6-7.** During the 2003 blackout, a 345 kV transmission line sagged as it became hot from the summer heat and from the resistive heating of the high current it was carrying.

The transmission line sagged and shorted to ground when it came in contact with a tree that had not been properly trimmed. The transmission line was automatically removed from service and the load was shifted to other lines. The other transmission lines became overloaded too. Two more 345 kV lines sagged and shorted out within an hour. The utility's alarm system was temporarily offline and other utilities were not warned of the problems. The load shifted to other lines that shut down in turn from the overloads.

Transmission Line Clearance

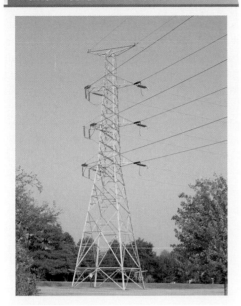

Figure 6-7. Large clearances are required around transmission lines to protect the lines from vegetative growth.

The overcurrent protective devices functioned as they were supposed to do to protect the equipment. The utility failed to properly shed loads to balance power requirements with power available. The communications systems that were intended to isolate problems and keep them confined to local areas failed and the problem quickly cascaded across the network. Within 7 minutes of a major transmission line failure, large sections of the Eastern Connect power grid were shutting down. Over a period of only 9 seconds, the grid collapsed and more than 250 electric power plants were shut down. These power plants represented over 61,800 MW of generating capacity, and 50 million people were without power.

System Startup

Starting up an electric generating power plant is much more complex than starting up a machine in an industrial plant. All the generators must be connected in parallel and the power from each generator must be exactly the right phase, frequency, and voltage before it can be connected to the grid. If there is a frequency mismatch, a generator turning slower is a load to the other generators. If the voltages are not equal, one of the generators could become a reactive load to the other generators. This causes high currents to circulate between the generators and can cause extensive damage. If the voltage phases are not matched, large opposing voltages are developed. This causes high currents to flow that can damage the generators.

Parallel Operations. When referring to placing generators in parallel, the term "operating source" refers to the existing power grid or a generator that is already running. This is typically represented in a drawing as busbars connected to the power grid. The term "incoming source" refers to a new generator being brought on-line. This may be a generator being brought on-line in a power plant or it may be a backup generator being started up in a plant. The incoming source should be running at a slightly higher frequency than the operating source. This is to allow the incoming source to slow down when a load is applied. This puts less stress on the generator than having to speed up when a load is applied.

For a power plant with several generators in parallel, the generators must be started and synchronized sequentially. The first generator is started to excite the step-up transformer that feeds the transmission lines. This transformer is sometimes called a power transmission transformer. When the second unit is switched on-line, the speed and the potential of each phase must be of the same polarity and level at precisely the same time.

As in a synchronous motor, the rotating DC field of the rotor of the second generator is locked up, or synched, with the first generator or the grid. If the second generator starts to slow, the locked field with the rotating field set up by the first generator keeps the second running at the same speed as the first. The same thing applies to the first unit if it tries to slow, as the rotating AC field provided by the second unit keeps the first generator at synchronous speed. Now that two generators are synchronized in parallel, any remaining generators can be brought on-line sequentially.

Synchroscopes. The speed regulators of a generator do a good, but not perfect, job of controlling the speed to generate exactly 60 Hz. Whenever a regulator tries to maintain a certain speed, it will hunt and wander slightly. As this is happening, it controls close to the required speed, but not exactly correctly. A synchroscope can be used to help match the frequency of a generator to the busbar, or to match the frequencies of two generating units. **See Figure 6-8.** A *synchroscope* is a device that indicates whether two AC sources to be connected in parallel are in the correct phase relationship. The synchroscope is connected to both potentials. When the needle on the synchroscope is stationary at the twelve o'clock position, the frequencies and phases are matched. When the needle turns clockwise, the incoming frequency is higher than the operating frequency. When the synchroscope shows that the units are operating at the correct frequency, a switch can be closed that brings the second unit on-line at the right time.

Lights Out Methods. Other methods of synchronizing generators are the various lights out methods. These methods connect a lamp across each of the phases of two generators or between a generator and the power busbars. The voltage across the lamps is the potential difference between the phases of the generators. When the phase rotation between the generators is not matched, the lamps flicker on and off, but not in unison. This means that phase connections on the incoming source are incorrect and must be reversed. When the peak voltages of the two generators match, the lamps go out because there is no longer any potential difference between the two generators. Two common lights out methods are the three-lights-out method and the one-light-out method.

In the three-lights-out method, a lamp is connected as a load across each phase of the two sources, A to A, B to B, and C to C. When all three lamps are dark, the potentials of both sources are the same and the second source can be brought on-line. The phase connection is incorrect if the lights do not all turn on and off at the same time. **See Figure 6-9.** The one-light-out method is very similar except that the phases are connected A to B, B to A, and C to C. When the C to C lamp is dark and the other two are brightly lit with no flickering, the generators are synchronized and the second source can be brought on-line.

DEFINITION

*A **synchroscope** is a device that indicates whether two AC sources to be connected in parallel are in the correct phase relationship.*

Synchroscopes

Figure 6-8. *Synchroscopes are used to synchronize the frequency of generators when parallel connections are made.*

TECH FACT

Most electric power is produced by alternating current (AC) generators that are driven by rotating prime movers. Many of the prime movers are steam turbines whose thermal energy comes from steam generators that use either fossil or nuclear fuel. Combustion turbines are often used for smaller units and in applications where gas, diesel, or oil is the only available fuel. After the power is produced, step-up transformers are used to raise the voltage to the level required for transmission lines.

Lights Out Methods

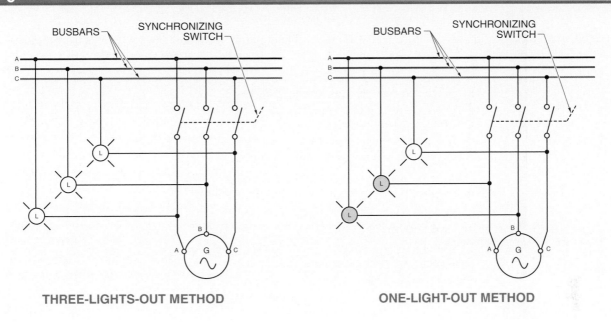

THREE-LIGHTS-OUT METHOD   **ONE-LIGHT-OUT METHOD**

Figure 6-9. Lights out methods are used to synchronize the frequency of generators when parallel connections are made.

SUMMARY

- Generators use a variety of power sources to generate force to turn a turbine or rotor.

- A transformer is used to step up the voltage from the generator resulting in lower tranmission currents and lower line loss.

- A transformer is used to step down the voltage at substations and the point of use.

- Three-phase power is commonly used for utility power distribution.

- The power grid consists of many generators connected in parallel.

- All generators must be synchronized before being connected to the power grid.

DEFINITIONS

- A *generator* is a machine that converts mechanical energy into electrical energy by means of electromagnetic induction.

- The *phase voltage* is the voltage measured across the windings.

- The *line voltage* is the voltage measured line-to-line.

- The *line current* is the current flow through the lines to a load.

- The *phase current* is the current flow through the individual windings.

- A *synchroscope* is a device that indicates whether two AC sources to be connected in parallel are in the correct phase relationship.

REVIEW QUESTIONS

1. Explain what will be seen on a synchroscope when the frequency of the incoming source is higher than the operating source.

2. Explain why the most efficient way to send electric power over transmission lines is at the lowest current possible.

3. List at least four sources of energy that are used to generate electric power.

4. Explain how a synchroscope can be used to help match the frequency of a generator to a busbar.

5. Explain how the three-lights-out method is used to help match the frequency of a generator to a busbar.

Transformer Principles and Applications

Reactors and Isolation Transformers

CHAPTER

7

Many types of sensitive electronic equipment, such as uninterruptible power supplies, computers, process controllers, data communication systems, and electronic ballasts, can be exposed to voltage and current distortion levels that exceed the power condition they were designed for. Reactors and isolation transformers can be used to improve the power quality available to the electronic equipment. Reactors limit inrush current and reduce notching. Chokes are a special type of reactor used to reduce short-circuit current and common-mode noise.

POWER QUALITY

There are many types of problems associated with with the use of electrical power. Many involve distortions to the voltage and current sine waveforms. Others involve unexpected currents and noise on the power lines. Two sources of power quality problems are rectifiers and drive-induced ground currents.

Rectifiers

One common source of power quality problems is nonlinear rectified power supplies. Motor drives of all types rectify incoming power to form a DC voltage. Common types of nonlinear rectified power supplies are the switched-mode power supply (SMPS), the insulated gate bipolar transistor (IGBT), and the silicon controlled rectifier (SCR). For a DC motor drive, the DC power is connected directly to a DC motor. For a variable-frequency AC motor drive, the DC power is used to power an inverter, which supplies adjustable frequency and voltage to AC motors.

Integrated Gate Bipolar Transistors. The widespread use of the integrated gate bipolar transistor (IGBT) as the switching element in inverters is a relatively recent development. The main benefits of an IGBT are the very fast switching time and high current-carrying capacity. This results in very efficient circuits.

Silicon Controlled Rectifiers. The process of rectifying DC voltage causes the current waveform to be nonlinear. A silicon controlled rectifier (SCR) is a semiconductor similar in construction to a transistor or a diode. An SCR is used as a switch where a small triggering current controls the flow of a larger current. A common application

SCR Power Supplies

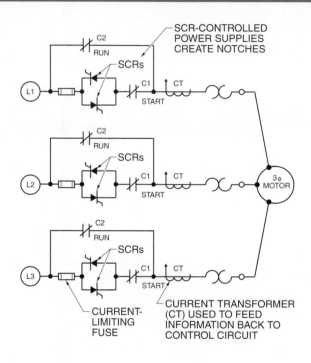

Figure 7-1. An SCR-controlled power supply for 3-phase motors creates notches in the voltage waveform.

DEFINITION

*A **notch** is a distortion in a voltage waveform where the voltage quickly drops toward zero and then returns to the correct value.*

***Common-mode noise** is a type of electromagnetic interference induced on power or communications lines.*

uses an SCR power supply as part of a motor drive. **See Figure 7-1.** In the special case of 3-phase motor drives using SCR six-pulse (six-diode) rectifiers, there are short intervals of time when more than one SCR in ON. This causes a transient short circuit to flow six times per line power cycle. This short-circuit current flow causes nonlinear voltage to drop across the system impedance and results in distortion in the voltage waveform called a notch.

Notches. A *notch* is a distortion in a voltage waveform where the voltage quickly drops

toward zero and then returns to the correct value. **See Figure 7-2.** The number of notches in a cycle is equal to the number of rectifiers, or pulses, in the power supply. The notch can appear at any time during the voltage cycle, depending on the load. In the worst case, the voltage can be reduced to zero. This creates extra zero crossovers that can cause problems with devices that switch on the crossover.

Zero crossovers are often used for timing signals, such as when clocks and alarms in a school must be synchronized throughout the building. DC drives use the zero crossover to determine when to fire the SCR. When there are extra zero crossovers caused by notches, clocks may run fast and switches may activate more often than desired.

In other cases, the zero crossovers are used to activate equipment without the high inrush current that can occur when activation occurs at other points of the cycle. A zero-switching relay is a solid state relay that turns a load ON when the control voltage is applied and the voltage crosses zero. The relay turns the load OFF when the control voltage is removed and the current in the load crosses zero. **See Figure 7-3.** Zero-switching relays are used to control the switching of many types of lamps and for PLC interfacing.

Common-Mode Noise

Common-mode noise is a type of electromagnetic interference induced on power or communications lines. Problems created by induced currents include nuisance trips of ground fault relays or breakers, disruption of data communication, telephone interference, and computer malfunctions.

Variable-frequency drives are a common source of common-mode noise. Unwanted pulses of current are caused by fast-changing voltages from the drive due to the stray capacitance and inductance through the system. These pulses of current are a source of noise that can interfere with proper device functioning. **See Figure 7-4.** The effect of common-mode noise can be reduced by providing a low-impedance return path for the current to return to the source.

Notching

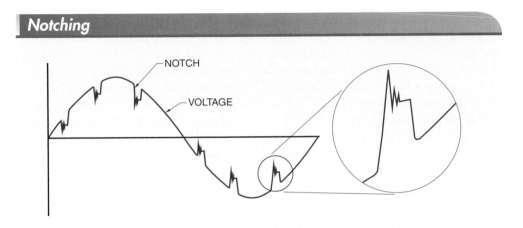

Figure 7-2. Notches are distortions in the voltage waveform.

Zero-Switching Solid State Relays

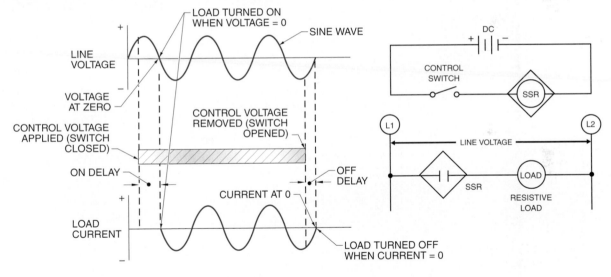

Figure 7-3. Zero-switching solid state relays have a delay to start and stop a device at a zero crossing.

REACTORS

Reactance is the opposition to current flow in an AC circuit. A *reactor*, or *line reactor*, is a coil added in series with a load to reduce inrush current, voltage notching effects, and voltage spikes. Reactors may be tapped so that the voltage in series can be changed to compensate for a change in the load that the motor is starting. Reactors are rated by the ohms of impedance that they provide at a given frequency and current. Reactors may also be rated by the I^2R drop across the device at a certain frequency at a rated current.

Two common types of reactors are the dry-type and the oil-immersed. The dry-type is open and relies on the air to circulate and dissipate the heat. Dry-type reactors are common in low-voltage applications.

 **DEFINITION**

A *reactor*, or *line reactor*, is a coil added in series with a load to reduce inrush current, voltage notching effects, and voltage spikes.

Oil-immersed reactors are common in high-voltage applications. Oil-immersed reactors are placed in tanks and require a magnetic shield to prevent eddy currents from circulating in the tank. The shield is made from laminated steel sheets like the transformer core and motor stators.

Reactors may be used as line input or load output reactors. **See Figure 7-5.** Line input reactors are used when low line impedance allows high inrush current, when power factor correction capacitors are used, or when a motor drive causes notching. Load output reactors are installed at the output of a motor drive. Output reactors help eliminate voltage spikes or reflected wave noise by slowing down the rate of change in the drive output voltage. However, output reactors have a tendency to overheat due to the harmonic content of the output waveform from the motor drive. The reactor must be designed to allow for harmonics.

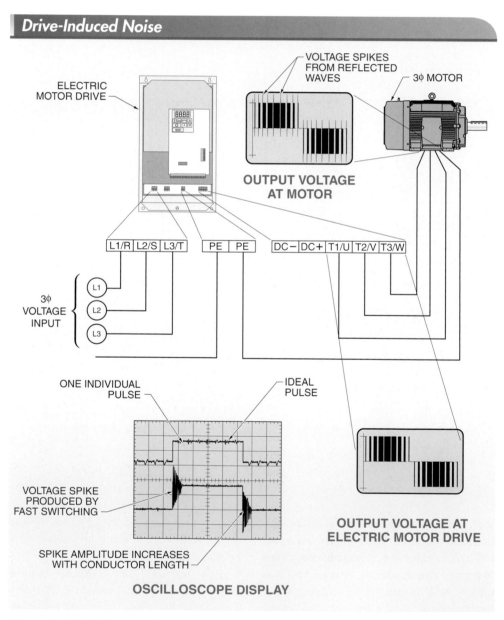

Figure 7-4. *Variable-speed motor drives generate noise that can interfere with proper drive functioning.*

Reactors

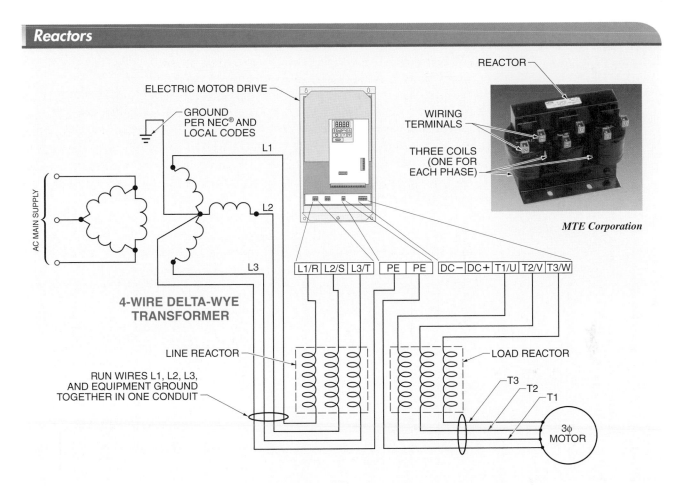

REACTOR

WIRING
TERMINALS

THREE COILS
(ONE FOR
EACH PHASE)

ELECTRIC MOTOR DRIVE

GROUND
PER NEC® AND
LOCAL CODES

L1

L2

L3

AC MAIN SUPPLY

4-WIRE DELTA-WYE
TRANSFORMER

LINE REACTOR

RUN WIRES L1, L2, L3,
AND EQUIPMENT GROUND
TOGETHER IN ONE CONDUIT

L1/R L2/S L3/T   PE PE   DC− DC+ T1/U T2/V T3/W

LOAD REACTOR

T3
T2
T1

3ɸ
MOTOR

MTE Corporation

Figure 7-5. Reactors are used as line input or load output reactors.

Inrush Current

Many electrical devices draw high currents at startup or have very low impedance to the flow of current. For example, electric motors typically draw many times their full-load current during startup. This inrush current can cause voltage sags that trip out other equipment. Many full-voltage motor starters use reactors to increase the impedance and limit the inrush current.

Large capacitor banks used to correct for low power factor have very low impedance when the capacitor bank is first switched ON and the capacitors begin charging. Low impedance means that the flow of current is very high. A reactor can be added in series to increase the reactance. The increased reactance increases the impedance and reduces the inrush current. **See Figure 7-6.**

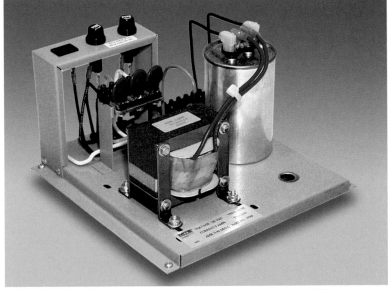

MTE Corporation

Reactors are used as harmonic filters in electronic power supplies.

Inrush Current

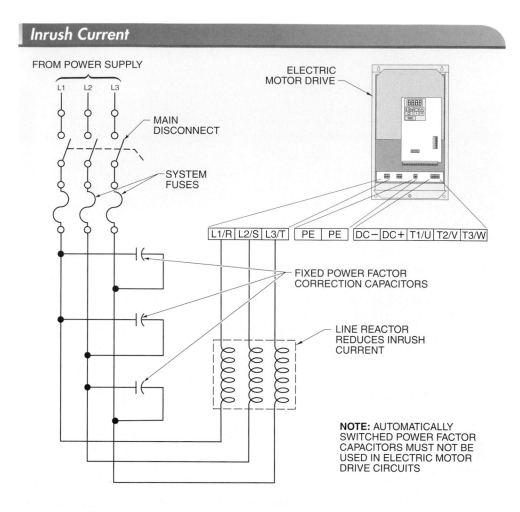

FROM POWER SUPPLY

L1 L2 L3

MAIN DISCONNECT

SYSTEM FUSES

ELECTRIC MOTOR DRIVE

| L1/R | L2/S | L3/T | PE | PE | DC− | DC+ | T1/U | T2/V | T3/W |

FIXED POWER FACTOR CORRECTION CAPACITORS

LINE REACTOR REDUCES INRUSH CURRENT

NOTE: AUTOMATICALLY SWITCHED POWER FACTOR CAPACITORS MUST NOT BE USED IN ELECTRIC MOTOR DRIVE CIRCUITS

Figure 7-6. Line reactors are used to reduce inrush current.

Reduced Notching

In order to reduce notching, the source of the notching needs to be isolated or buffered from other equipment that uses the same power distribution system. A relatively simple method to reduce notching is to create a voltage divider. **See Figure 7-7.** When impedance in the form of a reactor is added in series with an SCR controller, the notch voltage is distributed across the new impedance and the impedance already existing in the feeder lines. The added impedance decreases the notch depth and widens the notch width.

Experience has shown that the reactor should have about 3% impedance to reduce the notch depth by about 50%.

This is enough to eliminate the extra zero crossovers that cause problems. Higher impedance may cause problems with sensitive equipment because the wider notch may be seen as a loss of voltage. Lower impedance may not reduce the notch depth enough to eliminate the problems.

TECH FACT

Transients on a line can cause electronic equipment to generate errors. Digital electronic circuits operate on low-level digital signals that can be corrupted by a false signal induced by the transient voltage.

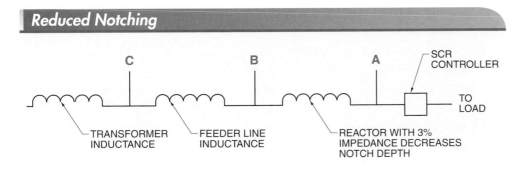

Reduced Notching

Figure 7-7. *A reactor can be added in series with an SCR power source to reduce notching.*

Saturable-Core Reactors

When an iron core is saturated, substantially all the magnetic domains are aligned with the applied magnetic field. Increases in the applied magnetic field do not result in increases in magnetic flux. Therefore, there is no increase in the voltage induced in opposition to the change in current. In other words, an inductor loses its ability to oppose changes in current when its core becomes saturated.

A *saturable-core reactor* is an inductor whose inductance is controlled through the use of a magnetic field created by a second winding wound around the same iron core as the primary winding. The "power" winding of a saturable-core reactor is the winding carrying the AC load current. The "control" winding of a saturable-core reactor is the winding carrying the DC control current. The control winding is carrying DC strong enough to create a magnetic field that saturates the core.

An increase in DC through the control winding produces an increased magnetic flux in the reactor core. An increase in the magnetic flux moves the reactor core closer to saturation and decreases the inductance of the power winding. A decrease in inductance in the power winding increases the current delivered to the load through the power winding. Therefore, a saturable-core reactor can be used as an amplifier where a relatively small DC through a control winding can control a relatively large AC through the power winding.

In actual practice, a saturable-core reactor consists of two pairs of windings. **See Figure 7-8.** The small dots near the saturable-core reactor coils indicate polarity. The power windings are in phase with each other and the control windings are out of phase with each other. This allows the saturable-core reactor to saturate the core equally in both positive and negative alternations of the AC cycle.

Saturable-core reactors were very popular in the plating industry before the advent of DC drives to control the current in the plating solution. In the case of a plater, the part being plated is the load. If no DC is flowing in the control coil, then the IR drop will be controlled by the amount of current in the reactor. With DC current in the control coil, the DC flux will flow in the core and limit the amount of AC flux in the core. Lower AC flux means less reactance and less impedance to the circuit current. Large amounts of AC current can be controlled by a small amount of DC current. This control is very linear and very reliable. Saturable-core reactors have fallen out of use in this type of application because the cost to build a reactor is much higher than building a DC drive.

DEFINITION

A *saturable-core reactor* is an inductor whose inductance is controlled through the use of a magnetic field created by a second winding wound around the same iron core as the primary winding.

TECH FACT

Saturable-core reactor power supplies used at high power levels are extremely reliable devices because there are no moving parts.

Saturable-Core Reactors

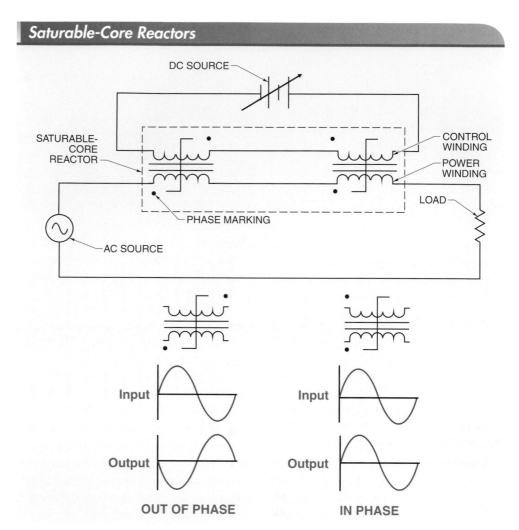

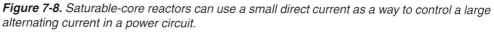

OUT OF PHASE   **IN PHASE**

Figure 7-8. Saturable-core reactors can use a small direct current as a way to control a large alternating current in a power circuit.

Chokes

A *choke*, or *line choke*, is a reactor used to restrict the current to AC or DC drives in the case of short circuits in the drive itself. When large short-circuit currents are drawn from the source, the choke starts to build a countervoltage, and the voltage available to the drive is reduced. The reduced voltage causes the instantaneous electronic trip (IET) circuit to trip the drive off-line to avoid damage. Chokes have large conductors with fewer turns and offer low impedance to the line into the drive.

A *common-mode choke* is a reactor used to reduce common-mode noise current

generated by the rapid switching of a motor drive or a signaling device. **See Figure 7-9.** Load current flows through one winding of the common-mode choke to the load and then flows through the other winding away from the load. This results in two opposite magnetomotive forces that cancel and result in zero inductance. Any other current, such as common-mode noise current, typically flows to ground and does not cancel in the choke. This results in a magnetomotive force that opposes the flow of the noise current. The net result is that a common-mode choke allows the load current to flow almost unimpeded while blocking the flow of noise current.

Common-Mode Chokes

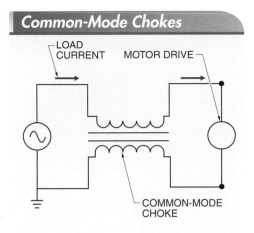

Figure 7-9. Common-mode chokes are used to reduce the severity of drive-induced common-mode noise.

Common-mode chokes are often used to reduce drive-induced common-mode noise. A common-mode choke provides high inductance and impedance generated during drive switching. The magnitude and rise time of the noise current are reduced to the point where they are below the noise threshold of affected equipment.

Common-mode chokes can also be designed for differential-mode filtering in digital communications. Digital communication protocols often use differential-mode signals where the difference between two signals must be measured. Differential-mode impedance causes the waveforms to lose shape. This puts limits on communication speeds.

Resonance

Capacitor banks are often used to correct low-power-factor situations. In systems with large amounts of capacitance used to correct power factor, high-voltage distortion can cause resonance at system harmonic frequencies. This results in series- or parallel-resonant currents, which can be very damaging to the electrical system. To reduce the effects of motor drives on other loads in the electrical system, one common technique is to add a reactor to the incoming power line to the motor drives. **See Figure 7-10.** The added

reactance ahead of a motor drive changes the resonance frequency and reduces the amount of distortion present in the input current to the motor drive.

Synchronous Motor Starting

Reactors are used in the start circuit of large synchronous motors to provide a voltage to the AC coil of the polarized-frequency relay (PFR). A reactor is in series with the rotor circuit and provides power for the PFR coil, which is in parallel with the reactor. The rotor current starts out at 60 Hz at locked rotor, but decreases in frequency as the rotor speeds up. This changes the inductive reactance, and the voltage dropped across the reactor is lowered. At a point where the AC frequency from the reactor is low enough, the relay drops out and the DC rotor supply is connected to the rotor.

ISOLATION TRANSFORMERS

Isolation transformers act as a buffer between a nonlinear power supply, the load, and the power source. Isolation transformers are used to provide cleaner power to nonlinear loads and to prevent harmonics from moving upstream into the power source. Isolation transformers, as well as reactors, can be used to reduce distortion of the motor drive current.

Isolation transformers provide reactive control of current harmonics for both AC and DC motor drives, whether they are the static bridge or the silicon controlled rectifier (SCR) switch input type. Reduction of current harmonics improves line current waveform distortion, thereby improving the power factor of the drive load and reducing voltage waveform distortion effects in the feeders ahead of the transformer.

 TECH FACT

The addition of a reactor or a choke can create a series- or parallel-resonant circuit, especially where power factor correction capacitors are used.

Resonance

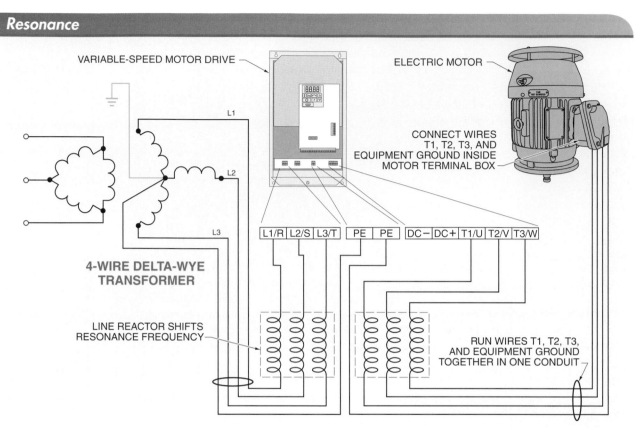

VARIABLE-SPEED MOTOR DRIVE

ELECTRIC MOTOR

CONNECT WIRES
T1, T2, T3, AND
EQUIPMENT GROUND INSIDE
MOTOR TERMINAL BOX

L1

L2

L3

| L1/R | L2/S | L3/T | PE | PE | DC− | DC+ | T1/U | T2/V | T3/W |

**4-WIRE DELTA-WYE
TRANSFORMER**

LINE REACTOR SHIFTS
RESONANCE FREQUENCY

RUN WIRES T1, T2, T3,
AND EQUIPMENT GROUND
TOGETHER IN ONE CONDUIT

Figure 7-10. A reactor in series with a variable-speed motor drive shifts the resonance frequency away from any harmonics on the line.

Drive Isolation Transformers

Drive isolation transformers are specially designed isolation transformers that are placed between the system supply power and an electronic motor drive or group of drives. **See Figure 7-11.** The three basic functions performed by drive isolation transformers are voltage change, reduction of drive-induced ground currents, and reduction of common-mode noise. They provide similar benefits for both DC drives and variable-frequency AC drive applications.

All two-winding transformers are isolation transformers. However, general-purpose transformers are not fully rated for motor drive applications because they cannot supply the required distorted current at full load without exceeding their design temperature rise. In addition, many standard transformers have shortened life expectancy due to mechanical stress when these transformers

supply DC motor drive current transients. Another contributor to shortened life expectancy is the severe cyclic nature of motor drive process control applications.

Drive Isolation Transformers

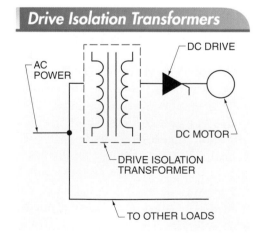

AC POWER

DC DRIVE

DC MOTOR

DRIVE ISOLATION
TRANSFORMER

TO OTHER LOADS

Figure 7-11. Drive isolation transformers are placed between a power source and a motor drive.

Design. Special designs are needed for transformers used as drive isolation transformers. A drive isolation transformer is designed with additional reactance to reduce power quality problems associated with the effects of nonlinear loads. **See Figure 7-12.** A drive isolation transformer must be able to withstand the heating effects of the nonlinear drive loads and operate well above the ambient temperature.

Drive Isolation Transformer Design

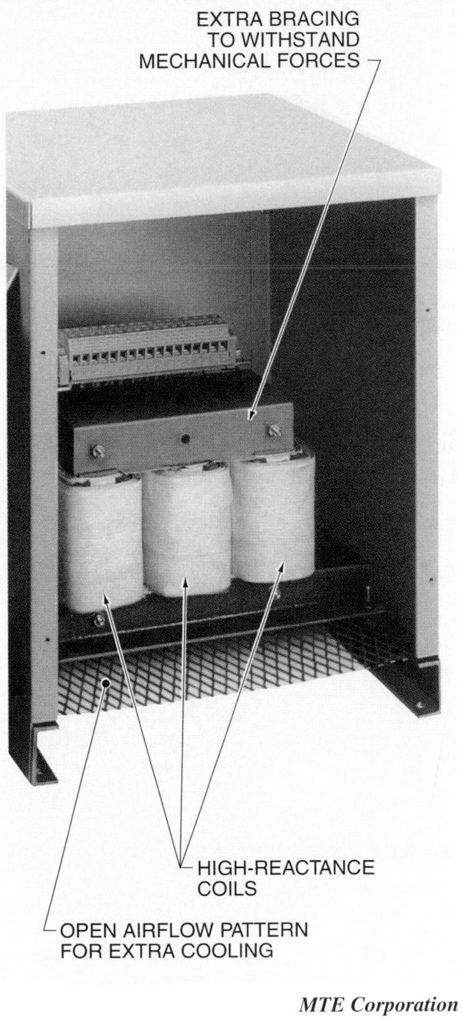

EXTRA BRACING
TO WITHSTAND
MECHANICAL FORCES

HIGH-REACTANCE
COILS

OPEN AIRFLOW PATTERN
FOR EXTRA COOLING

MTE Corporation

Figure 7-12. Drive isolation transformers have extra bracing to withstand mechanical forces, high-reactance coils to control current flow, and increased airflow for extra cooling.

Drive isolation transformers must also withstand the thermal and mechanical stresses caused by the highly cyclic load demands of both AC and DC motor drive process applications. Many loads have characteristics similar to AC and DC motor drives. This is because the load inputs are either designed with 3-phase static diodes or 3-phase, six-pulse SCR bridge rectifier circuits. Drive isolation transformers can also control the effect of these loads on the primary power system. General equipment applications for drive isolation transformers include SCR-controlled heating of furnaces, 3-phase rectifier input DC power supplies, and 3-phase switched-mode power supplies.

The benefits provided by drive isolation transformers can vary and are system-related. For example, some problems with harmonic distortion are the result of drives which are too large for the user's supply capacity, or the result of the drive and affected loads sharing long service lines with a large source impedance. Under these conditions, the additional reactance of drive isolation transformers may provide little or no improvement in related problems. Solutions for such installations typically are to reroute distribution lines and loads, or to install harmonic filters at appropriate locations in the electrical system.

Grounding and Shields

Proper grounding practices along with drive isolation transformers are used to reduce the effects of common-mode noise. If the secondary of a drive isolation transformer is wye-connected, it can be grounded. **See Figure 7-13.** In general, grounding the secondary provides three valuable benefits. Grounding provides superior transient and impulse immunity to the load side of the transformer. Grounding reduces common-mode noise coupling from primary to secondary. Grounding prevents induced ground current from transferring upstream into the primary system. However, a few motor drive designs do not function properly if the secondary neutral is grounded. In this case, the motor drive manufacturer's recommendations must be followed.

Separately Derived Systems. Unlike simple line reactors, the secondary of a drive isolation transformer represents a separately derived system that is electrically isolated from the primary source. With the secondary grounded, a transformer wired as a separately derived system acts to prevent the transfer of common-mode noise and transients from primary source to the drive, as well as from the drive to the primary system.

Electrostatic Shields. Electrostatic shields can prevent common-mode noise from transferring from the transformer's primary to the secondary only if the secondary is ungrounded. Some tests have shown that the typical noise and transients in industrial systems are actually transferred better through shielded transformers than those without shields. If the drive transformer has an ungrounded secondary, or if a high-resistance ground system is used, an electrostatic shield may be beneficial. However, if the

drive isolation transformer's secondary is grounded, primary source common-mode noise cannot be induced between the secondary neutral and ground. Under these conditions, an electrostatic shield serves no useful purpose.

Reactors and Drive Isolation Transformers

Drive isolation transformers and line reactors both provide reactive control of distorted current harmonics and line notching. However, that is the extent of their similarity. **See Figure 7-14.** Drive isolation transformers provide additional benefits over reactors because their wye-connected secondary can be grounded. This protects the motor drive system from common-mode transients that originate with the primary source. Drive isolation transformers also protect the primary source from common-mode energies that originate with the motor drive system. In addition, separately grounding the transformer secondary prevents ground current from transferring back through the primary ground system.

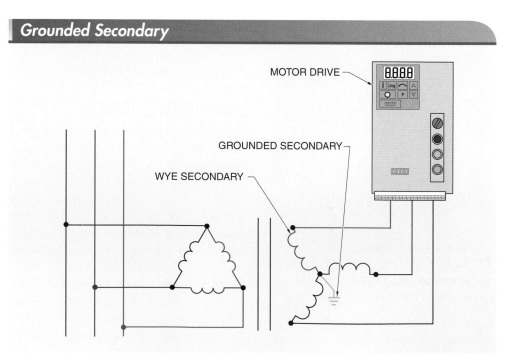

Figure 7-13. A grounded secondary helps reduce noise and improve power quality.

Reactors and Drive Transformers

Function	Capacitive Load (Most Adjustable-Frequency Drives)		Inductive Load (Most DC Drives)	
	Reactor	Drive Transformer	Reactor	Drive Transformer
Change voltage	N/A	Best	N/A	Best
Provide isolation	N/A	Best	N/A	Best
Correct line voltage imbalance	N/A	Best	N/A	Best
Reduce drive-induced currents	N/A	Best	N/A	Best
Reduce common-mode noise	Poor	Best	Poor	Best
Reduce line current harmonic injection into primary source	Good	Good	Poor	Poor
Reduce source voltage notching	N/A	N/A	Good	Good
Reduce feeder short-circuit capability	Best	Good	Best	Good
Meet line impedance requirement	Best	Good	Best	Good

Figure 7-14. A reactor and a drive isolation transformer perform different functions. A reactor is best at controlling line reactance and controlling short-circuit current flow. A drive isolation transformer is best at isolation, voltage control, and noise reduction.

SUMMARY

- Nonlinear rectified power supplies are a common source of power quality problems.

- Rectified power supplies can generate notches in the voltage waveform.

- Drive-induced ground currents are a type of noise generated by variable-speed drives.

- Line input reactors are used when power factor correction capacitors are used or when a motor drive causes notching.

- Output reactors are used to eliminate voltage spikes by slowing down the rate of change of the drive output voltage.

- A saturable-core reactor uses a small direct current to control a larger alternating current.

- A choke protects a circuit from short circuits or from common-mode noise.

- Drive isolation transformers have higher reactance and reinforced bracing to provide extra protection from overheating and mechanical forces.

- Drive isolation transformers can be grounded to reduce voltage transients, reduce common-mode noise, and prevent drive-induced ground current from transferring upstream into the primary system.

DEFINITIONS

- A *notch* is a distortion in a voltage waveform where the voltage quickly drops toward zero and then returns to the correct value.

- *Common-mode noise* is a type of electromagnetic interference induced on power or communications lines.

- A *reactor*, or *line reactor*, is a coil added in series with a load to reduce inrush current, voltage notching effects, and voltage spikes.

- A *saturable-core reactor* is an inductor whose inductance is controlled through the use of a magnetic field created by a second winding wound around the same iron core as the primary winding.

- A *choke*, or *line choke*, is a reactor used to restrict the current to AC or DC drives in the case of short circuits in the drive itself.

- A *common-mode choke* is a reactor used to reduce common-mode noise current generated by the rapid switching of a motor drive or a signaling device.

REVIEW QUESTIONS

1. What are the three basic functions of drive isolation transformers?

2. What are three benefits of grounding the secondary?

3. Explain how a voltage notch is formed.

4. Explain how power factor correction capacitors can cause a large inrush current.

5. Explain the difference between a general-purpose transformer and a drive isolation transformer.

CHAPTER

8

Autotransformers

Autotransformers are wired so that the primary and secondary circuits have a portion of their two windings in common. The power rating of an autotransformer is higher than that of an equivalent two-winding transformer. Autotransformers are limited by their low impedance and lack of electrical isolation between the primary and secondary circuits. Autotransformers are commonly used for reduced-voltage motor starting, in voltage regulators, and in variable transformers.

OPERATION

Transformers normally contain two independent circuits. The circuits are the primary circuit, which receives the energy from the source, and the secondary circuit, which delivers the energy to the load. An *autotransformer* is any transformer in which the primary and secondary circuits have a portion of their two windings in common. The power rating of an autotransformer is higher than that of an equivalent two-winding transformer.

Construction

An autotransformer is usually constructed from a single winding with multiple taps. **See Figure 8-1.** For a step-down application, the source is applied across the entire coil, which acts as the primary. The load is connected across only the part of the coil that acts as the secondary.

The electromagnetic field cuts across the entire winding. Therefore, the induced countervoltage is across the entire coil, and the design volts per turn is obtained across each turn of the entire coil. The voltage available to the load is less than the source voltage because there are fewer turns available to the secondary than the primary.

 DEFINITION

*An **autotransformer** is any transformer in which the primary and secondary circuits have a portion of their two windings in common.*

 **TECH FACT**

According to the NEC®, Article 411, Lighting Systems Operating at 30 Volts or Less, a secondary circuit shall be insulated from the branch circuit by an isolating transformer. Autotransformers cannot be used.

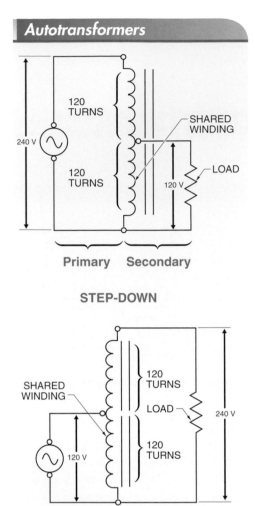

STEP-DOWN

STEP-UP

Figure 8-1. The primary and secondary of an autotransformer share part of the winding

For example, if a step-down autotransformer has 240 turns with a source voltage of 240 V, the source is applied across the entire coil and the volts per turn is 1 V/turn. If the secondary is tapped to use 120 of the turns, the voltage available to the load is 120 V.

For a step-up application, the source is applied across only part of the coil. The load is connected across the entire coil. The design volts per turn is again obtained across each turn of the entire coil. In this

case, the secondary voltage is higher than the source voltage because there are more turns available to the secondary than to the primary.

For example, if a step-up autotransformer has 240 turns with a source voltage of 120 V, the source is applied across only part of the coil. If the source is applied across only 120 of the turns, the volts per turn of the coil is 1 V/turn. If the secondary is tapped to use all 240 of the turns, the voltage available to the load is 240 V.

Two-Winding Transformer Loading. The current in a standard two-winding transformer is determined by dividing the power rating, in VA, by the voltage, in V. For example, if a transformer primary is rated at 12 kVA (12,000 VA) at 480 V, the maximum current flow is 25 A (12,000 ÷ 480 = 25). **See Figure 8-2.** The secondary must be rated at the same 12 kVA. If the secondary is rated to deliver 120 V, the transformer is designed to step down the voltage from 480 V to 120 V and deliver a load current of 100 A (12,000 ÷ 120 = 100).

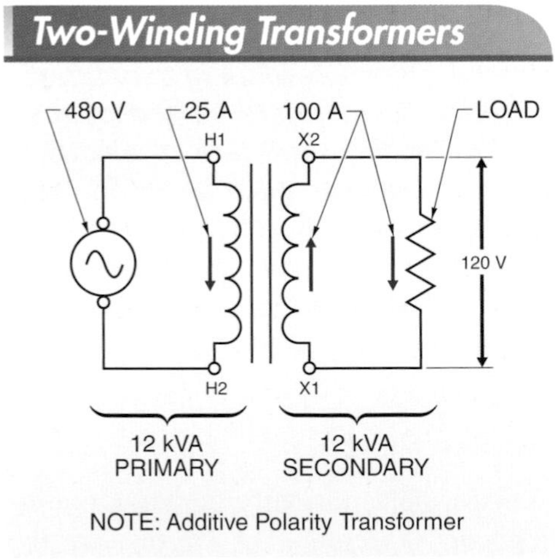

Figure 8-2. The primary and the secondary have the same power rating.

Two-Winding Autotransformer. An autotransformer can also be made from a two-winding transformer by connecting the two windings to form one continuous winding. When tap H2 is connected to tap X2, the windings are in series and the transformer operates as an autotransformer. The two windings are on the same core and connected in series. The primary and secondary windings can both be tapped to allow the autotransformer to operate as a step-up transformer or a step-down transformer.

Autotransformer Loading. The power rating of a single-coil autotransformer is higher than the rating of the individual windings of an equivalent two-winding transformer. When that same transformer is wired as an autotransformer by connecting H2 to X2, more power can be delivered to the load without exceeding the power rating of the windings.

The individual windings still have the same rating as they did originally, but they are now wired in series. The original primary still carries 25 A and the original secondary still carries 100 A. The original ratings are still being met.

However, the 25 A current flows through the new connection from the original primary to tap X2 of the original secondary. **See Figure 8-3.** The 100 A of induced current in the original secondary combines with the 25 A flowing through the primary, and 125 A flows to the load. As a result of converting the two-winding transformer to an autotransformer, the power available to the load increases from 12 kVA to 15 kVA. The current of the primary flows through the load and increases the kVA available for the load.

This shows that a two-winding transformer wired as an autotransformer can deliver more power to a load than when wired as a two-winding transformer. Alternatively, a single-coil autotransformer can be designed specifically for the application. This single-coil transformer can be smaller and less expensive than the two-winding transformer.

Limitations

Autotransformers have several limitations that prevent their widespread use in power circuits. A significant limitation of an autotransformer compared to a conventional transformer is the low impedance. Another limitation is a lack of isolation between the primary and the secondary.

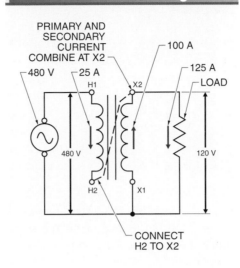

Autotransformer Loading

Figure 8-3. The current through the primary and the current induced in the secondary combine and flow through the load.

Low Impedance. The impedance of an autotransformer is less than the impedance of a two-winding transformer. For example, a two-winding transformer may have an impedance of about 3% of rated primary voltage. When wired as an autotransformer, the same windings have impedance that is equal to 3% of the ratio of the difference between the primary and secondary voltages and the primary voltage. The reduced impedance is calculated as follows:

$$Z_A = Z \times \frac{V_P - V_S}{V_P}$$

where

Z_A = autotransformer impedance (in %)

Z = two-winding transformer impedance (in %)

V_P = primary voltage (in V)

V_S = secondary voltage (in V)

For example, if the primary voltage is 240 V, the secondary voltage is 120 V, and the two-winding transformer impedance is 3%, the autotransformer impedance is calculated as follows:

$$Z_A = Z \times \frac{V_P - V_S}{V_P}$$

$$Z_A = 3\% \times \frac{240 - 120}{240}$$

$$Z_A = 3\% \times \frac{120}{240}$$

$$Z_A = 3\% \times 0.5$$

$$Z_A = \mathbf{1.5\%}$$

Short circuits are a significant concern. Because the impedance of an autotransformer is lower than that of a two-winding transformer, higher current flows when a short circuit fault occurs. **See Figure 8-4.** An autotransformer should be provided with external impedance in series in the circuit, such as from other transformers or reactors, or the impedance of the power lines.

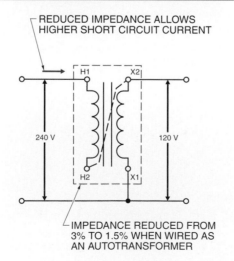

Reduced Impedance

REDUCED IMPEDANCE ALLOWS HIGHER SHORT CIRCUIT CURRENT

240 V   120 V

IMPEDANCE REDUCED FROM 3% TO 1.5% WHEN WIRED AS AN AUTOTRANSFORMER

Figure 8-4. The low impedance of an autotransformer allows higher short circuit current flow than a two-winding transformer.

Lack of Electrical Isolation. Autotransformers share part of the windings between the primary and secondary. The delta-wye or zigzag winding of standard transformers is no longer present. Therefore, autotransformers are not able to block the flow of transients and harmonics throughout a 3-phase power system.

During a fault condition, it is possible that part of the circuit may become open. In such a case, anyone coming in contact with the secondary is subject to the primary voltage. In addition, the primary may become grounded and establish a high voltage between one of the low-voltage conductors and ground. If the winding shorts to ground, it affects both windings and the load could be subjected to the high voltage. **See Figure 8-5.** The entire winding of an autotransformer should be designed with the insulation level required for the highest voltage expected.

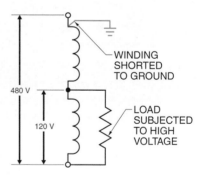

Lack of Isolation

WINDING SHORTED TO GROUND

480 V

120 V

LOAD SUBJECTED TO HIGH VOLTAGE

Figure 8-5. A short circuit from the primary winding to ground can subject the load to the full voltage.

For very large autotransformers, the cost of extra insulation increases the cost of the autotransformer faster than the reduction in copper reduces the cost. At a certain point, typically about 100 kVA or at a turns ratio of about 4:1, it becomes less expensive to use a two-winding transformer than an autotransformer.

APPLICATIONS

Autotransformers are used in many applications. A very common application is the use of autotransformers in reduced-voltage motor starting. Other applications include the use of autotransformers as voltage regulators and as variable transformers.

Reduced-Voltage Motor Starting

Reduced-voltage motor starting is a method of reducing the current through the motor windings and reducing the load on a power distribution circuit during motor starting. Autotransformer starting uses a tapped 3-phase autotransformer to provide reduced-voltage starting. Autotransformer starting is one of the most effective methods of reduced-voltage starting.

Induction motors and large synchronous motors use autotransformers to reduce the stator currents to a level that does not overtax the distribution systems that feed them. Synchronous motors have a two-part rotor. The rotor has a standard induction motor rotor section and a wound-rotor section. A synchronous motor is started as an induction motor and is accelerated to near synchronous speed using the induction part of the rotor. When the rotor approaches the synchronous speed, power is applied to the wound-rotor section and the motor pulls into synchronous speed. **See Figure 8-6.**

Load Current. Autotransformer starting is preferred over primary resistor starting when the starting current drawn from the line must be held below a maximum allowed value, yet the maximum starting torque per line ampere is required. Autotransformer reduced-voltage starting is used for starting blower, compressor, conveyor, and pump motors over about 10 HP.

Synchronous Motor Starting

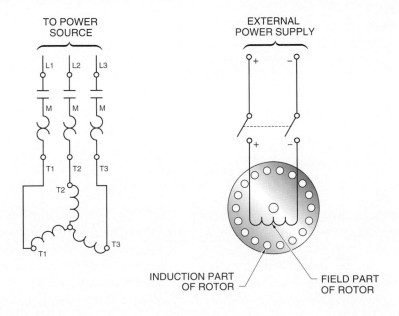

Figure 8-6. *Synchronous motors are started as induction motors to bring the rotor up to synchronous speed.*

Autotransformer reduced-voltage starting provides the highest possible starting torque per ampere of line current. However, because autotransformers are required, installation cost is higher than with other reduced-voltage starting methods. Autotransformer reduced-voltage starting can be used with any 3-phase motor.

In autotransformer starting, the motor terminal voltage does not depend on the load current. The current to the motor may change because of the motor's changing characteristics, but the voltage to the motor remains relatively constant.

Autotransformer starting may use the turns ratio advantage to provide more current on the load side of the transformer than on the line side. In autotransformer starting, transformer motor current and line current are not equal as they are in primary resistor starting.

For example, a motor may have a full-voltage starting torque of 120% and a full-voltage starting current of 600%. The electric utility commonly sets a limit of 400% current draw from the power line. This limitation is for the line side of the transformer. Because the transformer has a step-down ratio, the motor current on the transformer secondary is larger than the line current, even though the primary current of the transformer does not exceed 400%.

In this example, 80% voltage can be applied to the motor, generating 80% motor current. The motor draws only 64% of line current (80% of 80% = 64%) due to the 1:0.8 turns ratio of the transformer. **See Figure 8-7.**

A large electric motor may not be started very often. Therefore, the autotransformer can be overloaded during starting and then allowed to cool while the motor is running. Since autotransformers are only in the starting circuit for a short time, an autotransformer can be made smaller than a general-purpose transformer. It does not have to survive a temperature rise caused

by high sustained currents. This makes it essential for the technician to know how often a motor can be started in a certain time period. It is often possible that a motor cannot be restarted for a required length of time to allow cooling. A typical duty cycle is 10 sec ON and 10 min OFF.

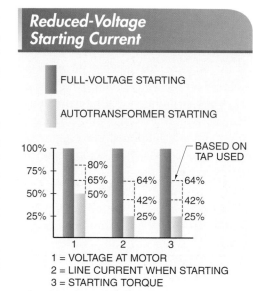

Figure 8-7. *Autotransformers are used in reduced-voltage motor starting to reduce the current drawn from the power line.*

Starting Circuits. Autotransformer reduced-voltage motor starting reduces the applied motor voltage to 50%, 65%, or 80% of the line voltage when starting. This is accomplished by placing a transformer coil in series with the motor for a given time period. After the time period, the motor is connected to full line voltage. The various windings of the transformer are added to and taken away from the motor circuit to provide reduced voltage when starting. The reduced voltage results in reduced current and torque.

Starting autotransformers are usually connected in an open delta or wye configuration. The open delta configuration is used for small motors and the wye configuration is used for large machines. The open delta configuration provides a simple means of voltage control, especially when more than one starting voltage is required. The wye configuration gives better voltage balance than the open delta configuration. Transformers connected in an open delta configuration require that the taps be changed on two coils only. Transformers connected in a wye configuration require that the taps be changed on three coils.

The different taps make an autotransformer very versatile by providing the option of different voltages for reducing the current in the stators of large motors. **See Figure 8-8.** Low stator current in a motor does not induce a strong pole in the rotor. If the rotor does not start, the next higher voltage can be selected for starting.

Reduced-Voltage Starting Circuit

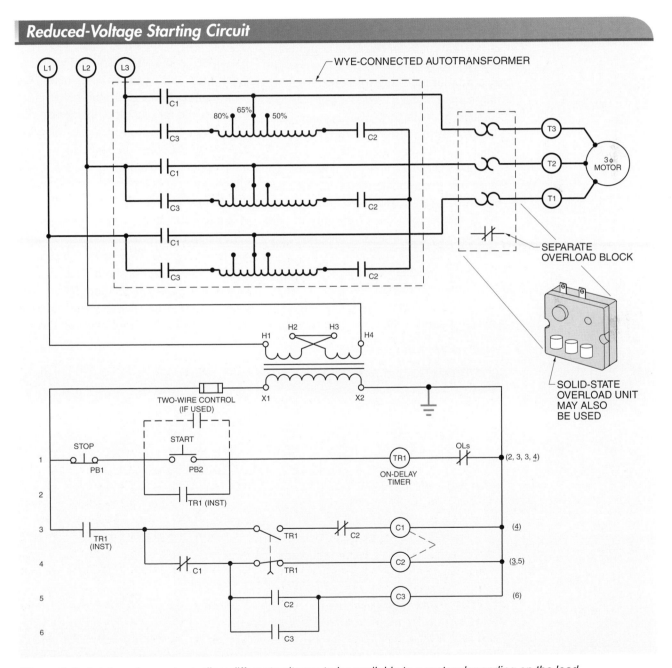

Figure 8-8. Autotransformer taps allow different voltages to be available to a motor depending on the load.

The control circuit consists of an ON-delay timer, TR1, and contactor coils C1, C2, and C3. Pressing start pushbutton PB2 energizes the timer, causing instantaneous contacts TR1 in line 2 and 3 of the line diagram to close. Closing the normally open (NO) timer contacts in line 2 provides memory for timer TR1, while closing NO timer contacts in line 3 completes an electrical path through line 4, energizing contactor coil C2. Energizing coil C2 causes NO contacts C2 in line 5 to close, energizing contactor coil C3. The normally closed (NC) contacts in line 3 also provide electrical interlocking for coil C1 so that they cannot be energized together. The NO contacts of contactor C2 close, connecting the ends of the autotransformers together when coil C2 energizes. When coil C3 energizes, the NO contacts of contactor C3 close and connect the motor through the transformer taps to the power line, starting the motor at reduced inrush current and starting torque. Memory is also provided to coil C3 by contacts C3 in line 6.

After a predetermined time, the ON-delay timer times out, and the NC timer contacts TR1 open in line 4, de-energizing contactor coil C2, and NO timer contacts

TR1 close in line 3, energizing coil C1. In addition, NC contacts C1 provide electrical interlock in line 4, and NC contacts C2 in line 3 return to their NC position. The net result of de-energizing C2 and energizing C1 is that the motor is connected to full line voltage.

Voltage Regulators

An autotransformer can be used to provide a slight boost (step up) or a slight buck (step down) to the line voltage to correct for small overvoltage or undervoltage conditions. The taps may be manually set, or automatic switchgear may be provided to adjust the taps in order to regulate the voltage at the desired value. This use allows changes in the voltage without interruption of the power to the loads, and allows for small changes in the voltage to the end user.

An autotransformer can be installed in combination with the tap changer to split the percentage of change as the changer is operated. **See Figure 8-9.** The position of the tap changer determines how much of the high-voltage winding is in the circuit. As the switches in the tap

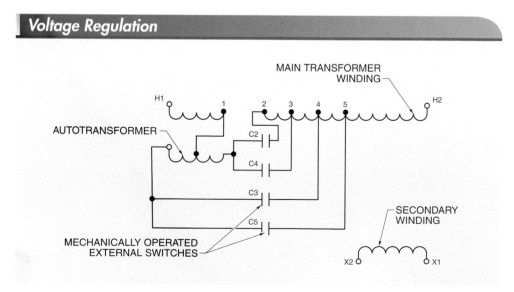

Figure 8-9. *Autotransformers are used with automatic switchgear to make small adjustments to the output of a two-winding transformer.*

changer close, varying amounts of the high-voltage winding provide flux for the secondary. The output can be adjusted to compensate for heavy loads. One end of a portion of the high-voltage winding can be connected to the middle of an auto-transformer. This method is used to take a 5% tap change and reduce it to a 2.5% change. This provides for closer control of the voltage in the circuit.

Variable Transformers

Autotransformers are commonly used as variable transformers. **See Figure 8-10.**

A variable transformer is a continuously adjustable autotransformer consisting of a single layer of wire wound on a toroidal core, and a carbon brush that traverses this winding. The brush track is made by removing a portion of the insulation from each turn of the winding, forming a series of commutator elements. The basic principle is that of a tap-changing trans-former. The brush is always in contact with one or more wires and continuously taps off any desired fraction of the wind-ing voltage. It is possible to remove the contact under load without interrupting the circuit.

Variable Transformers

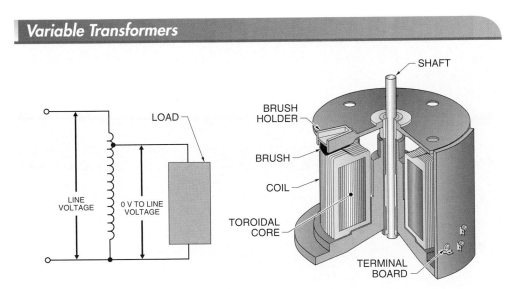

Figure 8-10. Variable autotransformers can be used whenever a variable voltage source is needed.

SUMMARY

- An autotransformer is constructed with a single winding that has multiple taps or with two windings that are coupled magnetically and electrically.

- An autotransformer can be used as a step-up transformer or as a step-down transformer.

- The power rating of an autotransformer is higher than the rating of an equivalent two-winding transformer.

- The impedance of an autotransformer is lower than an equivalent two-winding transformer.

- An autotransformer lacks electrical isolation between the primary and secondary.

- Autotransformers are often used in reduced-voltage motor starting.

- Autotransformers are often used to regulate voltage by boosting or bucking the line voltage.

- Variable autotransformers are used to vary the output when a changing level of AC voltage is needed.

DEFINITIONS

- An *autotransformer* is any transformer in which the primary and secondary circuits have a portion of their two windings in common.

REVIEW QUESTIONS

1. Explain how a two-winding transformer can be wired as an autotransformer.

2. Explain how an autotransformer can have a higher power rating than an equivalent two-winding transformer.

3. Why does an autotransformer have a limited duty cycle when used in reduced-voltage starting?

4. Why is it beneficial for an autotransformer to have several voltage taps when used in motor starting?

5. How can an autotransformer be used as a voltage regulator?

6. What are two limitations of autotransformers?

7. List several uses for autotransformers.

Transformer Principles and Applications

Buck-Boost
Transformers

CHAPTER

9

Buck-boost transformers are typically used to make relatively small changes to a source voltage. The source voltage may need to be boosted to raise it or may need to be bucked to lower it to the needed value. Buck-boost transformers are typically wired as autotransformers. Buck-boost transformers generally are not used to boost voltage where there is a fluctuating load because the voltage may become very high when the load is reduced. Common applications of buck-boost transformers are to boost voltage from 208 V to 230 V or to buck voltage from 240 V to 208 V to match the nameplate of existing equipment.

OPERATION

A *buck-boost transformer* is a small transformer designed to buck (lower) or boost (raise) line voltage. The major advantages of buck-boost transformers are their low cost and small size in comparison to general-purpose transformers.

Typical buck-boost transformers have dual-voltage primaries and secondaries. Buck-boost transformers are wired as autotransformers with the primary and secondary wired in series. **See Figure 9-1.** A buck-boost transformer is usually used to change the voltage by about 5% to 25%. Common primary voltages are 120 V, 240 V, or sometimes 480 V. Buck-boost transformers are typically available in sizes up to about 10 kVA.

DEFINITION

A buck-boost transformer is a small transformer designed to buck (lower) or boost (raise) line voltage.

 TECH FACT

Buck-boost transformers can be connected in parallel to supply a larger load than the rating of the individual transformers. The terminals must be connected in phase for like polarity, and they all must have the same turns ratio. The individual transformers share the load in inverse proportion to their impedances. The connection should be checked for current circulating between the individual transformers. Any significant current indicates a mismatch between the transformers. Individual overcurrent protection is required.

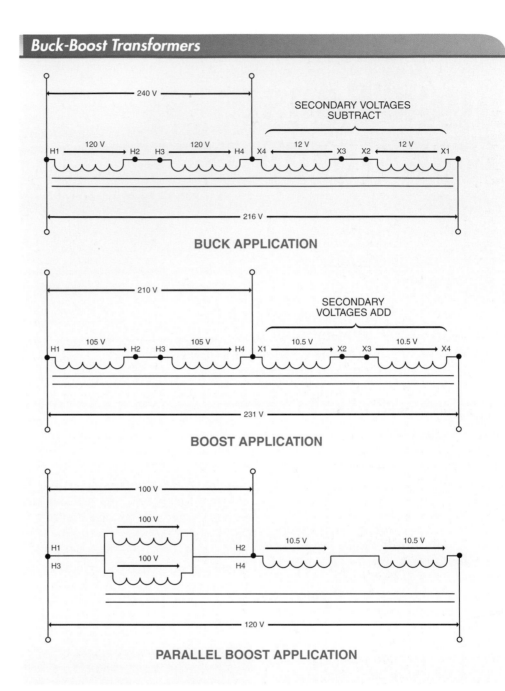

Figure 9-1. Buck-boost transformers reduce or increase the source voltage because the primary and secondary are wired in series with each other.

A common type of buck-boost transformer has the transformer primary rated at 120 V × 240 V and the secondary rated at 12 V × 24 V. This type of rating means that each of the two individual coils in the primary is rated at 120 V. When the primary coils are wired in series, a source of up to 240 V may be applied. When the coils are wound in series, this allows only 120 V to drop across each coil.

The voltage connected across the primary should never be allowed to exceed the rating of the coils. When the primary coils are wired in parallel, no more than 120 V may be applied across them.

An autotransformer may also be known as a compensator or balancing coil.

The amount of voltage available at the secondary to buck or boost the source depends on the voltage dropped across the primary. The polarity of the voltage that causes current to flow from H1 to H2 in the primary results in current flow from X1 to X2 in the secondary. Buck-boost transformers can be wired with subtractive polarity to buck the source voltage or with additive polarity to boost the source voltage.

For example, if the source is 240 V and the primary is wired in series, each primary coil drops 120 V. Each secondary coil then has 12 V induced in it. The secondary coils can be wired in subtractive polarity to buck a total of 24 V to bring the voltage down to 216 V.

The secondary coils can also be wired in additive polarity. If the source is 210 V and the primary coils are wired in series, each primary coil drops 105 V. Each secondary coil then has 10.5 V induced in it. The voltage across each secondary coil adds to the primary to boost a total of 21 V to bring the voltage up to 231 V.

Fusing

According to 2005 NEC® Section 450.4, *Autotransformers 600 Volts, Nominal, or Less*, each buck-boost autotransformer 600 volts, nominal, or less shall be protected by an individual overcurrent device installed in series with each ungrounded input conductor. The overcurrent protection device cannot be installed in series with the shared winding of the autotransformer. **See Figure 9-2.**

For autotransformers, the NEC® makes a distinction between circuits with a rated input current of less than 9 A and circuits with a rated input current of 9 A or more. When the current is less than 9 A, an overcurrent protective device is allowed to be rated at not

more than 167% of the input current. When the current is 9 A or more, the overcurrent protection device is allowed to be rated at not more than 125% of the full-load current of the autotransformer. When the current is 9 A or more and the calculated fuse rating does not correspond to a standard fuse rating, the next higher standard fuse rating is permitted. Standard fuse ratings are described in NEC® Section 240.6 *Standard Ampere Ratings*.

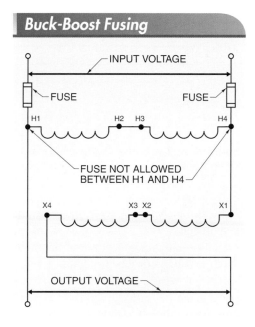

Figure 9-2. A fuse is not allowed between the H1 and H4 taps of the primary.

Grounding

NEC® Section 450.5, *Grounding Autotransformers,* covers the provisions for grounding autotransformers. Section 450.5(A) lists four general provisions for grounding autotransformers used to create 3-phase, 4-wire circuits from 3-phase, 3-wire distribution systems. **See Figure 9-3.** This section covers the provisions for point of connections, overcurrent protection, transformer fault sensing, and rating. A 2005 revision to the NEC® Section 450.5 clarified that zigzag connected transformers are not permitted to be installed on the load side of any system grounding connection.

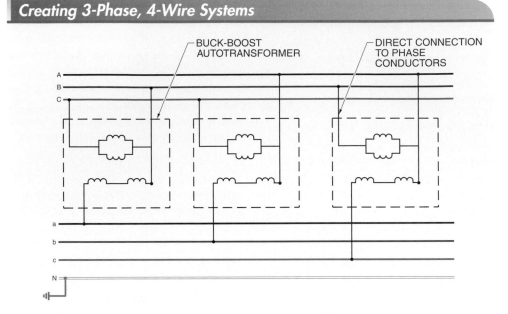

Figure 9-3. *Autotransformers must have a direct connection to the ungrounded phase conductors.*

Connections. An autotransformer may not be switched or provided with overcurrent protection that is independent of the main switch and overcurrent protection for the 3-phase, 4-wire system. The autotransformer must be directly connected to the ungrounded phase conductors.

Overcurrent Protection. An overcurrent protection device must be provided to cause the main switch or common overcurrent protection device to open if the load reaches 125% of its rating. However, delayed tripping is permitted to allow time for the common overcurrent protection device to open on branch or feeder circuits.

Fault Sensing. A fault-sensing system is required to guard against single-phasing or internal faults in 3-phase, 4-wire systems. The fault-sensing system can consist of current transformers wired to sense when an unbalance of 50% or more is present in the line current. **See Figure 9-4.**

Rating. The autotransformer needs to have a rating equal to the maximum possible unbalanced load current on the neutral of the 4-wire system.

 TECH FACT

A buck-boost transformer with two primary and two secondary windings can be connected in many different ways to provide a wide range of possible buck or boost voltages. This provides the flexibility required from a buck-boost transformer.

Limitations

There are several common limitations of buck-boost transformers. Buck-boost transformers should not be used to buck or boost a source feeding fluctuating loads, and should not be used to create a 240 V/120 V single-phase service from a 208Y/120 V source. NEC® limits the types of circuits where buck-boost transformers can be used. There are also limitations caused by the required overcurrent protection.

Fluctuating Loads. Buck-boost transformers are often used to adjust the voltage at the end of long transmission lines. However, buck-boost transformers should not be used to correct the voltage drop when the load fluctuates. The voltage drop on the line varies with load. When buck-boost transformers are designed and installed for high load conditions, unexpected high voltages may result during lightly loaded conditions.

240 V/120 V Service. Buck-boost transformers should not be used to create a 240 V/120 V single-phase service fed by a 208Y/120 V 3-phase, 4-wire source. If this is done, two neutrals would exist on the same circuit. This wiring arrangement creates unbalanced line-to-neutral voltages where one line is at 120 V and the other line is greater than 130 V. This connection is only permitted for creating a 240 V, 2-wire system. The impedance of the transformer is the only limit to the current of the unbalanced neutral.

Fault Sensing

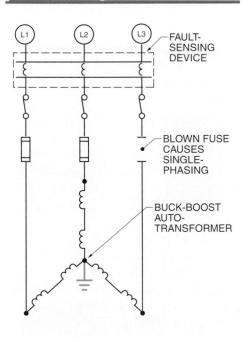

Figure 9-4. *NEC® Section 450.5 requires fault sensing in 3-phase, 4-wire systems to guard against single-phasing or internal faults.*

NEC® Section 210.9. Buck-boost transformers should not be used to buck or boost three units in a wye for creating a 3-phase, 4-wire wye circuit from a 3-phase, 3-wire delta circuit. The neutral created by this connection does not give proper line-to-neutral voltages under load. This application uses three buck-boost transformers in a 3-phase wye connection. This connection violates NEC® Section 210.9, *Circuits Derived from Autotransformers* Ex. No. 1. This wye connection can be used for 3-wire to 3-wire, 4-wire to 3-wire, and 4-wire to 4-wire applications.

Overcurrent Protection. If the fuse in the transformer primary blows, the secondary is still in series with the load across the supply. The secondary becomes a reactor in series with the load and drops the voltage at the equipment to a dangerously low level. If the overcurrent device on the feeder does not open, the motor or the secondary could be damaged. The voltage drop across the secondary could rise to a high level and the additional current could destroy the winding.

APPLICATIONS

A very common application of buck-boost transformers is where power line voltages are consistently above or below normal. Buck-boost transformers are used to make these small adjustments. Another common application is boosting 208 V to 230 V or bucking 240 V to 208 V. Many other buck-boost configurations are shown in the appendix.

Sizing

There are two common methods of sizing a transformer. The first method is to determine the load power and calculate the required size of the buck-boost transformer. The second method is to determine the load current and the buck or boost voltage and calculate the required size of the buck-boost transformer.

Load Power. One method of sizing a transformer is to calculate the load power, in kVA. This method can be used when the buck-boost transformer is wired as an isolation transformer and the entire load goes through the transformer secondary. **See Figure 9-5.** In the case of a single-phase motor, the power is the product of the current and the line voltage. The power is calculated as follows:

$$P = \frac{I \times E}{1000}$$

where

P = power (in kVA)
I = current (in A)
E = voltage (in V)

Motor Load Sizing

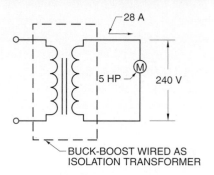

Figure 9-5. *Standard two-winding transformers can be sized by determining the load power.*

A motor nameplate gives the current at full load. For example, a 5 HP, 240 V single-phase motor may draw full load current of 28 A. The load power is calculated as follows:

$$P = \frac{I \times E}{1000}$$

$$P = \frac{28 \times 240}{1000}$$

$$P = \frac{6720}{1000}$$

$$P = \mathbf{6.72\,kVA}$$

TECH FACT

Buck-boost transformers rated for 60 Hz should be derated to operate at 50 Hz.

The load power is 6.72 kVA and the transformer secondary needs to be rated to carry that load. The user should select a transformer slightly larger than the expected maximum load. Manufacturers have selection tables to help determine appropriate transformer size.

Load Current. The alternative sizing method for buck-boost transformers is to calculate the power, in kVA, from the load current. This method can be used when the buck-boost transformer is wired as an autotransformer. **See Figure 9-6.** In the case of a single-phase motor, the power is the product of the current and the voltage buck or boost of the transformer. The power is calculated as follows:

$$P = \frac{I \times E}{1000}$$

where

P = power (in kVA)
I = current (in A)
E = voltage boost or buck (in V)

A motor nameplate gives the current at full load. For example, a 5 HP, 240 V, single-phase motor may draw full load current of 28 A. When the line voltage is 208 V, the buck-boost transformer can be used to raise the voltage to 236 V. This is a value that can be used by the 240 V motor. With the addition of the boost transformer on the line, the primary of the boost is across the line and the secondary is in series with the load across the line. The circuit current is now the total of the load and the transformer primary. The load power is calculated as follows:

$$P = \frac{I \times E}{1000}$$

$$P = \frac{28 \times 32}{1000}$$

$$P = \frac{896}{1000}$$

$$P = \mathbf{0.896\ kVA}$$

This calculation indicates that a 1 kVA buck-boost transformer can be added to a 208 V system to boost the voltage to 236 V to supply 28 A to a 5 HP motor.

Boosting at the End of a Line

Many loads at the end of a long line suffer from low voltage during peak load periods. The resistance of the lines drops the line voltage when the current flow is high at peak load.

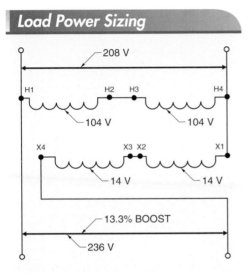

Load Power Sizing

Figure 9-6. Buck-boost transformers wired as autotransformers can be sized by determining the load current and multiplying by the buck or boost voltage.

Shelter. A buck-boost transformer can be used to boost the voltage at the end of a power line where there is voltage drop because of the length of the line. For example, a shelter (a small structure used for protection from the weather) is located a sizable distance from the service and contains a 120 V, 2 kW heater and a 100 W lamp. **See Figure 9-7.** With the light and heater OFF, the measured voltage at the shelter is near the nominal 120 V service.

There is a significant voltage drop as the current flows through the long line to the loads. With the light and heater ON, the measured voltage at the shelter drops considerably, perhaps to 95 V.

This is a drop of 25 V across the power line resistance. The heater output and the candlepower of the light are significantly reduced. At 95 V, the voltage is only 79% of nominal. Therefore, the current is also only 79% of nominal. The power dissipated is proportional to the square of the current. Therefore, the heater output is reduced to about 1250 W, greatly reducing its effectiveness on a cold winter day. The lamp load is reduced to about 63 W.

Sizing. A buck-boost transformer can be used to boost the voltage at the end of the line. In order to size the buck-boost transformer, the power required for the transformer can be calculated from the load current. The current can be calculated by dividing the power by the applied voltage as follows:

$$I = \frac{P}{E}$$

where
I = current (in A)
P = power (in W)
E = voltage (in V)

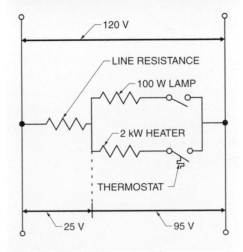

Loads on a Long Line

Figure 9-7. The voltage at the end of a long line drops because of the line resistance when a large load is connected.

For the 100 W lamp, the current is calculated as follows:

$$I = \frac{P}{E}$$

$$I = \frac{100}{120}$$

$$I = \textbf{0.833 A}$$

For the 2000 W heater, the current is calculated as follows:

$$I = \frac{P}{E}$$

$$I = \frac{2000}{120}$$

$$I = \textbf{16.667 A}$$

The total current, 17.5 A, is the sum of the individual load currents. The power required in the secondary of the buck-boost transformer is calculated by multiplying the current by the voltage rise in the transformer as follows:

$$P = \frac{I \times E}{1000}$$

where

P = power (in kVA)

I = current (in A)

E = voltage boost or buck (in V)

If the buck-boost transformer has a 7.5:1 turns ratio, the transformer can be wired to boost the voltage by 26.7%. At 95 V this is 25.3 V, and at 120 V this is 32 V. This is accomplished by placing the primary coils in parallel and across the line, and by placing the secondary 16 V coils in series with the load across the line. **See Figure 9-8.** The buck-boost transformer size can be calculated as follows:

$$P = \frac{I \times E}{1000}$$

$$P = \frac{17.5 \times 32}{1000}$$

$$P = \frac{560}{1000}$$

$$P = \textbf{0.560 kVA}$$

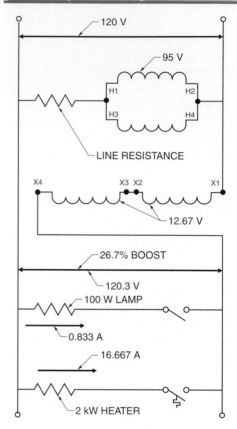

Figure 9-8. *Buck-boost transformers can be used to boost the voltage at the end of a long line.*

A buck-boost transformer larger than 0.560 kVA can be installed to boost the voltage to the required level. When both loads are turned on, the voltage at the loads is 120.3 V. The lamp and the heater operate normally.

Low Load. After the buck-boost transformer is installed at the shelter, the voltage can be measured. With the loads turned off, there is no current flow and no voltage drop on the power line to the shelter. The voltage will be boosted by the same percentage as before.

The lamp is a small part of the total load at the shelter. If the lamp is turned on, it draws only about 5% of the current that the heater draws, so the voltage drop across the line resistance is only about 5% of the 25 V drop shown previously. **See Figure 9-9.** Therefore, the lamp will see about 150 V instead of the 120 V it needs to operate efficiently. The current rises from 0.833 A to just over 1 A and the power dissipated by the lamp rises from 100 W to about 150 W. This increase in voltage results in a much shorter life for the lamp as the filament will burn out faster. This is why buck-boost transformers are not recommended for boosting when there is a varying load.

Wiring Alternatives. There are two common wiring alternatives that can reduce the problems with shortened lamp life. First, the lamp can be connected ahead of the buck-boost transformer. When the heater is not operating, the lamp operates normally. When the heater is operating, the lamp operates with a lowered voltage. The shelter will not have normal illumination, but the lamp will have a normal life.

A second alternative is to center tap the buck-boost transformer between the two 16 V coils and use a 130 V 100 W bulb. **See Figure 9-10.** The current drawn by this lamp on 130 V is 0.769 A. By tying the lamp circuit between the coils, the voltage available to the bulb is 107 V when the circuit is fully loaded, which provides 68 W. With the heater OFF, the lamp sees approximately 134 V and provides 109 W. This is within the design parameters of the lamp bulb.

Boosting Low Line Voltage

Many small plants in the middle of an industrial park or at the end of a long line suffer from low voltage during peak load periods. A 208Y/120 V 3-phase, 4-wire system seems to be especially vulnerable to low power during these peak periods. The phase-to-phase voltage drop can sometimes go as low as 185 V between lines, and the phase-to-neutral can go down to 107 V.

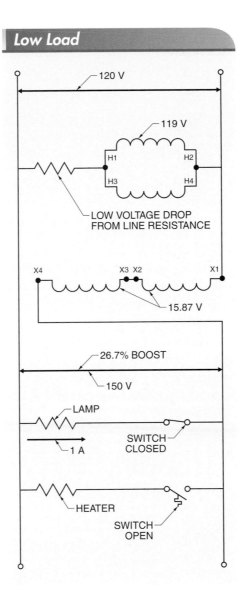

Low Load

Figure 9-9. Under low-load conditions, the voltage drop because of the line resistance is very small.

This situation describes a fluctuating load. As the loads are removed from the adjacent users, the supply voltage rises and the boosted voltage also rises from line to line and on the single-phase-to-neutral. The single-phase-to-neutral can rise to a level that is a problem for some equipment.

Motor Starting. Low line voltage reduces the available starting torque on 3-phase and single-phase motors, such as water pumps, water coolers, air conditioning compressors in window and rooftop units, and many other applications. Buck-boost transformers can be used to boost the line voltage up to the nominal level.

Serial Secondary. Starting with a 185 V source at times of peak load and a 120 × 240 V to 16 × 32 V buck-boost transformer, the 120 V coils can be put across the lines in parallel and the 16 V coils wired in series. This results in 107 V per coil with an output of 14 V on each coil in the secondary. **See Figure 9-11.** The two secondary coils give a boost of 28 V, which boosts the line-to-line voltage from 185 V to 213 V. This gives 123 V from phase to neutral, which is very near the nominal 120 V expected.

In off-peak times, the input to the buck-boost transformer rises back up to 208 V and the amount of boost rises in direct proportion to the increase in primary voltage. With 120 V on each of the primary coils, the output of each secondary coil rises to 16 V. With the secondary coils in series, a line-to-line value of 240 V is present along with a line-to-neutral value of 139.2 V. This is probably too high and can cause problems with electrical equipment designed for 120 V.

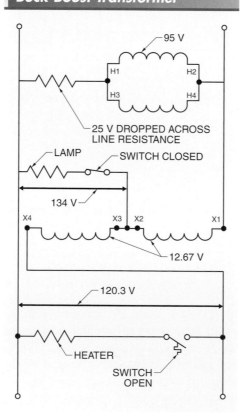

Center-Tapped Buck-Boost Transformer

95 V

H1   H2
H3   H4

25 V DROPPED ACROSS LINE RESISTANCE

LAMP   SWITCH CLOSED

134 V

X4   X3 X2   X1

12.67 V

120.3 V

HEATER

SWITCH OPEN

Figure 9-10. *A buck-boost transformer can be wired with a center-tapped secondary to provide intermediate voltages for some of the loads.*

GE Motors & Industrial Systems
Electric motors have a nameplate that gives the required operating voltage. Buck-boost transformers can be used to adjust the source voltage to create the required nameplate voltage.

Series Secondary

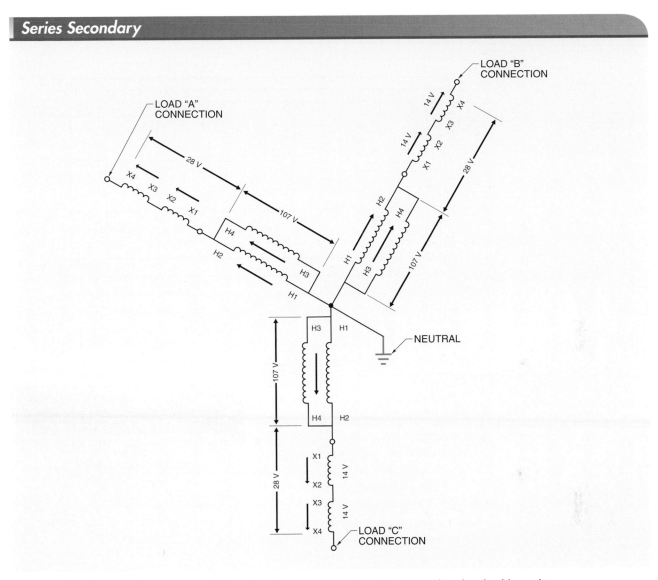

Figure 9-11. The maximum amount of buck or boost occurs when the secondary is wired in series.

Parallel Secondary. Starting with a 185 V source at times of peak load and a 120 × 240 V to 16 × 32 V buck-boost transformer, the 120 V coils can be put across the lines in parallel and the 16 V coils wired in parallel. This results in 107.3 V per coil with an output of 14.3 V on each coil in the secondary. **See Figure 9-12.** The two secondary coils give a boost of 14.3 V, which boosts the line-to-line voltage from 185 V to 199.3 V. This gives 115.6 V from phase to neutral, which is very near the nominal 120 V expected.

In off-peak times, the input to the buck-boost transformer rises back up to 208 V and the amount of boost rises proportionally. With 120 V on each of the primary coils, the output of each secondary coil rises again to 16 V. With the secondary coils in parallel, a line-to-line value of 224 V is present along with a line-to-neutral value of 129.2 V.

The user can select either the series or the parallel secondary, depending on need. If a change is needed between the two, it is simple to rewire the secondary to provide the solution.

Parallel Secondary

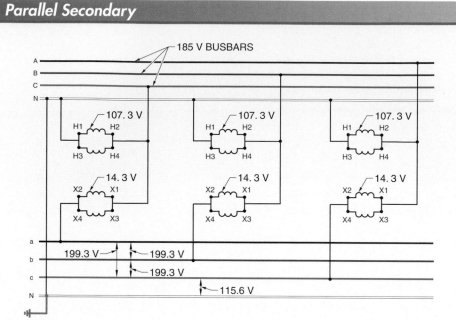

Figure 9-12. *The minimum amount of buck or boost occurs when the secondary is wired in parallel.*

Boosting 208 V to 230 V

There are applications where 208Y/120 V power is available but 230 V power is needed. This type of application uses a 230 V motor or a 230 V air conditioning system powered from a 208 V supply power line.

This required boost can be accomplished by wiring both the primary and secondary coils of a buck-boost transformer in series. The coils of the primary in series have 104 V across each one. With the turns ratio of the transformer being 7.5:1, the secondary coils have 13.9 V across each coil. This gives a boost of 27.8 V, giving just under 236 V line-to-line.

Parts Heater. A small industrial facility is fed with a 208Y/120 V service. **See Figure 9-13.** Small parts are heated in a small cabinet. The facility is concerned about the amount of time to heat the parts because it slows down production.

TECH FACT

AC voltage in power systems should be within −10% and +5% of rated voltage.

Parts Heating

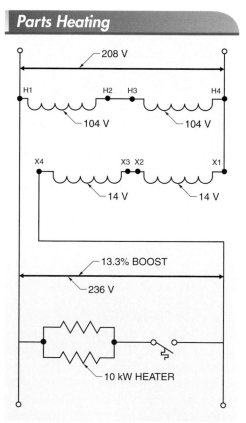

Figure 9-13. *Voltage can be boosted from 208 V to 236 V to supply a 230 V parts heater.*

The cabinet has two 5 kW, 230 V heating elements. This would give the cabinet 10 kW of heat at 230 V. Since only 208 V is available, the voltage is only 90% of the nominal 230 V required by the heaters. Therefore, the power is only 81% of the nominal 10 kW, or about 8.1 kW.

A voltage boost of about 10% is needed for the heaters to work at nominal load. The wiring configuration described above gives a 13.3% boost from 208 V to just under 236 V. This is within the voltage tolerances of most loads. Since this not a situation where the load is fluctuating, this is a good application for a buck-boost transformer.

Extruder Heater. A manufacturing plant is fed with a 208 V, 3-phase service. **See Figure 9-14.** The plant uses an extruder to heat plastic material in a 230 V machine. The material is forced through a barrel, similar to a cannon, by a large screw located in the bore. The barrel is surrounded by band heaters that provide heat to melt the plastic granules so that the material can be forced through a die to make various products.

Extruder Heater

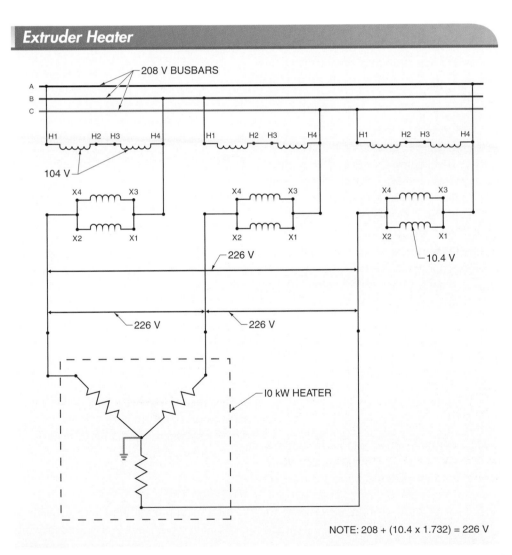

NOTE: 208 + (10.4 x 1.732) = 226 V

Figure 9-14. Voltage can be boosted from 208 V to 226 V to supply a 230 V extruder band heater.

The heated barrel, combined with the friction of the material being forced through the bore, brings the plastic material up to the desired temperature. Different plastics and different colors require different amounts of heat to melt, so each color has its own heat setting.

A 10 kW 230 V band heater on 208 V only supplies about 8.1 kW instead of the design 10 kW. Materials that have the highest heat settings are not melting properly because of the low voltage to the heaters. This type of extruder may have eight band heaters for each extruder. Therefore, the extruder heaters only provide about 65 kW of heat instead of the 80 kW of heat required.

The load on the extruder is calculated by checking the nameplate for the minimum circuit. In this case, the total load current is 227 A. A boost of 10% is enough to raise the 208 V source to nearly 230 V for the extruder heaters. A buck-boost transformer with a 10:1 winding ratio can be used to achieve this amount of boost. The primary is sized for 120 × 240 V while the secondary is sized for 12 × 24 V. The buck-boost transformer size for each phase is calculated as follows:

$$P = \frac{I \times E}{1000}$$
$$P = \frac{227 \times 24}{1000}$$
$$P = \frac{5448}{1000}$$
$$P = \textbf{5.448 kVA}$$

The actual voltage rise is only about 18 V (10.4 × 1.732 = 18). Therefore, the manufacturing plant should be able to install three 5 kW buck-boost transformers to bring the extruder heaters up to about 226 V (208 + 18 = 226). Larger buck-boost transformers may be better to minimize problems with heat buildup but can lower the power factor.

Bucking 240 V to 208 V

There are applications where it may be required to buck a supply voltage for special equipment. Motors are rated to operate at ±10% of the nominal voltage. A piece of equipment may be rated at 208 V in a location where 240 V is available. The voltage difference is enough that the source needs to be reduced in order to prevent an overcurrent situation.

Conveyor Motor. In the case of a 208 V rated motor turning a gear for a conveyor drive, a 240 V supply is too high. Since this is a nonfluctuating load, a buck-boost transformer is a good choice to reduce the voltage.

The voltage must be reduced by 32 V. A buck-boost transformer with a 7.5:1 winding ratio wired with subtractive polarity can be used to achieve this amount of buck. The buck-boost transformer can be wired in an open delta with only two transformers supplying the load. **See Figure 9-15.** The primary is sized for 120 × 240 V while the secondary is sized for 16 × 32 V. If the motor pulls 24 A, the size of the buck-boost transformer can be calculated as follows:

$$P = \frac{I \times E}{1000}$$
$$P = \frac{24 \times 32}{1000}$$
$$P = \frac{768}{1000}$$
$$P = \textbf{0.768 kVA}$$

The manufacturer can install two 1 kVA buck-boost transformers to supply the load. The open delta is popular because only two transformers are needed.

Bucking 240 V to 200 V

There are applications where a piece of equipment operates at a nonstandard voltage. This may occur when the equipment is modified locally or when it is purchased from another country.

Conveyor Motor

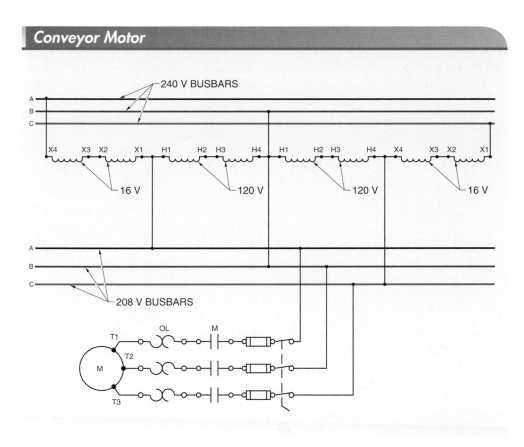

Figure 9-15. *A buck-boost transformer can be wired in an open delta when the load is relatively low to buck the voltage from 240 V to 208 V.*

Oil Heater. A manufacturing plant has 240 V service available. A piece of equipment that was built overseas has a built-in oil heater that is rated at 10 kW at 200 V on 50 Hz. The frequency is not a factor since the only opposition to current flow is the resistive heater. If the heater is operated at 240 V instead of the nominal 200 V, the power is 14.4 kW instead of the nominal 10 kW. This extra power is well above the design power and could cause overheating and a possible fire.

The 240 V source needs to be reduced to a value near the nominal 200 V required by the oil heater. A buck-boost transformer with a 7.5:1 winding ratio wired with subtractive polarity can be used to buck the source down to 208 V. **See Figure 9-16.** The primary is sized for 120 × 240 V while the secondary is sized for 16 × 32 V.

Oil Heater

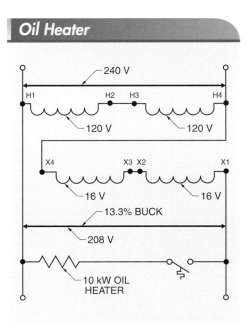

Figure 9-16. *Voltage can be bucked from 240 V to 208 V to supply a 200 V heater.*

At 208 V, the heater provides 10.8 kW of heat and draws 52 A. The size of the buck-boost transformer can be calculated as follows:

$$P = \frac{I \times E}{1000}$$

$$P = \frac{52 \times 32}{1000}$$

$$P = \frac{1664}{1000}$$

$$P = \textbf{1.664 kVA}$$

A buck-boost transformer larger than 1.664 kVA and wired in subtractive polarity needs to be installed. This bucks the voltage down to 208 V.

Other Applications

There are many applications where a technician may need to determine the proper wiring connections to develop a particular voltage.

This requires knowledge of the available source voltage, the desired output voltage, and the type of buck-boost transformer available. There are three very common transformers available for use as buck-boost transformers, all with dual-wound primaries and secondaries.

These common transformers are a 120 V × 240 V with a 10:1 turns ratio that results in either 12 V or 24 V from the secondary; a 120 V × 240 V with a 7.5:1 turns ratio that results in either 16 V or 24 V from the secondary; and a 240 V × 480 V with a 10:1 turns ratio that results in either 24 V or 48 V from the secondary.

The most common case is where a known source voltage is available, but it is not correct for the particular application. A technician can use a table of common voltages to determine the correct wiring arrangement. **See Figure 9-17.**

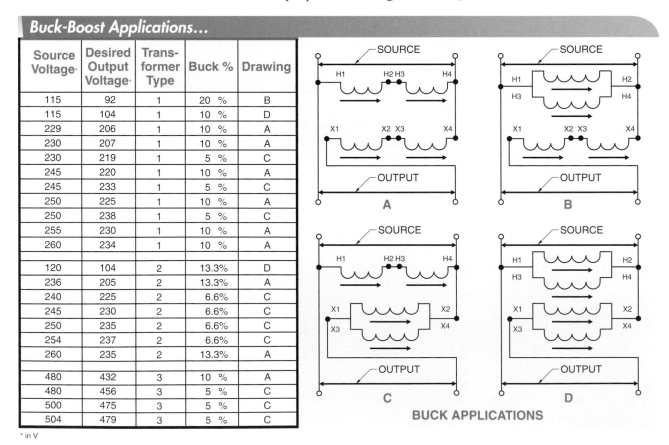

Buck-Boost Applications...

Source Voltage*	Desired Output Voltage*	Transformer Type	Buck %	Drawing
115	92	1	20 %	B
115	104	1	10 %	D
229	206	1	10 %	A
230	207	1	10 %	A
230	219	1	5 %	C
245	220	1	10 %	A
245	233	1	5 %	C
250	225	1	10 %	A
250	238	1	5 %	C
255	230	1	10 %	A
260	234	1	10 %	A
120	104	2	13.3%	D
236	205	2	13.3%	A
240	225	2	6.6%	C
245	230	2	6.6%	C
250	235	2	6.6%	C
254	237	2	6.6%	C
260	235	2	13.3%	A
480	432	3	10 %	A
480	456	3	5 %	C
500	475	3	5 %	C
504	479	3	5 %	C

* in V

BUCK APPLICATIONS

Figure 9-17. *A table of common input and desired output voltages can be used to determine the correct wiring arrangement for a buck-boost transformer.*

...Buck-Boost Applications

Source Voltage*	Desired Output Voltage*	Transformer Type	Boost %	Drawing
100	110	1	10 %	H
100	120	1	20 %	F
105	116	1	10 %	H
105	126	1	20 %	F
110	121	1	10 %	H
115	127	1	10 %	H
189	208	1	10 %	E
208	229	1	10 %	E
215	237	1	10 %	E
220	242	1	10 %	E
225	248	1	10 %	E
230	242	1	5 %	G
230	253	1	10 %	E
95	120	2	26.6%	F
100	126	2	26.6%	F
105	119	2	13.3%	H
110	125	2	13.3%	H
120	136	2	13.3%	H
120	136	2	13.3%	H
120	152	2	26.6%	F
208	236	2	13.3%	E
210	224	2	6.6%	G
215	229	2	6.6%	G
215	244	2	13.3%	E
220	235	2	6.6%	G
220	250	2	13.3%	E
225	240	2	6.6%	G
230	246	2	6.6%	G
231	277	3	20 %	F
385	424	3	10 %	E
416	458	3	10 %	E
425	468	3	10 %	E
430	473	3	10 %	E
435	457	3	5 %	G
440	462	3	5 %	G
440	484	3	10 %	G
460	483	3	5 %	E

* in V

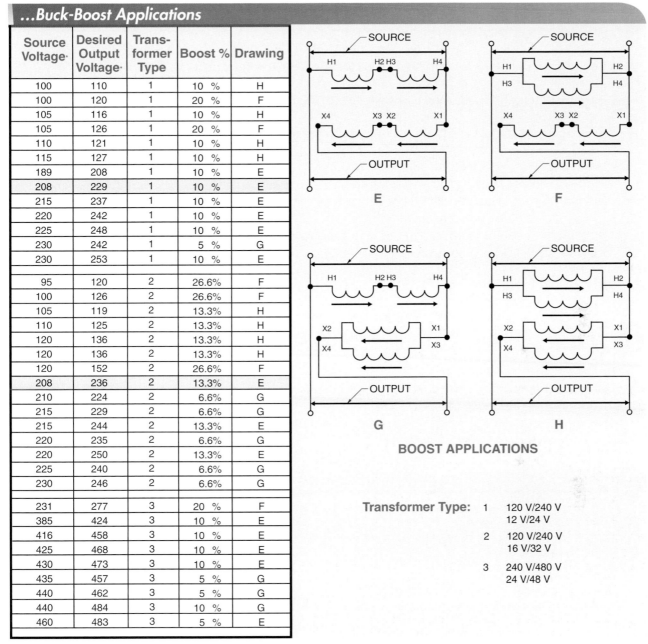

BOOST APPLICATIONS

Transformer Type:
1 120 V/240 V 12 V/24 V
2 120 V/240 V 16 V/32 V
3 240 V/480 V 24 V/48 V

The first step in using this table is to decide whether a buck or a boost is needed to adjust the source to the desired output voltage. This establishes which table must be used. The next step is to determine the source voltage available and the desired output voltage. A set of numbers near those voltages is probably in the table. After the correct row in the table is identified, the transformer type and the drawing are selected.

For example, if a 208 V source is available and 230 V is desired, the Boost Application table must be used. Consulting the table, there are two rows that are appropriate. A transformer with a 120 V × 240 V primary and a 12 V × 24 V secondary (Type 1) can be used to boost the voltage to 229 V. A transformer with a 120 V × 240 V primary and a 16 V × 32 V secondary (Type 2) can be used to boost the voltage to 236 V. In both cases, Drawing E gives the correct wiring connections.

SUMMARY

- Buck-boost transformers are typically used to buck (reduce) or boost (raise) a source voltage.

- Buck-boost transformers usually have dual-voltage primaries and secondaries.

- Buck-boost transformers generally should not be used to boost the voltage when there are fluctuating loads.

- The two methods of sizing buck-boost transformers are matching the load power requirement and determining the load current and the buck or boost voltage.

- Buck-boost transformers can have the primaries and secondaries wired in series or in parallel to give a wide range of buck and boost options.

DEFINITIONS

- A ***buck-boost transformer*** is a small transformer designed to buck (lower) or boost (raise) line voltage.

REVIEW QUESTIONS

1. Why should caution be exercised when a buck-boost transformer is used to raise voltage that has dropped due to peak loads?

2. What are the two methods used when sizing single-phase buck-boost transformers?

3. If a load of 1500 W on 120 V has a line loss of 10%, calculate the buck-boost transformer size, in kVA, required to raise the voltage back to nominal.

Transformer Principles and Applications

Special Transformers

There are several special types of transformers that are similar to the more common types of transformers, but have special features for particular applications. Control transformers are used to provide low-voltage power to control systems and devices. Potential transformers are used to measure high voltage present on power lines. Current transformers are used to measure high current present on high-power circuits. Variable transformers provide any voltage within the design range of the transformer.

CONTROL TRANSFORMERS

A *control transformer* is a transformer that is used to step down the voltage to the controls and control devices of a circuit or machine. The reduced voltage provides a much safer environment for technicians working on the equipment. Residential heating, ventilating, and air conditioning systems and other machines use reduced voltage for the controls, thermostats, relays, and contactors. Control transformers are often referred to as power supplies.

Operation

The most common control transformers have two primary coils and one secondary coil. The primary coils of a control transformer are crossed so that the metal links can be used to connect the primaries for either 240 V or 480 V operation. **See Figure 10-1.** Many control transformers are designed to provide 12 V or 24 V power from the secondary. However, some control transformers provide 120 V power. For example, a control transformer with a 10:1 turns ratio can have the two primary coils wired in parallel. This steps down the primary voltage from 240 V to 24 V on the secondary. When the control transformer has the two primary coils wired in series, the voltage is stepped down from 480 V to 24 V.

A control transformer is a transformer that is used to step down the voltage to the controls and control devices of a circuit or machine.

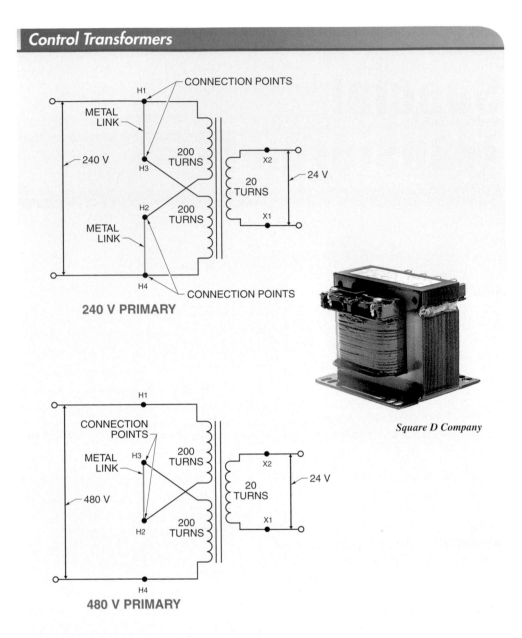

Square D Company

Figure 10-1. *In order to deliver 120 V, the control transformer primary coils can be connected in parallel for a 240 V source or in series for a 480 V source.*

Some plants have a 600 V system that requires the control transformer primary windings to be rated for 600 V. This type of transformer typically has a dual-voltage secondary, which makes this unit highly versatile. **See Figure 10-2.** This type of transformer typically delivers 120 V or 240 V power from the 600 V source. Control transformer sizes typically range from 50 VA to 3 kVA.

Sizing

There are three characteristics of a circuit that must be determined before selecting a control transformer. The three characteristics are sealed VA, inrush VA, and inrush load power factor. The sealed VA is the power, in VA, that the transformer must deliver to the load for an extended period of time. The inrush VA is the power, in VA, that the transformer must deliver to the load upon startup. The inrush VA can often be 10 times the sealed VA. The inrush load power factor is the power factor of the inrush current. The inrush load power factor is difficult to determine, but 40% is a reasonable estimate for most situations. Manufacturers provide tables to help in selection.

The size of a control transformer can be determined by adding the power requirements of the components in the control circuit. For example, a heating, ventilating, and air conditioning (HVAC) building controller may require 10 VA of 24 V power. The controller also provides power to two humidity sensors and two electronic/pneumatic transducers. The two sensors and two transducers require 3 VA each for a total of 12 VA. Therefore, the control transformer must be sized to provide at least 22 VA (10 + 12 = 22) of sealed VA.

Undersized Control Transformers. Processes often change over time, with the result that control systems are modified to match the changes. Additional relays and contactors may be installed without considering the size of the control transformer. If the control transformer is not replaced, it may be undersized for the new circuits. When the total load connected to a control transformer exceeds the rating of the transformer, the voltage output of the transformer decreases.

An undersized control transformer can make it difficult to troubleshoot a circuit. A technician needs to understand how the other equipment in the circuit should operate. For example, a relay, contactor, or starter has two power ratings. The lower rating is the hold-in VA rating. The higher rating is the pull-in VA rating. The modified control system may have enough power to hold in an armature, but it may not have enough power to pull in the armature. Therefore, the relay, contactor, or starter cannot close and complete the circuit. However, if the armature is manually pushed in, the circuit will operate normally.

It can also be difficult to determine when a control transformer is undersized because most troubleshooting measurements are taken when all or almost all loads in a circuit are OFF. A digital multimeter set to record the minimum and maximum voltage can be used to detect a low-voltage condition caused by circuit overloading. **See Figure 10-3.**

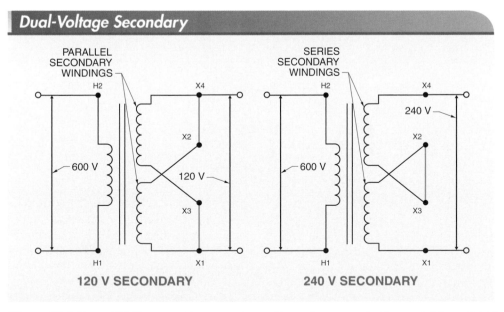

Dual-Voltage Secondary

PARALLEL SECONDARY WINDINGS

SERIES SECONDARY WINDINGS

120 V SECONDARY

240 V SECONDARY

Figure 10-2. For a 600 V source, the secondary coils can be connected in parallel for a 120 V secondary or in series for a 240 V secondary.

Undersized Control Transformers

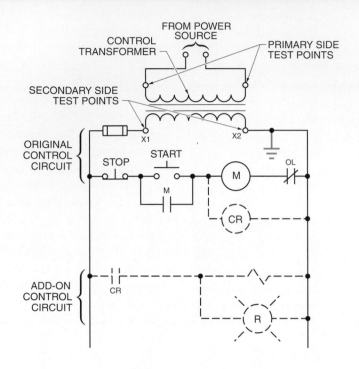

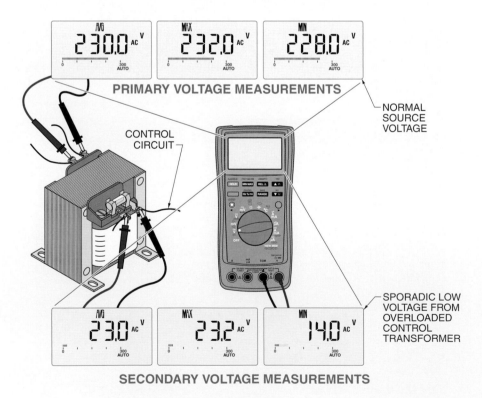

Figure 10-3. A DMM set in MIN MAX recording mode can be used to identify a circuit with an overloaded control transformer.

Overcurrent Protection

NEC® Section 430.72(C) lists requirements for transformers used in motor control circuits. This section says that control transformers with a current in the primary winding of less than 2 A shall be protected by an overcurrent device of 500% or less of the rated primary current. Accessories such as fuse kits that mount on the unit and covers that isolate the front and terminal strips are readily available.

It is common industry practice to protect a control transformer on both the primary and secondary sides. The reason for fusing both sides of a control transformer is to protect the primary if there is a short in the secondary. For example, a heating element contactor controlled by a temperature switch is used to control a 3-phase heater. **See Figure 10-4.** The control circuit has a fuse between the control transformer and the switch. The power circuit has fuses between the source and the control transformer. If either circuit has an overload, a fuse will open the circuit.

The phase difference between the primary and the secondary is 180°. This tells us that, as the current in the primary flows from H1 to H4 in the primary, the current in the secondary flows in the opposite direction from X1 to X2. As current in the secondary increases, the flux increases. This flux has opposite polarity to the flux in the primary and the fluxes cancel. This lowers the reactance and the current increases.

Control Transformer Fusing

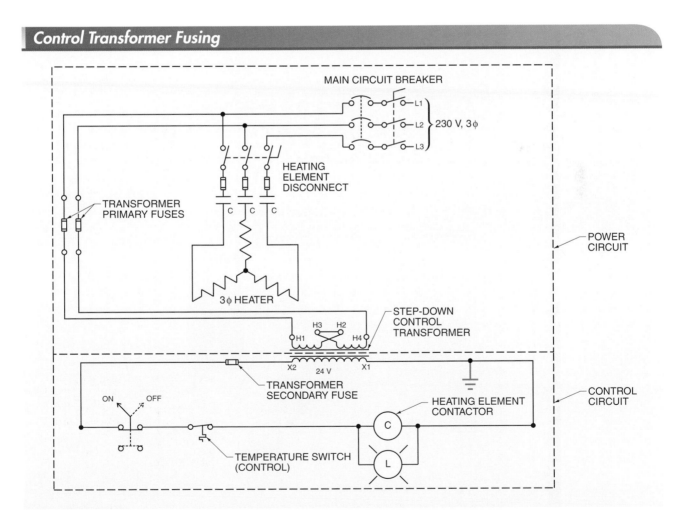

Figure 10-4. *A control transformer is normally fused on the primary and secondary sides for maximum protection.*

A short in the secondary results in a high current in the secondary. The high current increases the flux that opposes the flux induced in the primary. This results in higher current in the primary. As the current increases in the primary, the reactance decreases and the primary becomes more resistive than reactive. The reduced reactance allows too much current to flow in the primary. This change in current happens very quickly, and a short in the secondary can induce enough current in the primary to cause damage. Therefore, both sides of a control transformer should be fused.

Grounding

A control transformer may have one side of the secondary grounded. A *grounded system* is a control system with a grounded secondary in the control transformer. When one side of the transformer is grounded, a voltmeter can easily be used to test the system. One lead of the voltmeter can be connected to ground and the other lead can be used to measure voltages in the control circuit.

A control transformer may have the secondary ungrounded. A *floating system* is a control system without a grounded secondary in the control transformer. When the secondary is not grounded, a voltmeter cannot be used to measure voltage to ground to troubleshoot the control circuit. Voltages in the control circuit must be measured relative to one side of the control transformer. **See Figure 10-5.**

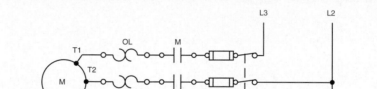

DEFINITION

*A **grounded system** is a control system with a grounded secondary in the control transformer.*

*A **floating system** is a control system without a grounded secondary in the control transformer.*

Floating System Voltages

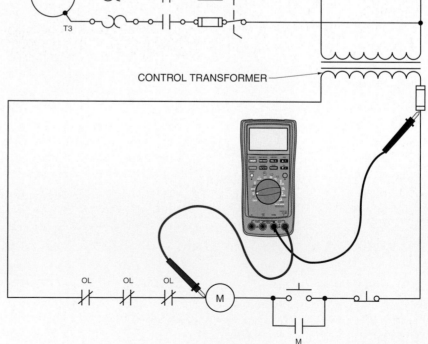

CONTROL TRANSFORMER

Figure 10-5. *Voltages in a floating system must be measured relative to one side of the control transformer instead of to ground.*

Power supply isolation from ground may be necessary to prevent ground loops. A *ground loop* is a circuit that has more than one point connected to earth ground, with a voltage potential difference between the two ground points. The voltage potential difference can produce a circulating current in the ground system. Isolation transformers are used to prevent a ground loop from causing problems in a circuit. **See Figure 10-6.**

HVAC Control Systems. Control transformers for HVAC systems are usually rated at 120 V or 240 V on the primary and 24 V on the secondary. **See Figure 10-7.** This allows a low-voltage jacketed cable to be run from the control panel to the thermostat. Many commercial applications running HVAC units at 240 V or 480 V 3-phase systems use a dual-voltage control transformer with a 24 V secondary.

DEFINITION

*A **ground loop** is a circuit that has more than one point connected to earth ground, with a voltage potential difference between the two ground points.*

Isolation Transformer

Figure 10-6. Isolation transformers can be used in controllers to prevent a ground loop from causing problems in the circuit.

Applications

Most machines in industry contain a transformer in the panel to provide power for logic inputs, relays and solenoids, panel lighting, and numerous other loads in industrial process control. When a disconnect is pulled to remove power at the control panel, the control power is also turned off.

Most control transformers used in control panels are the dual-voltage type, just like the motors that power the equipment. Common applications of control transformers are to provide power for HVAC control systems and to provide power to control a heating element.

HVAC Control Systems

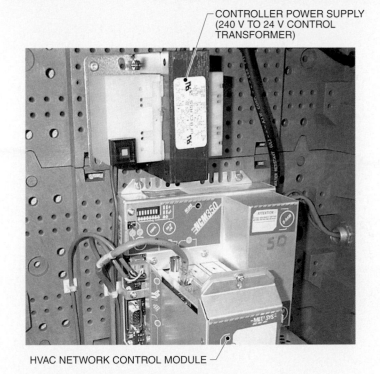

Figure 10-7. HVAC control systems often use control transformers to step down the 240 V supply to 24 V for the control circuit.

TECH FACT

Small control transformers operating at 24 V and supplied from 120 V or 240 V sources are common in small commercial buildings and residential complexes. These transformers are used to supply control voltage for HVAC units.

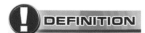

DEFINITION

*A **bell transformer** is a transformer used to supply low-voltage, low-power circuits, such as for doorbells, annunciators, and similar systems.*

Heating Element Control. A control transformer may be used to supply power to contactors for a heating element. **See Figure 10-8.** The control transformer provides low-voltage power to the contactors. A control device, such as a float switch, pressure switch, thermostat, or other device, closes a contact. This allows power to flow to the heaters. For safety reasons, heater controls are usually designed to be normally open and only operate when needed.

Heating Element Control

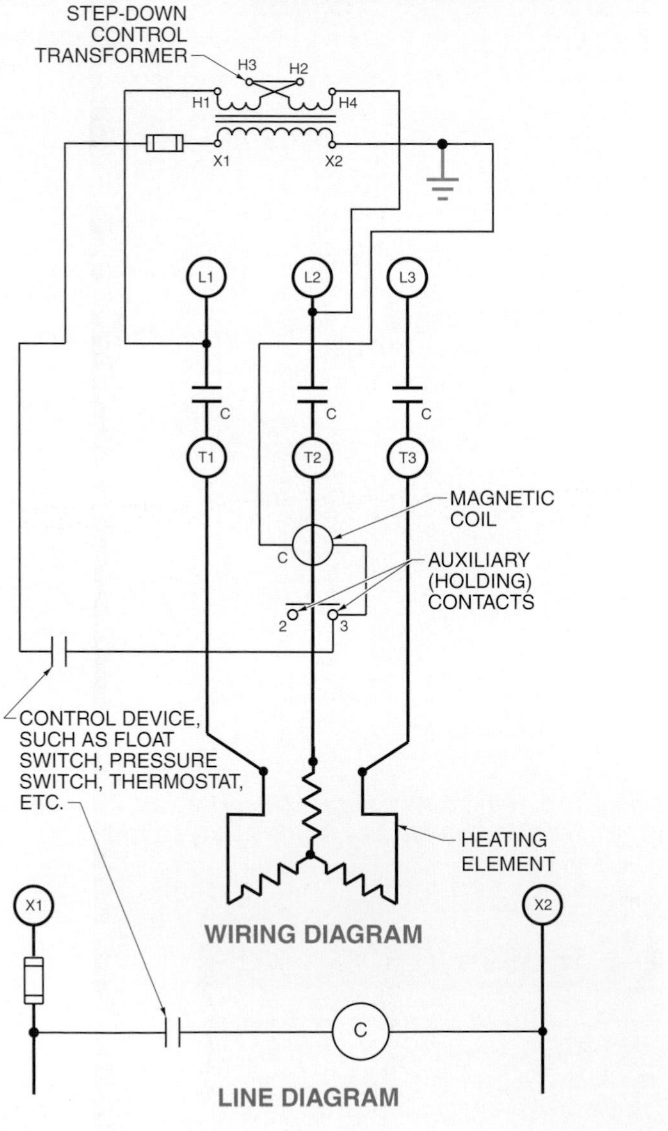
STEP-DOWN CONTROL TRANSFORMER
H3
H2
H1
H4
X1
X2
L1
L2
L3
C
C
C
T1
T2
T3
MAGNETIC COIL
C
AUXILIARY (HOLDING) CONTACTS
2
3
CONTROL DEVICE, SUCH AS FLOAT SWITCH, PRESSURE SWITCH, THERMOSTAT, ETC.
HEATING ELEMENT
WIRING DIAGRAM
X1
X2
C
LINE DIAGRAM

Figure 10-8. *A heating element is controlled by contacts powered by the low-voltage supply from a control transformer.*

Bell Transformers. A *bell transformer* is a transformer used to supply low-voltage, low-power circuits, such as for doorbells, annunciators, and similar systems. The primary application for bell transformers is as power supplies for doorbells and annunciators. Other common applications include thermostat circuits, music or intercom systems, burglar alarms, and hard-wired water sprinkler systems.

Bell transformers operate like a standard step-down control transformer. They usually have an insulated panel with four screws for connections. One of these screws is the common and the other three are taps that allow different voltages to be taken off the coil. These taps are usually 6 V, 12 V, 16 V, and sometimes 24 V. **See Figure 10-9.** The primary is usually 120 V, but may be 240 V.

Small bell transformers are usually rated for about 20 VA to 40 VA and can be mounted in a number of fashions. The primary side may have a threaded nipple that allows it to be attached to a box cover with a locknut through a knockout. It may also have a tab with a setscrew at the end that also fits through a knockout for mounting. These transformers also come equipped with mounting feet for control panel applications.

POTENTIAL TRANSFORMERS

The high voltages typically seen on power lines are a hazard to technicians working on or near the power lines. It is a difficult task to design a voltmeter to measure these high voltages. A *potential transformer* is a precision two-winding transformer used to step down high voltage to allow safe voltage measurement.

Operation

The stepped-down voltage from a potential transformer can be measured directly with a voltmeter. The line voltage can be calculated by multiplying the measured voltage by the turns ratio of the transformer. However, a better solution is to use a modified voltmeter.

Bell Transformers

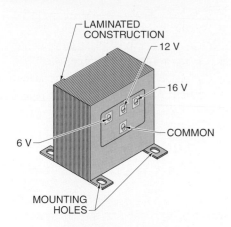

SECONDARY SIDE

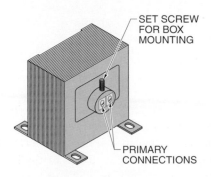

PRIMARY SIDE

Figure 10-9. A bell transformer has taps for 6 V, 12 V, and 16 V to power low-power devices such as doorbells and annunciators.

The display of the meter can be modified with a new meter face or programmed to show a value corresponding to the actual line voltage, even though a stepped-down value was actually measured. The meter can multiply the measured value by the turns ratio to display the actual line voltage.

The primary side of the potential transformer is connected across the power lines. **See Figure 10-10.** Fuses can be added for safety and to make it easy to remove the potential transformer from the circuit for maintenance. To ensure that the voltage measurement is as precise as possible, the load on the potential transformer must be

kept to a minimum. The voltmeter should be a high-impedance model to draw as little current as possible from the transformer. This nonchanging load keeps the voltage ratio constant and as the primary voltage changes, the secondary voltage changes proportionally.

Potential transformers can be made with almost any turns ratio so that the voltage can always be reduced to 120 V. This allows standard voltage meters to be used. For example, a potential transformer with a turns ratio of 60:1 can be used to measure 7200 V. The same meter can be used to measure 34.5 kV if a potential transformer with a turns ratio of 287.5:1 is used to step down the line voltage to 120 V. In this case, a different multiplication factor is used in the display or a different face installed on the meter.

Potential transformers are usually fairly small. They are typically rated at 500 VA or less. Most of the size of a potential transformer is the heavy insulation on the primary winding required to withstand the high voltages present on power lines.

*A **potential transformer** is a precision two-winding transformer used to step down high voltage to allow safe voltage measurement.*

Control tranformers step down the line voltage for use in control systems.

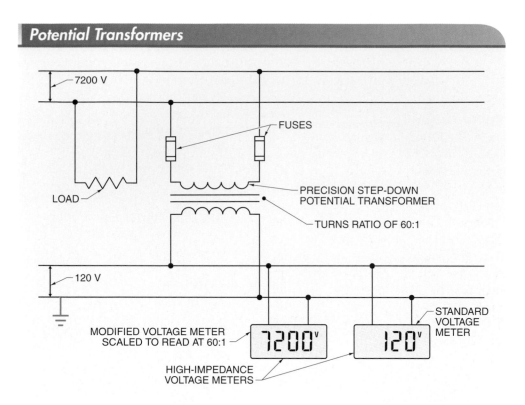

Figure 10-10. *A potential transformer is used to step down the high voltage of a power line in order to make it easier to measure.*

Accuracy. Potential transformers are often used for metering and billing. Therefore, the accuracy of potential transformers is critical. ANSI has established standard methods of classifying potential transformers for accuracy and load. The accuracy classification includes the standard load as well as the maximum percent error allowed.

The design, construction, and installation of the transformer all affect the accuracy. The load rating must include the total of all loads including the circuit wiring connected to the secondary of the transformer. The total load must be calculated and the proper transformer selected from a table provided by the manufacturer.

The transformer correction factor is a number provided by the manufacturer that is used as a multiplier to correct for inaccuracies. The correction factor corrects for effects of magnetizing current or internal phase angle shift created by the internal inductance of the transformer.

The transformer correction factor is used to define an accuracy class. Typical values of the accuracy class are 0.3, 0.6, and 1.2. A lower accuracy class number means a more accurate current transformer.

Applications

Potential transformers have several common uses. Important uses for potential transformers are as voltage meters, to feed voltage relays, and for load shedding during peak load periods.

Voltage Relays. Potential transformers are often used as part of a system to monitor voltage on power lines. A sudden fall or rise in the voltage activates an undervoltage or an overvoltage relay. An undervoltage relay is switched when the voltage drops below a setpoint. An overvoltage relay is switched when the voltage rises above a setpoint. Under- and overvoltage relays are used to protect equipment from undervoltage or overvoltage conditions. For example, a relay can signal a tap changer to step up or step down or an undervoltage relay can start the transfer of a load from one supply to another in the event of a power failure. **See Figure 10-11.**

Load Shedding. Voltage relays are also used in load-shedding applications. A potential transformer can be used to monitor a power line. When a power line is overloaded and the voltage drops below a setpoint, the relay switches and removes some of the load from the power line. For example, a large industrial facility with its own generating equipment may be designed to automatically remove loads when the generating system is overloaded. The loads to be shed are specified in advance.

CURRENT TRANSFORMERS

Very high currents can be present on power lines and in high-power circuits. It can be difficult to measure these currents directly. A *current transformer* is a transformer used to step down line current to make it easier to measure. The total power in a transformer is the same on the primary and secondary sides. The only way to step down the current is by

stepping up the voltage. Therefore, a current transformer is a modified step-up voltage transformer.

Operation

Current transformers are unique because they usually have only one winding. The primary is connected in series with the line load. When the primary has a large current rating, the primary winding may consist of a straight conductor passing through the center of the magnetic circuit. This straight conductor represents a one-turn winding. **See Figure 10-12.**

When the primary has a low current rating, the primary winding may consist of several turns wrapped around the core. This provides the flux on low-current applications or to compensate for line drop to a power meter.

The secondary consists of many turns of wire wrapped around a core. The number of turns is determined by the desired turns ratio of the current transformer. The primary current of a current transformer is not controlled by the secondary, as it would be in a two-winding potential transformer. The secondary of a current transformer cannot affect the current in the primary, as the load on the feeder determines the primary current.

*A **current transformer** is a transformer used to step down line current to make it easier to measure.*

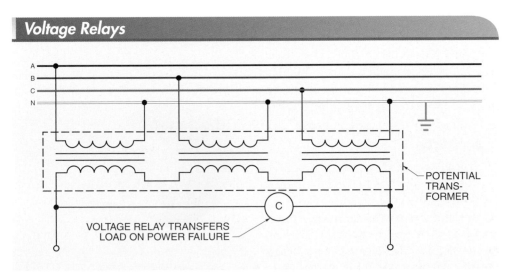

Figure 10-11. *A potential transformer can be used to feed a voltage relay that is used to transfer a load in the event of a power failure.*

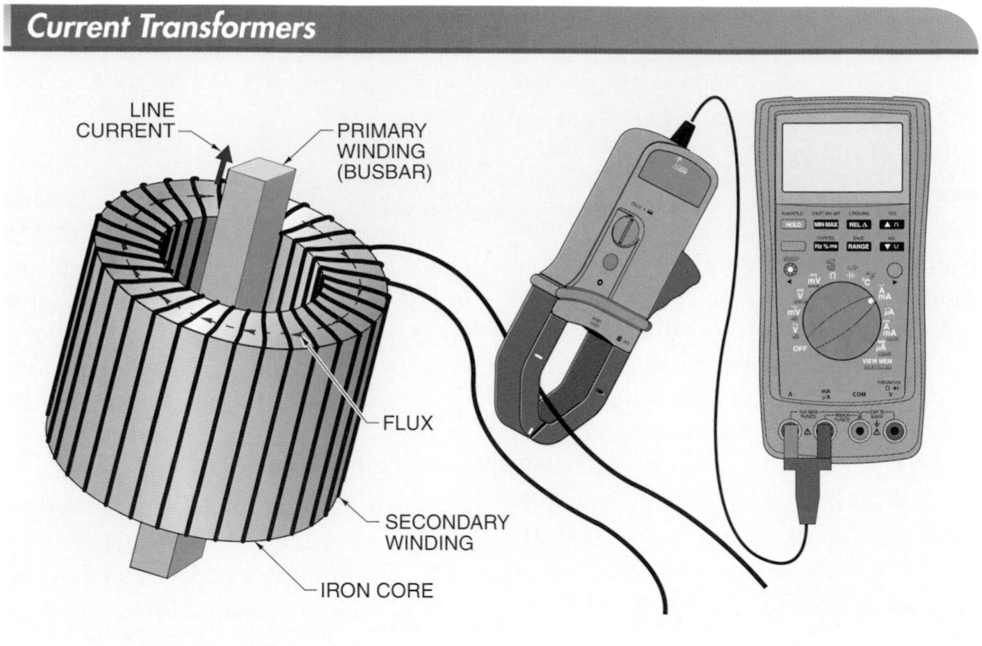

Current Transformers

Figure 10-12. A current transformer typically has one pass of a conductor as the primary and many turns of wire for the secondary.

The secondary of a current transformer must never be allowed to be open when the primary circuit is energized. When the circuit is operational, the load on the secondary keeps the magnetizing currents low, keeping the turn-to-turn potentials low. When the secondary becomes an open circuit, the magnetizing currents rise and the current transformer acts as a step-up potential transformer. The voltage can rise to a destructive level and cause a short between the turns as the result of the degradation of the insulation. Therefore, a current transformer should always have its secondary shorted when not connected to an external load. **See Figure 10-13.**

Construction

All transformers have losses in power transfer from resistance, magnetizing current, hysteresis, and other factors. These factors must be compensated for in the design of the transformer in order to ensure an accurate measurement.

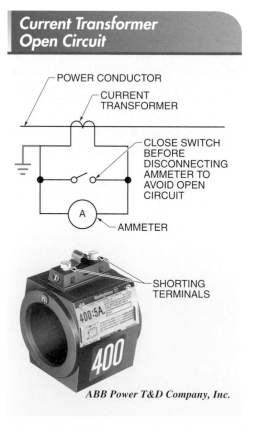

Current Transformer Open Circuit

ABB Power T&D Company, Inc.

Figure 10-13. The secondary of a current transformer must never be allowed to be part of an open circuit.

A current transformer is constructed of high-permeability steel at the flux density at which the transformer operates. The flux density is kept to a low value so that the magnetizing current is low. The circular coil of high-silicon steel provides the low-reluctance magnetic circuit needed to provide the necessary field strength for the secondary winding. The three types of current transformers in general use are the window, bar, and wound.

Window. A *window current transformer* is a transformer that consists of a secondary winding wrapped around a core and the primary sent through the opening in the core. After the secondary is wound around the core, the assembly is placed into a mold and an insulating material is injected around the transformer. Taps are brought out from the winding. **See Figure 10-14.** A power line is passed through the window and acts as the primary. This completed assembly is referred to as a window current transformer.

Bar. A *bar current transformer* is a special type of window current transformer with a solid bar placed permanently through the window. A bar current transformer can withstand the stresses of heavy overcurrent. To avoid magnetic stresses that could destroy the bus and damage the transformer, care must be taken to properly mount these transformers with respect to adjacent conductors. This type of transformer is typically found on installations where the potential is 25 kV or less. **See Figure 10-15.**

Wound. A *wound current transformer* is a transformer with separate primary and secondary windings wrapped around a laminated core. A wound current transformer is designed so that the primary winding consists of one or more turns of large cross-section wire connected in series with the circuit to be measured. This type of current transformer is found in the high-voltage side of substations and has a primary conductor that carries the current and a low-turn secondary for the output current. **See Figure 10-16.**

Window Current Transformers

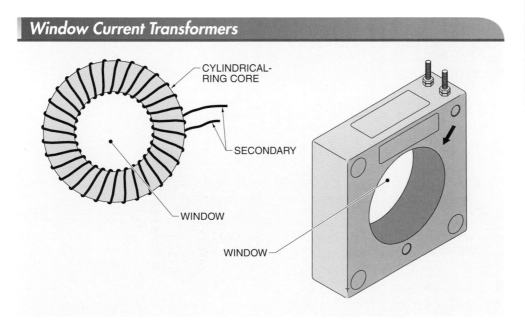

Figure 10-14. A window current transformer has an open area in the center for a power line to be passed through as the primary.

Bar Current Transformers

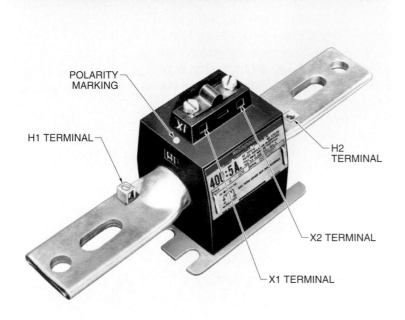

POLARITY MARKING

H1 TERMINAL

X1

400:5A

H2 TERMINAL

X2 TERMINAL

X1 TERMINAL

ABB Power T&D Company, Inc.

Figure 10-15. *A bar current transformer has a bar permanently placed in the window. Primary connections are made on the bar.*

Current Rating

The current rating of the primary winding of a current transformer is determined by the maximum value of the load current to be measured. For example, if the current rating is 400 A with a secondary rating of 5 A, the ratio between the primary and secondary is 400:5, or 80:1. This means that the secondary winding has 80 times as many windings as the primary and the current transformer can be used to measure a line load of 400 A. The primary must be rated to withstand 400 A.

TECH FACT

Manufacturers often provide manuals and troubleshooting charts with their products to assist in locating the cause of a problem. These manuals should be consulted when a problem occurs because they provide detailed information on remedying a specific problem and finding a permanent solution. These charts and manuals are often found on the Internet.

 **DEFINITION**

*A **wound current transformer** is a transformer with separate primary and secondary windings wrapped around a laminated core.*

Wound Current Transformers

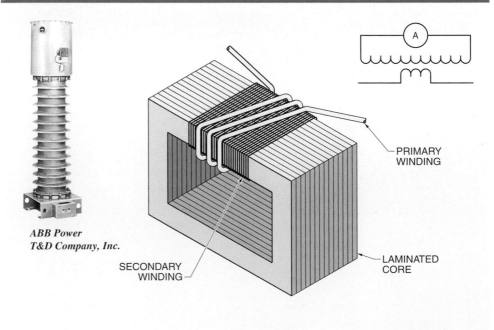

ABB Power T&D Company, Inc.

PRIMARY WINDING

SECONDARY WINDING

LAMINATED CORE

Figure 10-16. *A wound current transformer has several turns of wire for the primary.*

The output of the secondary is current proportional to the primary current. The output is used to measure the primary current and used to provide power to the instruments used to make the measurement. The secondary of a current transformer is always rated at 5 A, regardless of the current rating of the primary. This enables the production of standardized current devices rated at 5 A. The nameplate would commonly have a rating like 400:5 to show that the secondary is designed to carry 5 A.

Applications

Current transformers are used in many applications in industry, both for metering and for feedback. All of the loads that could possibly be connected to the secondary are designed to accept a maximum current of 5 A. Common applications of current transformers include power metering, motor current monitoring, and variable-speed-drive monitoring.

Power Metering. Current transformers are used with potential transformers for power metering from a utility to a customer. Power, or wattage, is found by multiplying the voltage and the current. A potential transformer provides a way to measure the voltage. A current transformer provides a way to measure

the current. **See Figure 10-17.** The watt meter is in series with the ammeter and must operate on 5 A or less.

Motor Current Monitoring. Large motor starters have current transformers on the lines to the starter to monitor the motor currents. **See Figure 10-18.** The outputs of the current transformers are connected across the terminals of an overload relay. The maximum current on the secondary of the current transformer is 5 A. The overload heater is in series with the current transformer and must operate on 5A or less.

A 200 HP motor draws approximately 280 A at full load on 480 V. Current transformers rated at 300:5 (60:1) can be used to monitor the current. Overload heaters rated at 280 A are not readily available. An overload relay can be used to monitor the output of the current transformer. Since the current transformer ratio is 60:1, the heaters can be reduced by the same ratio. The load current of 280 A divided by 60 is 4.67 A. Therefore, heaters rated at 4.67 A can be used.

Variable-Speed-Drive Monitoring. Variable-speed drives use current transformers to monitor the incoming current. This is part of the instantaneous electronic trip (IET) circuit that takes the drive off-line if a sudden rise in current occurs that exceeds the rating of the drive.

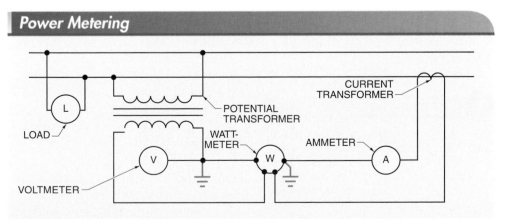

Power Metering

Figure 10-17. Current transformers are used with potential transformers to meter the power delivered from a utility to a customer.

Motor Current Monitoring

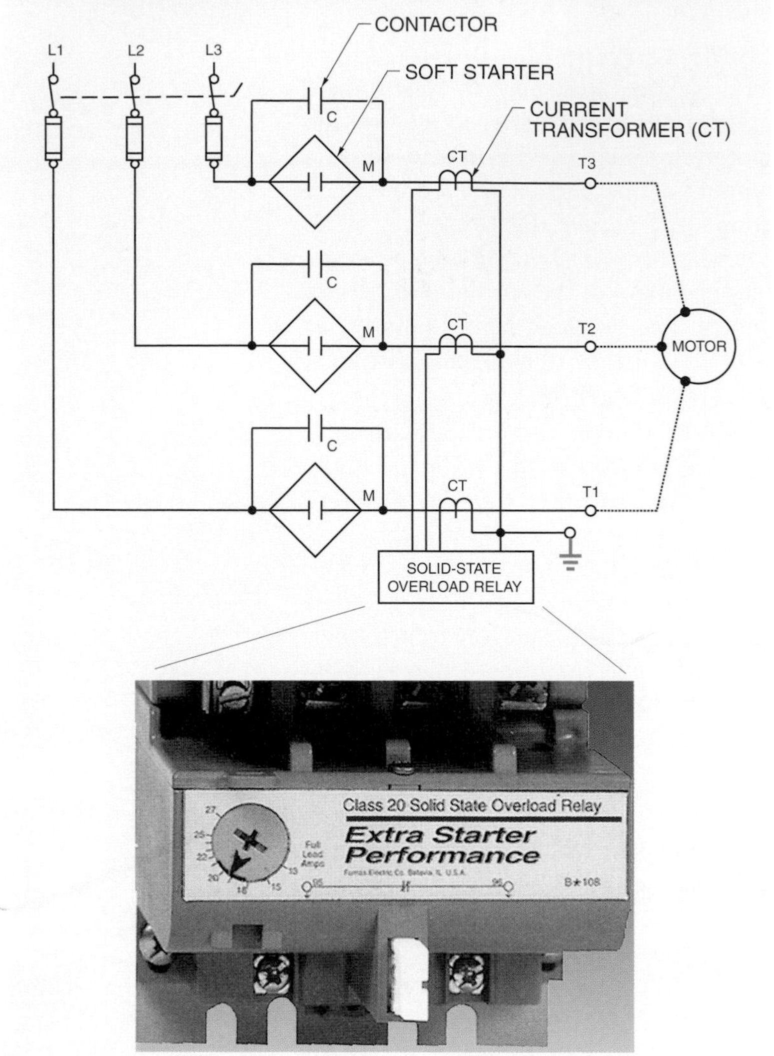

CONTACTOR

SOFT STARTER

CURRENT TRANSFORMER (CT)

Furnas Electric Co.

Figure 10-18. *Current transformers are used to feed an overload relay in motor starter applications.*

DEFINITION

*A **variable transformer** is a transformer used to make fine adjustments to the output voltage.*

TECH FACT

For current transformers, the accuracy rating is determined at full rated load. The full load includes the impedance of the secondary winding itself, the impedance of the leads from the transformer to the load, and the load itself. At lower loads, the accuracy may be only half the stated accuracy.

VARIABLE TRANSFORMERS

A *variable transformer* is a transformer used to make fine adjustments to the output voltage. A typical variable transformer can be adjusted from 0% to about 117% of the input voltage.

Operation

A variable transformer usually consists of a wiper or brush that can be rotated across the windings to create a variable turns ratio. Variable transformers are usually wired as autotransformers with no electrical isolation between the source and the output.

Variable transformers are often built with less than 1 V/turn, permitting fine voltage adjustment. The winding turns are evenly spaced so the output voltage is proportional to the position of the knob. Almost all variable transformers use manual control to adjust voltages. Some models have motorized controls to allow adjustment at a distance.

Construction

A typical variable transformer consists of a copper winding on a toroidal core. This core consists of laminated, grain-oriented silicon steel to create low reluctance. This makes a variable transformer very efficient. A brush, usually made of carbon, rotates across the winding to take off voltage at any value from 0 V to the maximum design output voltage. **See Figure 10-19.**

Variable transformers can have extremely low power losses with efficiencies as high as about 98%. Variable transformers can be built for either bench or panel mount with an adjustable shaft to allow for varying panel thicknesses.

Most 240 V models also have a tap to allow for 120 V input. This allows for normal output with only half of the normal input voltage. The output current must be reduced with the output voltage when it exceeds 125% of the rated input voltage.

Variable Transformers

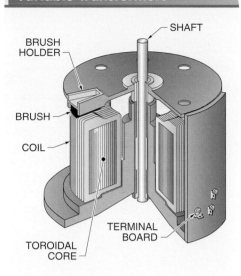

Figure 10-19. *Variable transformers have a brush that rotates across the winding to take off any voltage from 0 V to the maximum design voltage.*

Variable transformers are constructed to minimize losses. Therefore, these transformers can be operated at full current rating at ambient temperatures up to 122° F (50° C). In installations above 122° F, the output current must be reduced. On single transient loads and loads that are cycled on and off, much higher output current may be carried for brief intervals. The specifications provided by the manufacturer should be consulted for temperature ratings.

Combinations

Most variable transformers are rated at 120 V or 240 V. Units with higher voltage requirements may be created by combining, or ganging, the standard units in series. Units with increased current requirements may be created by ganging the standard units in parallel.

For example, two units can be ganged in a 3-phase open delta. Three 120 V units can be ganged in a wye connection to result in a 240 V line-to-line 3-phase assembly. **See Figure 10-20.** In a system that typically has a common neutral or ground between source

and load, the neutral or ground must be connected to the variable transformer common terminals. If the system has no neutral, the loads must be balanced or the transformers will be damaged from circulating currents.

In other applications of variable transformers, three 240 V units can be ganged in a wye connection, creating a 380 V or a 480 V line-to-line 3-phase assembly. A 480 V single-phase application of variable transformers can be met with two 240 V units ganged in a series connection.

Three-Phase Connections

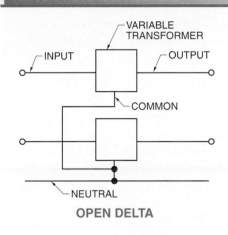

OPEN DELTA

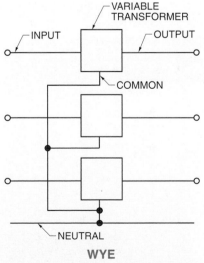

WYE

NOTE: IF NO NEUTRAL IS PRESENT, THE LOADS MUST BE BALANCED TO AVOID DAMAGE FROM CIRCULATING CURRENTS

Figure 10-20. *Variable transformers can be ganged together to develop the desired voltage output.*

Applications

Variable transformers are used in many applications requiring precise voltage control. A very common application of variable transformers is in voltage regulation. Other applications of variable transformers include temperature control, control lighting, and laboratory equipment testing for overvoltage and undervoltage conditions.

Voltage Regulation. The input voltage to a process can vary by the normal ±10% allowed by the electric utility. There are many applications that require more precise voltage control. Variable transformers can be used with buck-boost transformers to create a voltage regulation system. **See Figure 10-21.** The variable transformer is center-tapped to allow a buck-boost transformer to apply a buck or boost as necessary.

A stepper motor is used to adjust the location of the output tap from the variable transformer. A controller measures the output voltage and adjusts the stepper motor to provide the desired amount of buck or boost to provide the required output regulated voltage.

For example, a plating application requires a stable voltage to ensure consistent plating thickness. A voltage regulator using a variable transformer is part of the power source to the plater.

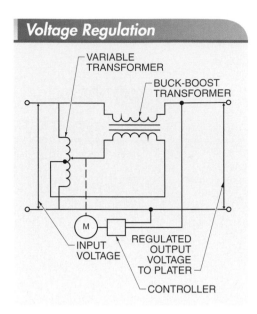

Figure 10-21. Variable transformers can be used as part of a voltage regulation system.

SUMMARY

- Machine and system controls often use low-voltage power to provide a safer environment for technicians working on the equipment.

- A control transformer is used to provide low-voltage power to control systems and devices.

- A control transformer usually has two primary coils to allow for 240 V or 480 V operation.

- The two most common control transformer secondary voltages are 24 V and 120 V.

- When control circuits are modified, the control transformer should be evaluated to make sure it can still provide enough power to operate the circuits.

- It is common industry practice to protect a control transformer on both the primary and secondary sides.

- Control systems can be wired as grounded systems or floating systems.

- A bell transformer is commonly used for doorbells and annunciators.

- A potential transformer is a step-down transformer used to measure very high voltages.

- A potential transformer is most accurate when the load on the secondary is constant and very small.

- The secondary output of potential transformers is 120 V.

- Important uses for potential transformers are as voltage meters, to feed voltage relays, and for load shedding during peak load periods.

- A current transformer is used to step down a high current to make it easier to measure.

- The primary winding of most current transformers consists of a single pass of a power line.

- The secondary of a current transformer must never be allowed to be open. Dangerous voltage results.

- The three common types of current transformers are window, bar, and wound current transformers.

- The maximum current output of a current transformer is 5 A.

- Current and potential transformers are used together to measure the power delivered from a utility to the customer.

- A current transformer can be used to monitor motor and drive currents.

- A variable transformer uses a wiper or brush that rotates across the winding to provide any desired output voltage within the design range.

- A variable transformer can be combined, or ganged, to produce a wide range of voltage and current outputs.

DEFINITIONS

- A *control transformer* is a transformer that is used to step down the voltage to the control circuit of a circuit or machine.

- A *grounded system* is a control system with a grounded secondary in the control transformer.

- A *floating system* is a control system without a grounded secondary in the control transformer.

- A *ground loop* is a circuit that has more than one point connected to earth ground, with a voltage potential difference between the two ground points.

- A *bell transformer* is a transformer used to supply low-voltage, low-power circuits, such as for doorbells, annunciators, and similar systems.

- A *potential transformer* is a precision two-winding transformer used to step down high voltage to allow safe voltage measurement.

- A *current transformer* is a transformer used to step down line current to make it easier to measure.

- A *window current transformer* is a transformer that consists of a secondary winding wrapped around a core and the primary sent through the opening in the core.

- A *bar current transformer* is a special type of window current transformer with a solid bar placed permanently through the window.

- A *wound current transformer* is a transformer with separate primary and secondary windings wrapped around a laminated core.

- A *variable transformer* is a transformer used to make fine adjustments to the output voltage.

REVIEW QUESTIONS

1. Explain how overcurrent protection should be installed for a control transformer.

2. Why do most control transformers have two primary coils?

3. Why should a control transformer be evaluated when control circuit systems are modified?

4. What are the most common control transformer secondary voltages?

5. What is the secondary output of a potential transformer?

6. List a common application of potential transformers.

7. List three common types of current transformers.

8. What will happen if the secondary of a current transformer is opened during load?

9. Explain the difference between a variable transformer and a transformer with multiple output taps.

10. List a common application of variable transformers.

Transformer Principles and Applications

CHAPTER

Special Connections

11

Many special connections are used in the field. Transformers connected in parallel can add power to a system as long as they are connected properly to give the same polarity to the coils. In some instances, a 277 V system can be set up using a dual-wound transformer rated for 480 V/240 V that is connected in a zigzag. There are 2-phase systems and low-frequency systems still in use today.

PARALLEL CONNECTIONS

Industrial and commercial processes change frequently, and power requirements change with them. If a load changes, the transformer that powered the load may not be large enough to power the new load. If a larger replacement is not readily available, then a new transformer of the same size can be connected in parallel with the existing one. A *parallel transformer con-* *nection* is two or more transformers wired in parallel that carry a common load. **See Figure 11-1.**

In general, a single transformer is more efficient than two or more transformers wired in parallel. However, there are many situations where transformers may need to be connected in parallel. For example, transformers may be paralleled if a larger transformer is not readily available or if the larger load is only temporary.

System Requirements

The requirements for satisfactory operation of two or more transformers connected in parallel include having the same polarity, angular phase displacement and phase sequence, voltage ratio, and percent impedance. The ratio of the resistance component to the reactance component should also be the same. In addition, the

DEFINITION

A *parallel transformer connection* is two or more transformers wired in parallel that carry a common load.

voltage rating must be the same for each transformer to ensure that each can handle the power requirements.

Polarity. When paralleling two or more transformers, the polarity of each must match. This means that the same windings of each transformer must be connected to the same line. This matched polarity ensures that both units are at the same potential at the same time, and that no potential difference exists between any of the transformers. If there is any doubt about the polarity, the transformers must be tested to identify the correct polarity.

Angular Phase Displacement. When two transformers are connected in parallel with each other, the connection of the two windings makes up a closed series circuit. In a single-phase circuit, the secondary windings of two parallel transformers are either in phase with each other or they are 180° out of phase with each other.

If the two voltages are in phase with each other, the induced voltages are such that no current flows in the series circuit. If the two voltages are out of phase with each other, the induced voltages are such that they cause current to circulate between the transformers.

To simplify the parallel connection of single-phase transformers into 3-phase circuits without having to test for polarity, angular phase displacement, and phase rotation, the lead markings have been standardized and the various 3-phase connections without taps have been placed in three different vector groups, depending on their angular displacement. **See Figure 11-2.** For example, all delta-delta or wye-wye connections have either 0° or 180° angular displacement. All transformers of any one vector group may be connected in parallel with each other, as long as the other requirements are met.

A transformer of one group cannot be connected in parallel with a transformer of another group because of the difference in angular displacement. Transformers from different groups cannot be connected in parallel by interchanging the external leads. To change the angular phase displacement from 0° to 180° or vice versa, the internal connections of the coils forming the wye or delta connection must be changed.

TECH FACT

Group 3 transformers may be labeled 1 for +30° and 11 for −30°.

Parallel Connections

Figure 11-1. *A parallel transformer connection has two transformers wired to the same power lines and carrying the same load.*

Transformer Connection Vector Groups

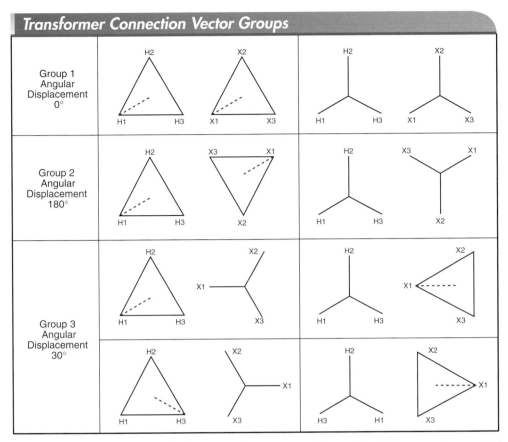

Figure 11-2. All 3-phase to 3-phase transformer connections without taps can be grouped into one of three groups with angular displacements of 0°, 180°, or 30°.

Voltage Ratio. The voltage ratio of any group of two or more transformers connected in parallel should be the same. If the voltage ratio is not the same, the difference in voltage between two windings causes a current to flow within the parallel circuit at all times.

It is also very important to ensure that the wire size is the same and any connections are made carefully to ensure that the voltage drop across the wires and terminations is equal for both transformers.

Percent Impedance. An external load divides between transformers connected in parallel. If the impedances of the two transformers are equal, the load divides equally between the transformers. If the transformer impedances are unequal, the load divides between them in inverse proportion to their impedances. The transformer with lower impedance takes

more current until its terminal voltage falls to a value equal to the terminal voltage of the other transformer.

The best case is to select transformers that have the same impedance to make sure that the load is split evenly. The impedance of a transformer may be found on the nameplate or in the manufacturer's literature. However, there may be circumstances where it becomes necessary to install two transformers in parallel even when they have unequal impedances.

Unequal Impedances. Transformers with unequal impedances can be wired to accomplish an equal division of the load. The method requires two transformers with dual-voltage primaries and secondaries. Each of the coils of the first transformer primary is wired in series with one of the coils of the second transformer primary. **See Figure 11-3.**

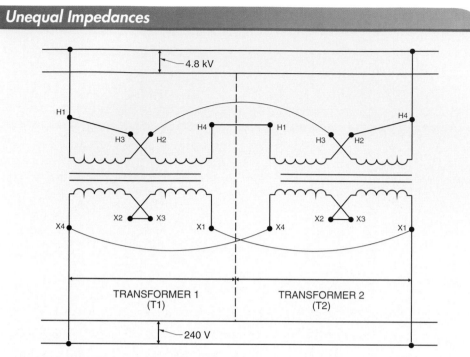

Unequal Impedances

NOTE: IMPEDANCE OF TRANSFORMER 1 IS NOT EQUAL TO IMPEDANCE OF TRANSFORMER 2

Figure 11-3. Transformers with unequal impedances can be wired in parallel by connecting one primary coil of each in series with a primary coil from the other.

This series winding connection results in a configuration where each new pair of coils is the same as the other new pair, but different from the original pairs of coils. For example, a transformer steps down a 4.8 kV source to 240 V as required by the load. The secondary windings are rated for 120 V and the primary windings are rated for 2.4 kV. Tap H2 of the first transformer (T1) is connected to tap H3 of the second transformer (T2). Tap H4 of T1 is connected to tap H1 of T2. The 240 V required by the load is obtained by connecting X2 and X3 to each other on each transformer.

Another method of using parallel transformers of unequal impedance uses a reactor in series with the transformer having the lower impedance. **See Figure 11-4.** Although this method is seldom used, it can be used to make quick repairs to a system. The reactance of this reactor should be such that the impedance of the reactor/transformer

pair is equal to that of the other transformer in the parallel circuit.

Wiring Connections

An additional transformer is added in parallel by connecting its primary side to the same source and its secondary to the same load as the first transformer. When making this connection, similarly marked leads of both transformers must be connected to the same sides of each circuit. Proper polarity is present if all "H" and "X" terminals with the same numbers are connected to the same line. This ensures that the primaries are energized at the same time and the secondaries have the same potential and phasing.

As a safety precaution, a transformer primary should be considered to be hot, even if the fuse is blown. It is a common industry practice to backfeed a transformer to avoid having an outage. Backfeeding occurs when a voltage is applied to the secondary. **See Figure 11-5.**

Reactor Usage

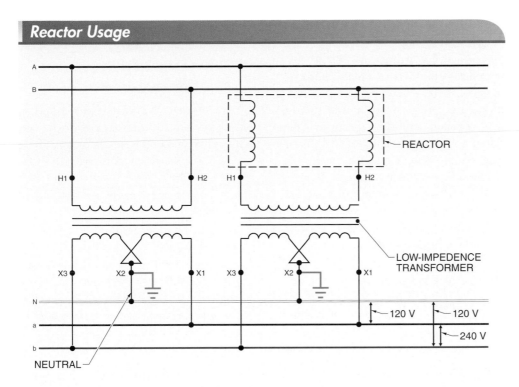

Figure 11-4. *A reactor can be placed in series with the transformer of lowest impedance to equalize the impedance between transformers.*

Transformer Backfeeding

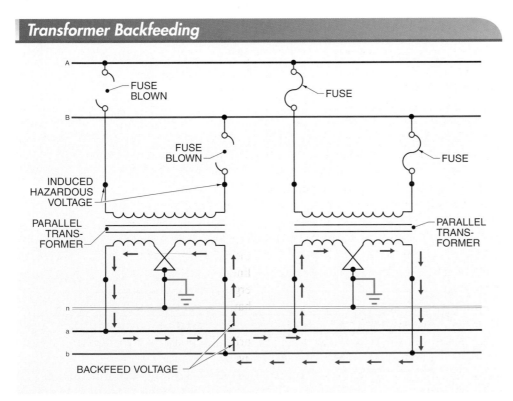

Figure 11-5. *A transformer primary should always be considered hot because it is common practice to backfeed a transformer.*

When testing the connections, it is best to have the transformers unloaded, but connected and hot, so that the potentials can be read on both transformers. For safety reasons, it is a good practice to install low-current fuses while doing this test, as a problem will blow the fuse without doing any damage. The potentials across each primary should match the source. The potentials across the secondaries should match each other. The potential between the two X1 leads should be 0 V, indicating that the two transformers are installed correctly. **See Figure 11-6.**

A simple test with an ammeter on the primary leads can be done to test for proper load balancing. When the current is equal in both transformers, the units are operating normally. If there is no current through one of the transformers, there may be a blown fuse. A reading of 0 V across the fuse indicates that the fuse is working properly. Any significant voltage reading across the fuse indicates a blown fuse.

T-CONNECTIONS

A *T-connection* is a low-cost method of wiring two transformers to provide 3-phase power. This method is an alternative to the open delta connection for providing low-cost power to a circuit with only two transformers. A T-connection maintains a balanced phase relation better than an open delta.

One of the transformers is called the main transformer. A *main transformer* is one of the transformers in a T-connected transformer bank and is provided with a 50% voltage tap. The other transformer is called the teaser transformer. A *teaser transformer* is one of the transformers in a T-connected transformer bank and is connected to the 50% tap of the main transformer.

 **DEFINITION**

*A **T-connection** is a low-cost method of wiring two transformers to provide 3-phase power.*

*A **main transformer** is one of the transformers in a T-connected transformer bank and is provided with a 50% voltage tap.*

*A **teaser transformer** is one of the transformers in a T-connected transformer bank and is connected to the 50% tap of the main transformer.*

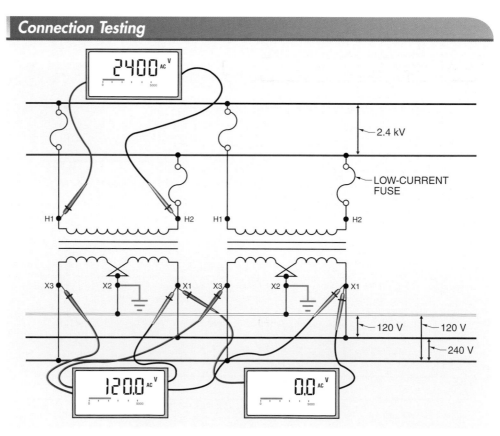

Figure 11-6. *When transformers are first wired in parallel, the potentials across the primaries should match line voltage, potentials across the secondaries should match each other, and potentials between the two X1 terminals should be 0 V.*

The capacity of the T-connected transformers is similar to two transformers in open delta. The advantage is that the T-connection with the teaser transformer operates at reduced flux and therefore operates with reduced iron loss and at a higher efficiency.

Wiring Connections

There are several common arrangements of T-connections that give useful output voltages. For example, if a source is 3-phase 2400 V, two transformers can be connected to give 3-phase 240 V at the secondaries. The teaser is connected to the 50% tap of the main transformer. The vector diagram shows 207.8 V across the teaser secondary. **See Figure 11-7.**

277 V and 208Y/120 V. A teaser transformer that is supplied with an 86.6% tap can also be used to develop 277 V from 480 V lines or to develop a 208Y/120 V source. **See Figure 11-8.** The transformer has 416 V from H2 to the center tap on the main transformer. The voltage between the H1, H2, and H3 taps is 480 V.

The voltage from H2 to the 86.6% tap is 277 V with 138.5 V to the center tap on the main transformer. The two secondaries are connected for 3-phase 208Y/120 V. The voltage between the X1, X2, and X3 taps is 208 V. The 86.6% tap is used as the zero point, and the voltage from X0 to the secondary taps is 120 V. Special T-connected transformers may be available for this purpose, and both transformers are contained in the same case.

T-Connections

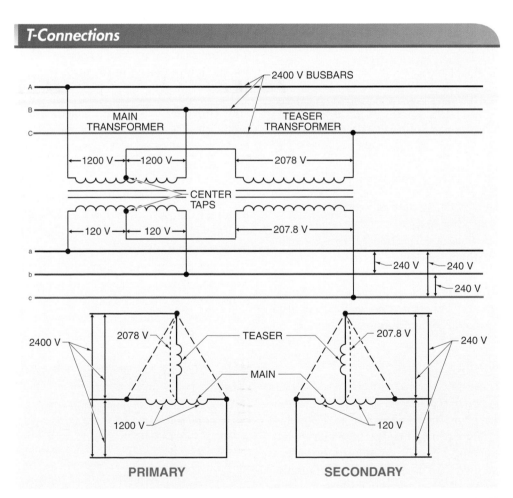

Figure 11-7. In a T-connection, one of the units is the main transformer and is provided with a 50% tap to which a teaser transformer is connected.

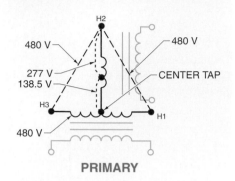

Developing 208Y/120 V

PRIMARY

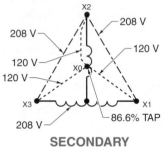

SECONDARY

Figure 11-8. A teaser transformer with an 86.6% tap can be used to develop 277 V from a 480 V source or to develop a 208Y/120 V source.

ZIGZAG CONNECTIONS

There are many applications where a zigzag connection is used. A *zigzag connection* is a method of wiring transformers where the windings are divided over several legs of the transformer core. Zigzag connections are commonly used to create a neutral or to cancel unwanted fluxes, such as from harmonics or synchronous converters.

Wiring Connections

There are several variations on zigzag wiring of transformers. In some situations, the primary windings are in a standard configuration while the secondary windings are wired in series from one leg to another. This is a common configuration for transformers used to mitigate harmonics. Another common configuration is to use a standard 480 V/240 V transformer

wired as a zigzag to create a neutral to derive 277 V for lighting systems from a 480 V source. Synchronous motors can also use a zigzag configuration to reduce or eliminate flux distortion.

277 V for Lighting. Industrial lighting systems often use 277 V. This is typically derived from the line-to-neutral voltage on a 480 V system. However, there is no fourth wire for a neutral in plants that have a 3-wire, 480 V ungrounded or corner-grounded delta.

When there is no neutral, there are two common ways to create a 277 V source for the lighting. The simplest is to purchase a transformer to step down the 480 V source to 277 V. This is not a common size and such a transformer would be a special order and would be expensive. Another method is to use a standard transformer with a 480 V primary and a 240 V secondary. The transformer can be wired in a zigzag to create a neutral that can be used to derive 277 V for the lighting. **See Figure 11-9.**

Synchronous Converters. A synchronous converter is a rotating machine that produces a DC current from an AC source or AC current from a DC source. A characteristic of synchronous converters is that the transformer feeding the converter can have a problem with flux distortion in the windings due to an imbalance of DC current flowing in the neutral. This neutral is the return path for the DC circuit.

In the case of a 3-phase synchronous converter, the input power is from a bank of transformers that are connected delta on the primary side and interconnected wye on the secondary side.

Interconnected Wye. In a 3-phase system, the flux distortion caused by a synchronous converter can be eliminated with an interconnected wye connection in the secondary. An *interconnected wye connection* is a method of wiring transformers where two separate windings are interconnected on each phase in a zigzag fashion. **See Figure 11-10.** The unbalanced current from the synchronous converter flows into the neutral of the interconnected wye windings and divides equally through each of the three legs. The unbalanced current flows in opposite directions in each secondary.

For example, the direction of the current flow in the half-section a-b is opposite to the flow of the current in the adjacent half-section c-d. The magnetizing action of the unbalanced current is neutralized because the two half-sections a-b and c-d form the two half-sections of the secondary winding of the first transformer.

The low-voltage side operates at only 86.6% of its normal capacity (if operated in straight wye connection) because the two half-sections of each transformer secondary winding are connected in different phases. Therefore, transformers wired for interconnected wye operation are larger than those wired for straight wye connection.

Double Delta. Some synchronous converters also operate on 6-phase power. A 6-phase synchronous converter uses a transformer with a dual secondary wired in a double-delta connection to feed the synchronous converter. A *double-delta connection* is a method of wiring transformers where both sets of secondary windings are connected in a delta connection, but one is reversed with respect to the other. As a result, two delta connections displaced 180° from each other are formed. **See Figure 11-11.** The high-voltage windings are usually wired in a delta connection.

The synchronous converter is tapped at 60° intervals to receive the 6 phases from the source. A synchronous converter with 6-phase input is more efficient than the 3-phase equivalent.

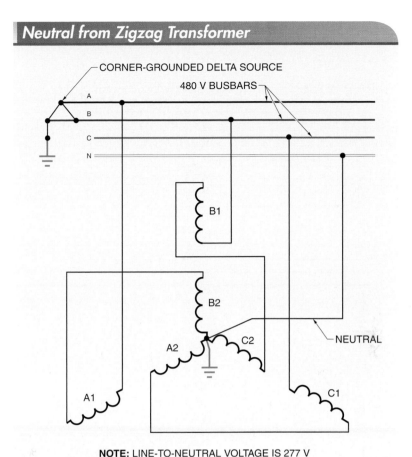

Neutral from Zigzag Transformer

NOTE: LINE-TO-NEUTRAL VOLTAGE IS 277 V

Figure 11-9. A zigzag transformer can be used to develop a neutral from a 3-wire, 480 V source.

TWO-PHASE SYSTEMS

There are a few places and applications where 2-phase systems are used. A *2-phase system* is a power supply consisting of two phases that are 90° out of phase. Two-phase systems are isolated and not connected to the grid networks that power most of industry today. Two-phase systems may be created at a generator or they may be created from 3-phase systems with special transformer connections.

Two-phase systems are usually found in older industrial areas and are being phased out and replaced as new facilities are built. A 2-phase motor can start on a 2-phase system without the need of a phase shift, as the 90° phase relationship sets up a rotating field similar to a 3-phase system.

 **DEFINITION**

An *interconnected wye connection* is a method of wiring transformers where two separate windings are interconnected on each phase in a zigzag fashion.

A *double-delta connection* is a method of wiring transformers where both sets of secondary windings are connected in a delta connection, but one is reversed with respect to the other.

Delta-to-Interconnected Wye Connections

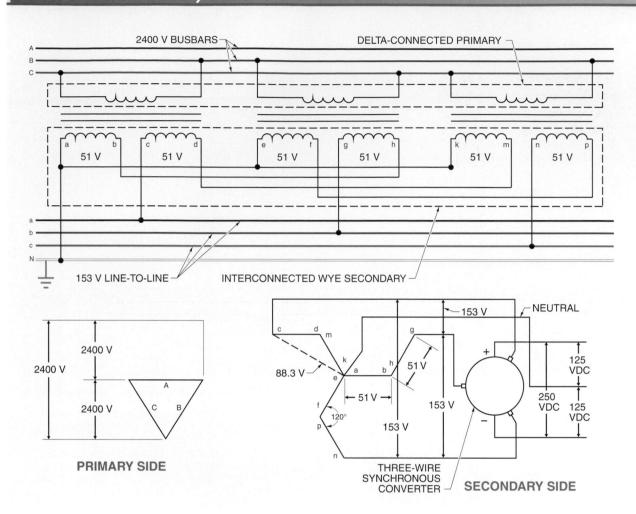

Figure 11-10. *An interconnected wye secondary can be used to eliminate any flux distortion caused by a synchronous converter.*

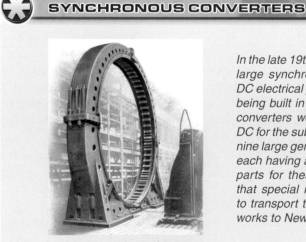

SYNCHRONOUS CONVERTERS

In the late 19th century, Westinghouse built large synchronous converters to supply DC electrical power for the subway system being built in New York. The synchronous converters were used to change AC into DC for the subway. Westinghouse supplied nine large generators to power this project, each having a capacity of 45,000 kW. The parts for these machines were so large that special railroad cars had to be built to transport them from the Westinghouse works to New York.

Library of Congress

Double-Delta Connections

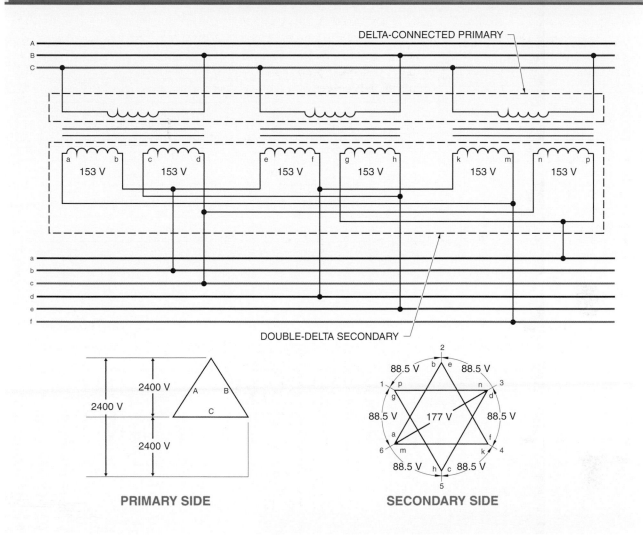

PRIMARY SIDE

SECONDARY SIDE

Figure 11-11. A double-delta connection can be used to create a 6-phase source for a synchronous converter.

Wiring Connections

A 4-wire, 2-phase system has no connection between the coils. **See Figure 11-12.** The most common 2-phase transformer voltage reduction for the end user is the use of two single-phase transformers with the primaries of each connected to each phase. The voltage between the two feeders on each phase is 2400 V. The vector diagram shows the voltage and phase relationship between the two phases.

A 3-wire, 2-phase system uses two single-phase transformers with the primary connected to a neutral conductor.

See Figure 11-13. Since the phases are 90° out of phase, the multiplier to use to find the phase-to-phase voltage is $\sqrt{2}$ or 1.414. If the circuit is 480 V phase-to-neutral, the phase-to-phase voltage is 678.8 V ($480 \times \sqrt{2} = 678.8$).

Scott Connections. A *Scott connection* is a method of wiring two transformers to transform from 3-phase to 2-phase or 2-phase to 3-phase. **See Figure 11-14.** A Scott connection is very similar to a T-connection. As with a T-connection, the two transformers are called the main and the teaser transformers.

*A **2-phase system** is a power supply consisting of two phases that are 90° out of phase.*

*A **Scott connection** is a method of wiring two transformers to transform from 3-phase to 2-phase or 2-phase to 3-phase.*

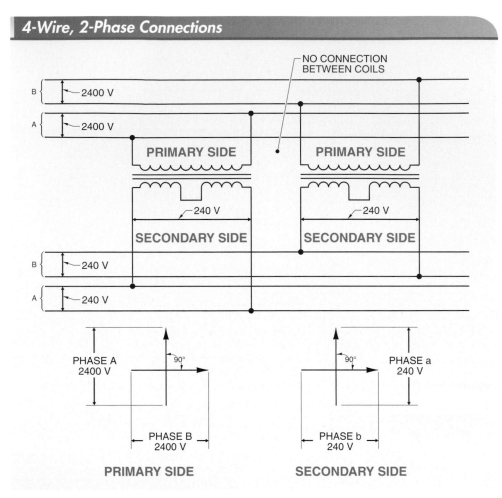

4-Wire, 2-Phase Connections

Figure 11-12. *A 4-wire, 2-phase connection has no connection between the coils.*

Transformers for commercial buildings are often mounted on a pad outside the building.

A Scott connection uses a tap at the 50% point on the main transformer and a tap at the 86.6% point on the teaser transformer. An 86.6% tap may be present on the main transformer and a 50% tap may be present on the teaser transformer, but they are not used.

TECH FACT

Residential wiring uses a center-grounded tap between two phases of a 3-phase system. This type of circuit can supply 120 V power for most household circuits and 240 V power for heavier loads. Two conductors and a ground are used. This leads people to mistakenly call this a 2-phase power system.

3-Wire 2-Phase Connections

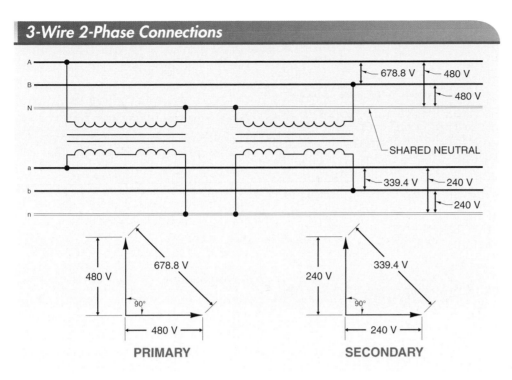

Figure 11-13. *A 3-wire, 2-phase connection has a neutral shared by both transformers.*

Scott Connection

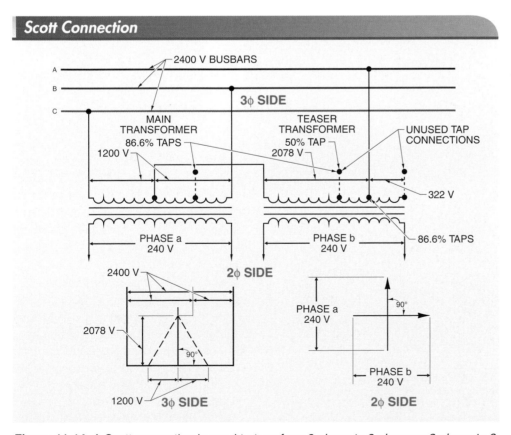

Figure 11-14. *A Scott connection is used to transform 3-phase to 2-phase or 2-phase to 3-phase.*

Two-Zone Furnace. A Scott connection draws balanced 2-phase power from a 3-phase source. This is useful when an application calls for power in two zones. This can happen in a furnace or other heating application where there are two zones that heat a product. For example, a furnace blows hot air into two separate zones. Each zone has separate resistance heating elements that heat the air. **See Figure 11-15.** Each zone heater can be connected to one of the two phases. As a result, the 3-phase power source is balanced even when only two loads are drawing power.

Two-Zone Furnace Heating

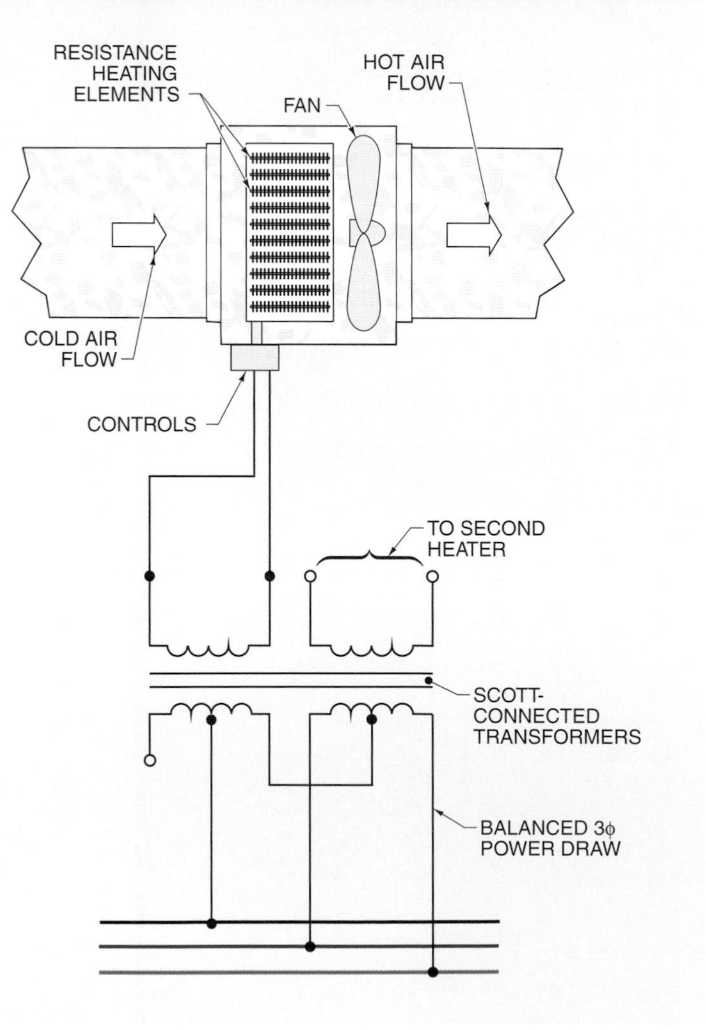

Figure 11-15. A two-zone furnace with a Scott connection in the power supply is an application of 2-phase power that draws balanced 3-phase power.

LOW-FREQUENCY SYSTEMS

In North America, virtually all power systems now operate at 60 Hz. Some early electric power systems were operated at 25 Hz. These older systems have been phased out, but there are still some 25 Hz systems in use today. There is one small 25 Hz electric power grid still in operation near Niagara Falls that is being phased out and other grids in Ontario, Canada. Everywhere else, customers using these systems are not supplied by a utility, but generate their own power.

Applications

Since 25 Hz power was common in the past, there are some large industrial customers that still have older equipment designed for 25 Hz. It is expensive to replace production equipment, so many of these older systems still exist. Low-frequency 25 Hz power is also occasionally used for low-speed electric motors. In addition, telephone circuits use low-frequency power for ringing signals.

Low-Speed Electric Motors. A 4-pole motor operating on 60 Hz has a synchronous speed of 1800 rpm, while the same motor on 25 Hz has a synchronous speed of 750 rpm. This allows a motor to operate at a slow speed with a smaller stator housing. **See Figure 11-16.** Standard motors rated at 60 Hz should not be used at 25 Hz without being derated. Since the rotor turns more slowly, the fan also turns more slowly and the motor no longer has adequate cooling.

In some older steel mills, 25 Hz motors are used to move the hot bar from the rolling dies and send it to the shear. Motors used for low speed applications like hoisting and crushing may also be 25 Hz motors. Traction motors, used to turn wheels on locomotives and electric trains, may also operate at 25 Hz. The use of 25 Hz motors is declining as high-efficiency AC induction motors are replacing these motors in many applications.

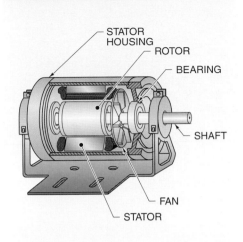

Low-Speed Electric Motors

- STATOR HOUSING
- ROTOR
- BEARING
- SHAFT
- FAN
- STATOR

Figure 11-16. Low-frequency power for electric motors allows for a smaller stator housing.

Telephone Ringing Signals. A telephone circuit uses a low-frequency ringing signal to operate telephone ringers. The actual frequency varies from country to country, but is typically about 20 Hz to 25 Hz. This application is growing as telephone systems are increasingly integrated into Internet-enabled systems and power is supplied over the Ethernet cables.

Transformers at 25 Hz

A transformer rated for operation at 25 Hz cannot be directly replaced with one rated at 60 Hz. A transformer has lower reactance at 25 Hz than at 60 Hz. The purpose of the iron core in a transformer is to concentrate the flux. At the lower frequency, the iron core has more time to saturate during each cycle. The low rate of change along with the saturation creates problems in the iron when the current changes polarity.

A transformer designed to operate at 60 Hz should not be used as a direct replacement for a transformer designed to operate at 25 Hz. The reactance of a transformer operated at 25 Hz is less than half the reactance of the same transformer operated at 60 Hz. Therefore, when the transformer is operated at 25 Hz, the current flow more than doubles compared to the current flow at 60 Hz. The increase in current flow is likely to burn out the windings.

Therefore, a 60 Hz transformer that has a voltage rating of more than twice the rating of a 25 Hz transformer can be used as a substitute. Alternatively, two transformers with a voltage rating matching the 25 Hz transformer can be connected in series. **See Figure 11-17.**

Transformers in Series

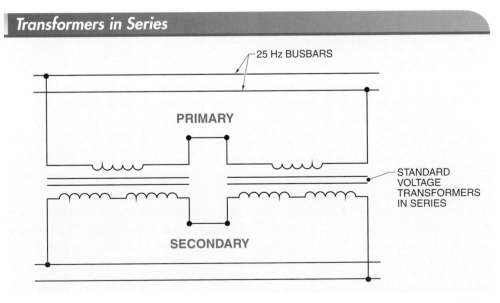

- 25 Hz BUSBARS
- PRIMARY
- STANDARD VOLTAGE TRANSFORMERS IN SERIES
- SECONDARY

Figure 11-17. Standard 60 Hz transformers can be wired in series when replacing a 25 Hz transformer.

4160 V/2400 V to 120 V/240 V

There are many industrial facilities that have 4-wire, 4160 V phase-to-phase power supplied by the utility. A system with 4160 V phase-to-phase is 2400 V phase-to-ground. Other facilities have a 3-wire, 2400 V phase-to-phase system. These circuits are often used to supply industrial loads. In some cases, industrial loads may cause power quality problems.

Because of the possibility of power quality problems with industrial power systems, industrial facilities may want to isolate their office loads from their industrial loads. The office loads are typically supplied with 3-wire, 120 V/240 V circuits. Both the 4160 V and the 2400 V systems can easily be stepped down to 120 V/240 V systems to power the office loads independently of the industrial loads. **See Figure 11-18.**

The 4-wire, 4160 V system can be stepped down by connecting the transformer across one of the phases to neutral. The 3-wire, 2400 V system can be stepped down by connecting the transformer across any two of the phases.

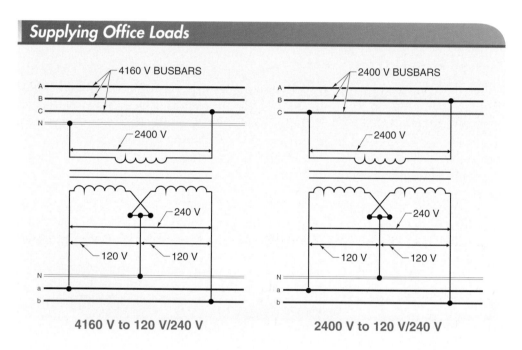

Supplying Office Loads

4160 V to 120 V/240 V

2400 V to 120 V/240 V

Figure 11-18. Office loads can be supplied separately from industrial loads by transforming 4160 V phase-to-phase or 2400 V phase-to-neutral to a 3-wire, 120 V/240 V system.

SUMMARY

- Transformers may be wired in parallel to increase the power available to a circuit.

- The requirements for satisfactory operation of two or more transformers connected in parallel include having the same polarity, angular phase displacement and phase sequence, voltage ratio, and percent impedance.

- All transformers of any one vector group may be connected in parallel with each other, as long as the other requirements are met.

- One method of using parallel transformers of unequal impedance has each of the coils of the first transformer primary wired in series with one of the coils of the second transformer primary.

- Another method of using parallel transformers of unequal impedance uses a reactor in series with the transformer having the lower impedance.

- As a safety precaution, a transformer primary should be considered to be hot, even if the fuse is blown.

- A T-connection is a low-cost method of wiring two transformers to provide 3-phase power.

- A T-connection uses a main transformer with a 50% tap and a teaser transformer connected to the 50% tap.

- A Scott connection uses two transformers to develop a 2-phase power source from a 3-phase source.

- A Scott connection uses a tap at the 50% point on the main transformer and a tap at the 86.6% point on the teaser transformer.

- A standard transformer operated at 25 Hz has lower reactance and can easily overheat.

- A standard transformer of at least double the voltage rating of a 25 Hz transformer can be substituted when replacing the 25 Hz transformer.

- Two standard transformers wired in series can be used when replacing a 25 Hz transformer.

DEFINITIONS

- A *parallel transformer connection* is two or more transformers wired in parallel that carry a common load.

- A *T-connection* is a low-cost method of wiring two transformers to provide 3-phase power.

- A *main transformer* is one of the transformers in a T-connected transformer bank and is provided with a 50% voltage tap. A main transformer is also used in a Scott connection.

- A *teaser transformer* is one of the transformers in a T-connected transformer bank and is connected to the 50% tap of the main transformer. A teaser transformer is also used in a Scott connection.

- A *zigzag connection* is a method of wiring transformers where the windings are divided over several legs of the transformer core.

- An *interconnected wye connection* is a method of wiring transformers where two separate windings are interconnected on each phase in a zigzag fashion.

- A *double-delta connection* is a method of wiring transformers where both sets of secondary windings are connected in a delta connection, but one is reversed with respect to the other.

- A *2-phase system* is a power supply consisting of two phases that are 90° out of phase.

- A *Scott connection* is a method of wiring transformers to transform from 3-phase to 2-phase or 2-phase to 3-phase.

REVIEW QUESTIONS

1. What system requirements must be met for satisfactory operation of transformers in parallel?

2. What is the meaning of a transformer group when connecting transformers in parallel?

3. What is the main transformer in a T-connection?

4. What is the teaser transformer in a T-connection?

5. What type of transformer connection can be used to transform from 3-phase to 2-phase or 2-phase to 3-phase?

6. What problems can come up when using a standard 60 Hz transformer on a 25 Hz system?

7. List two methods that can be used to replace a 25 Hz transformer with a 60 Hz transformer.

Transformer Principles and Applications

Selection and Installation

CHAPTER

12

The process of selecting and installing a transformer includes many factors. The selection process includes the load requirements for voltage, power consumption, and phase. The installation process includes the initial inspection and testing, guarding and enclosures, and overcurrent protection.

SELECTION

There are many considerations involved in selecting a transformer. These include the requirements for the primary and secondary voltage; whether single-phase or 3-phase; the present and future loads; whether it is to be used for step up, step down, or isolation; whether it is to be a two-winding transformer, autotransformer, or buck-boost; the permitted noise level; and the required and available cooling and ventilation.

Voltage

When selecting a transformer, the first decision that needs to be made concerns the required voltages at the primary and secondary. A facility may have several bus voltages present, and a transformer must be selected according to where and how it will be used. Once the bus voltage is known, the primary and secondary coil voltage can be selected to give the desired load voltage.

Primary. Once the voltage has been determined, there are usually several options for the transformer primary. For example, if the supply is 3-phase, 4160 V, there are several options for the primary coil voltage. **See Figure 12-1.** This source is 4160 V phase-to-phase and 2400 V phase-to-ground.

Single-wound 4160 V primary coils can be wired in a delta connection across the lines. In addition, dual-wound 2400 V primary coils can be wired in parallel and connected in a wye from phase to neutral.

A transformer with dual-wound 1200 V primary coils can also be selected. In this case, the two coils need to be wired in series and connected in a wye from phase to neutral in order to create the desired voltage. Having these three options available for one source can make the purchase of a transformer less expensive and delivery much faster.

Secondary. The transformer secondary offers many options for the different types of loads and voltage requirements of the equipment to be powered. For example, a common 3-phase transformer is rated at 480 V/277 V on the secondary and is used for power and lighting in many plants. **See Figure 12-2.** This type of transformer is equipped with three single-wound 277 V secondary coils connected in a wye with the center tap used as a neutral for the 277 V single-phase loads. The three phase conductors give 480 V phase-to-phase for equipment power. The center tap is grounded, making it possible for overload current protection devices to function under ground fault conditions.

If the needed secondary is a 3-wire, 240 V system, then a unit with dual-wound 240 V secondary coils can be used. The coils can be wired in parallel and connected in a delta. This system will usually have the B phase grounded for the operation of the overcurrent devices in case of a fault to ground in the system. This same transformer can be used for a 480 V, 3-wire system, with the 240 V coils wired in series and connected in a delta. Again, we usually find the B phase grounded.

Another common secondary connection is a 4-wire, 120 V/240 V service connected in a delta, with dual-wound 120 V coils connected in series. **See Figure 12-3.** The center connection between A and C phase is grounded. This arrangement gives 120 V phase-to-ground between each of these two phases and ground. This connection also gives 208 V phase-to-ground between the B phase and ground.

This same unit can be used for a 208Y/120 V service with the coils in parallel and connected in a wye. The center connection is used as a neutral and connected to ground. This connection gives 120 V phase-to-neutral or phase-to-ground, and 208 V phase-to-phase.

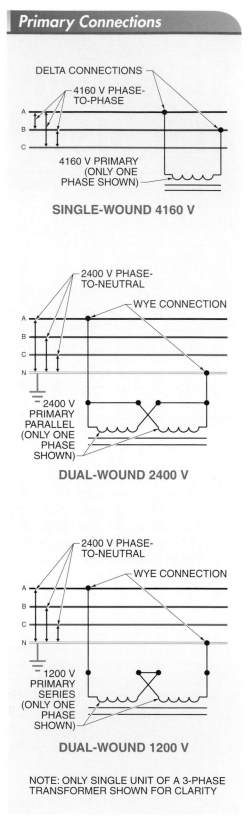

Primary Connections

SINGLE-WOUND 4160 V

DELTA CONNECTIONS
4160 V PHASE-TO-PHASE
4160 V PRIMARY (ONLY ONE PHASE SHOWN)

DUAL-WOUND 2400 V

2400 V PHASE-TO-NEUTRAL
WYE CONNECTION
2400 V PRIMARY PARALLEL (ONLY ONE PHASE SHOWN)

DUAL-WOUND 1200 V

2400 V PHASE-TO-NEUTRAL
WYE CONNECTION
1200 V PRIMARY SERIES (ONLY ONE PHASE SHOWN)

NOTE: ONLY SINGLE UNIT OF A 3-PHASE TRANSFORMER SHOWN FOR CLARITY

Figure 12-1. There are several options for the transformer primary wiring.

Secondary Connections – 480 V/240 V

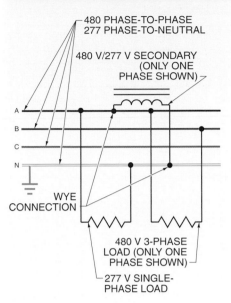

SINGLE-WOUND 480 V/277 V

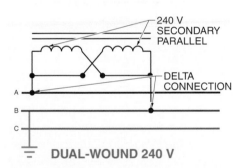

DUAL-WOUND 240 V

Figure 12-2. *There are several options for the transformer secondary wiring.*

Phase

The type of electrical system determines the type of transformer required. Single-phase transformers are used to transform single-phase power, and 3-phase transformers are used to transform 3-phase power. In addition, there are a few other types of phasing used in special applications.

Single-Phase. For single-phase transformers, almost any unit can be selected as long as the kVA rating matches the load requirements. A transformer should be sized so that it is equal to or larger than the load calculation. Single-phase transformers can be connected together into a 3-phase system.

Three-Phase. Individual 3-phase units with all the coils on a common core are very common. In addition, three single-phase transformers are often used to connect to a 3-phase circuit. **See Figure 12-4.** An advantage of using three single-phase units is that, in case of failure, it is much less expensive to replace one single-phase unit than a common-core 3-phase unit.

Secondary Connections – 120 V/208 V/240 V

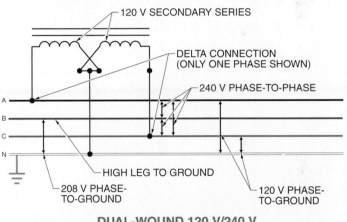

DUAL-WOUND 120 V/240 V

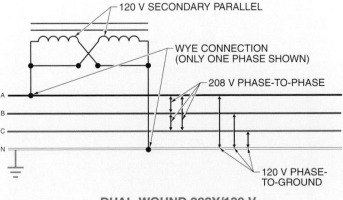

DUAL-WOUND 208Y/120 V

Figure 12-3. *A common secondary connection is a 4-wire, 120 V/208 V/240 V service.*

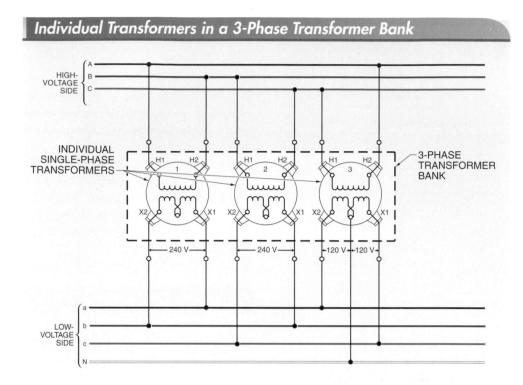

Individual Transformers in a 3-Phase Transformer Bank

Figure 12-4. Three single-phase transformers can be connected together to connect to a 3-phase circuit.

Other Phasing. In a few areas, 2-phase systems can be found. Synchronous converters use interconnected units to provide 6-phase as the input for the conversion to DC. These systems often use standard single-phase transformers with special connections to create these phase configurations.

Loads

When sizing a transformer, present loads and expected future loads should be considered. Large power transformers are expensive and future expansion should be taken into account when selecting a transformer. It may be more cost-effective to use an oversized transformer from the beginning than to have to replace a transformer in the near future. It may be more expensive than needed to use a transformer that is larger than needed. Power factor correction can be installed at the time of transformer installation to rectify the poor power factor of a large, unloaded transformer.

Open Delta. An open delta transformer configuration can be used when selecting a transformer to allow for future expansion. This connection allows the use of two transformers for the present load with the option of later ordering a third unit when the loads are increased. **See Figure 12-5.** This connection offers the user 58% of the capacity of a standard closed delta.

It is common to create an open delta with two transformers of different sizes. The larger unit is used to provide power to the single-phase loads. For example, a 25 kVA transformer can be connected to a 15 kVA transformer in an open delta. The 25 kVA transformer is used for the single-phase loads, while both transformers supply the 3-phase loads. When selecting the two units, they must be sized to operate at 86% or less of their rated capacity. The impedance of the two transformers should be matched in this installation. When a third unit is added to close the delta, each transformer can deliver 100% of its rating.

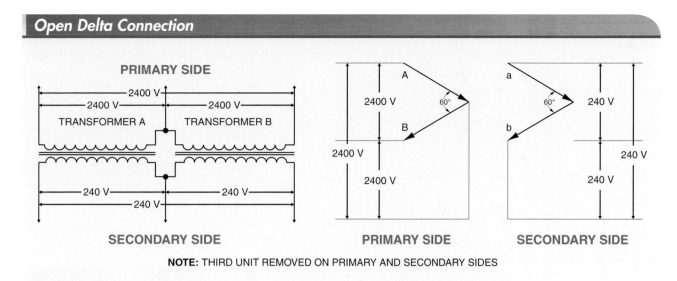

NOTE: THIRD UNIT REMOVED ON PRIMARY AND SECONDARY SIDES

Figure 12-5. An open delta configuration can be used when selecting a transformer to allow for future expansion.

Purpose and Type

An important factor in selecting a transformer is determining the purpose of the transformer. The particular application determines the type of transformer selected. The most common types of transformers are the two-winding types, buck-boost, and autotransformers. **See Figure 12-6.**

Two-Winding. A *two-winding transformer* is a transformer with no electrical connection between the primary and the secondary. A two-winding transformer is the most common type of transformer. All two-winding transformers have high-voltage leads and low-voltage leads connected to the windings.

A step-down transformer provides a decrease in voltage by powering the high-voltage leads and taking power off at the low-voltage leads. A step-up transformer provides an increase in voltage by powering the low-voltage leads and taking power off the high-voltage leads. An isolation transformer is used to reduce or eliminate the effect of voltage spikes, harmonics, and other line disturbances.

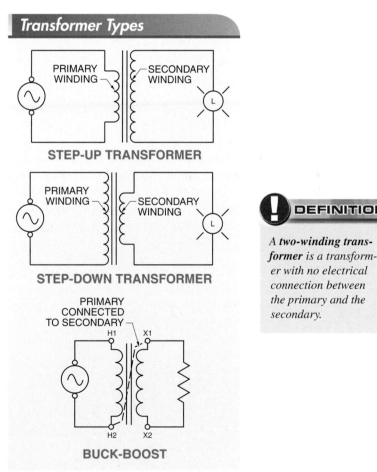

Figure 12-6. The most common types of transformers are two-winding transformers, wired as step-up or step-down, and auto-transformers.

TECH FACT

A two-winding transformer is the most common type of transformer.

DEFINITION

*A **two-winding transformer** is a transformer with no electrical connection between the primary and the secondary.*

Autotransformers. An autotransformer has all or part of the windings shared by the primary and secondary. When selecting a transformer, a two-winding transformer can be interconnected for use as an autotransformer by connecting the primary in series with the secondary. Autotransformers are used in reduced-voltage starting of large motors and must be sized to handle the locked rotor current at the reduced voltage.

A variable transformer is a special autotransformer that is able to make continuous adjustments in the output voltage. A buck-boost transformer is a special autotransformer used to make a small adjustment to the line voltage, typically in increments of 2½%. Buck-boost transformers are used to either buck (reduce) a high voltage or boost (increase) a low voltage to a level that can be used with the equipment involved. They are sized by using the circuit current multiplied by the amount of voltage change, either buck or boost.

Noise Level

Most transformers are placed in an area where transformer noise is not a concern. However, noise can be a critical factor when the transformer is placed in an area that is normally fairly quiet. Transformers have a noise rating based on the kVA rating.

Cooling and Ventilation

All transformers generate heat. If excessive heat is permitted to remain in or around a transformer, the insulation on the conductors that make up the windings would be subject to damage and perhaps failure. When selecting a transformer, the location where the transformer will be placed and the ambient temperature must be considered. NEC® Section 450.9 describes the requirements for adequate ventilation. **See Figure 12-7.**

Dry-Type. Dry-type transformers contain no oil and are cooled by circulating air. Dry-type transformers must be placed in a location where the natural circulation of air is sufficient to cool the transformer.

Dry-type transformers must be kept clear of debris and obstructions of the cooling vents so that air can cool the unit. Some dry-type transformers contain fins or radiators to aid in the dissipation of heat. In some cases, fans or blowers are used to blow air past a transformer for additional cooling.

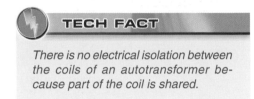

TECH FACT

There is no electrical isolation between the coils of an autotransformer because part of the coil is shared.

Transformer Cooling

COOLING VENT

COOLING VENT
DRY-TYPE

COOLING FANS

LIQUID-FILLED

Figure 12-7. Transformers need to be cooled. The location where a transformer will be placed and the ambient temperature must be considered.

Liquid-Filled. Liquid-filled transformers use oil or other liquid to surround the coils and conduct heat away from the core. If the natural circulation of the oil is not enough to do the job of cooling, fins or cooling fans may be required to remove the heat. Pumps with heat exchangers can be installed on transformers that operate in areas with high ambient temperatures. The oil is pumped out and cooled in the heat exchanger and then circulated back to the unit.

Some of the oils used as dielectric and for cooling of the transformer are flammable. These oils may require an expensive room or area where they are to be installed. Some common liquids used are mineral oil, askarels, less-flammable liquids, and nonflammable liquids.

Mineral Oil. Mineral oil does a good job of providing insulation between the transformer windings and the core. However, mineral oil is subject to oxidation that degrades the oil and results in sludge buildup. Certain types of gases are used to prevent oxygen and moisture from forming acids and sludge in the unit. These gases may require venting.

Askarels. Askarels are nonflammable liquids used in transformers. Askarels have been banned by the EPA because of concerns with polychlorinated biphenyls (PCBs), a class of poisonous chemicals present in askarels. NEC® Section 450.25 says that askarel-insulated transformers installed indoors and rated over 25 kVA shall be furnished with a pressure-relief vent. The EPA requires labeling of transformers and other equipment that contains PCBs. **See Figure 12-8.**

Less-Flammable Liquids. Because of the environmental and health problems associated with the use of askarels, liquids for use in transformers have been developed that are flammable but have a relatively high fire point. The NEC® Section 450.23 defines less-flammable liquids as those liquids which are listed and have a fire point of not less than 300°C.

Figure 12-8. Transformers containing askarels must be labeled to indicate that PCBs are present.

Listed liquids are those liquids that have been suitably evaluated and meet appropriate designated standards. These liquids do not contain any PCBs but they can burn. Therefore, the NEC® has special requirements for less-flammable liquid-insulated transformers. The installation requirements for these transformers depend on whether the transformer is located inside or outside.

Nonflammable Liquids. Because of the added benefit of having nonflammable liquid-filled transformers, the search continues for liquids that have the nonflammability of askarels without the use of PCBs. NEC® Section 450.24 recognizes nonflammable fluid-insulated transformers. These transformers are filled with a liquid that has no flash point or fire point and does not burn in air. A liquid confinement area and a pressure-relief vent must be furnished when these transformers are installed indoors.

INSPECTION AND TESTING

Installing a transformer is more than just connecting the wires according to the wiring diagram. The first part of the installation process includes initial inspection and testing of the transformer when it is received from the factory or warehouse. After a successful inspection, the installation can begin.

When a transformer arrives at a factory or job site, there are several things that should be inspected before accepting the shipment. For larger power transformers, there are some electrical tests that should be performed to verify that the unit was manufactured correctly and is in satisfactory condition. It is best to inspect and test a transformer before installation and before it is energized for the first time to ensure that it is in good working order.

A complete drawing of the coils and the regulator is found on the transformer nameplate. The nameplate gives the installer all of the data relating to the transformer, including the rating, impedance, primary and secondary voltage, the phasing, the allowable temperature rise, oil type (if used), weight, and connection diagrams. Also included are the name of the manufacturer, the model number, and the serial number.

Inspection

The first thing to do is to inspect for any damage that could have occurred during shipping. The bushings and the insulators should be inspected for cracks and chips in the porcelain. **See Figure 12-9.** The exterior finish should be inspected. If paint has been worn or scraped off, it must be repaired. A unit that sits outside or in a corrosive environment will corrode and leaks may develop. The cooling fins, if present, should be inspected for dents that may affect the ability of the cooling system to operate.

Insulation Resistance

A relatively common problem with transformers is insulation failure between the coils. An *insulation resistance test* is a test performed to measure the leakage current through the coil insulation. When the leakage current is too high, the insulation can fail and short out the coil. An insulation resistance test should be conducted with a

hipot tester or a megohmmeter. **See Figure 12-10.** Megger® is a company that manufactures megohmmeters and other test tools. The company name has become a common name for a megohmmeter.

There are several conditions that should be met for best results from insulation resistance testing. The transformer should not be in service and should not be connected to any circuits, switches, capacitors, etc. The temperature should be above the dewpoint of the ambient air to prevent a moisture coating from forming on the insulation surface.

Transformer Inspection

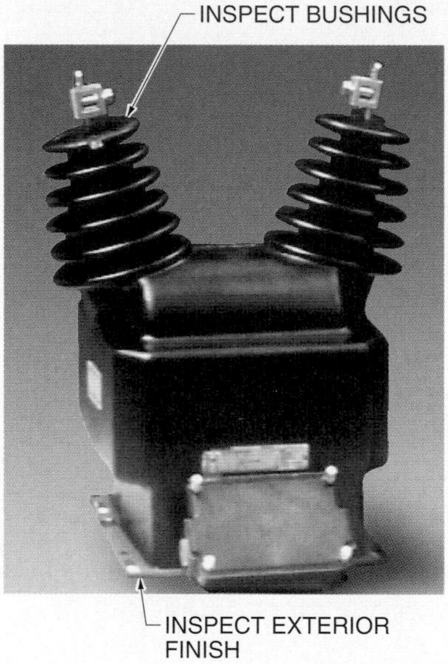

INSPECT BUSHINGS

INSPECT EXTERIOR FINISH

ABB Power T&D Company, Inc.

Figure 12-9. *When a new transformer arrives, the bushings and exterior finish must be inspected before installation.*

The test voltage should be at the correct level. Too much voltage can overstress or damage insulation. Each winding should be tested individually with all the other windings grounded. The transformer manual should give the allowable voltage and type of test required. A transformer rated at 600 V or less can typically be tested with a 500 V or 1000 V megohmmeter to look for any leakage to ground or between the primary and secondary.

Insulation Resistance Testing

MEGOHMMETER

MEASURE FROM
TAP TO GROUND

Fluke Corporation

Figure 12-10. *The transformer insulation should be tested before the transformer is installed.*

*A **winding resistance test** is a test performed to measure the electrical resistance of the transformer windings.*

Capacitors are sometimes used for power factor correction, and they must be discharged or disconnected before the transformer is tested. The temperature must be considered. The tests should be conducted at a temperature of 68°F (20°C).

As a rule of thumb, the resistance doubles for every 18°F (10°C) above the 68°F base temperature and halves for every 18°F (10°C) below the 68°F base temperature. For example, a resistance measured at 2 MΩ at 30°C translates to 1 MΩ at 20°C.

Winding Resistance

A *winding resistance test* is a test performed to measure the electrical resistance of the transformer windings. **See Figure 12-11.** If the resistance increases, extra heating of the wire making up the windings occurs. This can cause the coil to burn out

if the temperature gets hot enough to soften or melt the wire.

A precision ohmmeter is used for this test. All windings must be tested, so a load tap changer must be cycled through all its possible positions. For a very large transformer, such as a 20 MVA unit in a substation, this testing can take several hours.

Turns Ratio

A turns ratio test between the primary winding and secondary winding can be run to verify that all the windings are wired and operating correctly. Again, all windings should be tested, so a load tap changer must be cycled through all its possible positions since these devices effectively change the transformer turns ratio. Kits and test tools are available from test equipment suppliers to simplify this testing. **See Figure 12-12.**

Winding Resistance Testing

TRANSFORMER TAPS

PRECISION OHMMETER

Megger Group Limited

Figure 12-11. *The resistance of the transformer windings is measured with a precision ohmmeter.*

Turns Ratio Testing

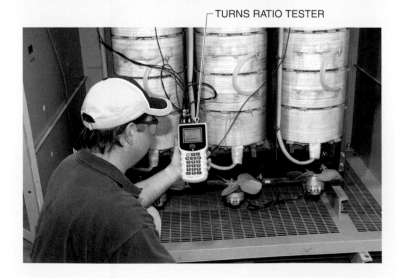

TURNS RATIO TESTER

Megger Group Limited

Figure 12-12. *The turns ratio of the transformer is measured with a turns ratio tester.*

GUARDING AND ENCLOSURES

Transformers are required to be protected from physical damage by NEC® Section 450.8(A). Physical protection of transformers can be provided by covering, shielding, fencing, enclosing, or otherwise prohibiting approach or contact by persons or objects.

Energized parts of transformers shall be guarded per NEC® Sections 110.27 and 110.34. Guarding transformers accomplishes two objectives. First, it protects the transformer from objects that could cause physical damage and threaten the reliability of the electrical system. Next, it protects workers and personnel in the vicinity of the transformer from accidental contact that could result in physical harm. All transformers are required to be guarded and marked by signs or other markings on the equipment indicating the operating voltage of the transformer.

Guarding can be accomplished by limiting access to the equipment, by use of partitions or barriers, or by elevation to restrict access to the energized parts. In the event that exposed live parts are present, NEC® Section 450.8(D) requires that the operating voltage of the live parts be clearly indicated by visible markings either on the equipment or on adjacent structures. **See Figure 12-13.**

In general, equipment such as switches shall not be installed within a transformer enclosure unless the equipment operates at 600 V, nominal, or less and only qualified persons have access to such equipment. NEC® Section 450.8(B) says that dry-type transformers shall be enclosed in a case or enclosure to protect against the insertion of foreign objects. In addition, NEC® Section 450.8(C) requires that all energized parts within the enclosure be suitably guarded.

TECH FACT

A turns ratio test, or turns-to-turns ratio test, is used to verify that the transformer is increasing or decreasing the voltage as expected.

Transformer Guarding

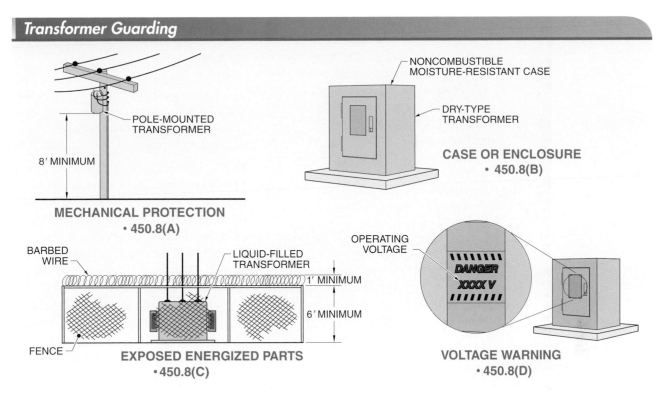

MECHANICAL PROTECTION
• 450.8(A)

POLE-MOUNTED TRANSFORMER

8′ MINIMUM

NONCOMBUSTIBLE MOISTURE-RESISTANT CASE

DRY-TYPE TRANSFORMER

CASE OR ENCLOSURE
• 450.8(B)

BARBED WIRE

LIQUID-FILLED TRANSFORMER

1′ MINIMUM

6′ MINIMUM

FENCE

EXPOSED ENERGIZED PARTS
•450.8(C)

OPERATING VOLTAGE

DANGER XXXX V

VOLTAGE WARNING
• 450.8(D)

Figure 12-13. *All transformers are required to be guarded.*

Dry-Type

The NEC® makes distinctions in the installation requirements between dry-type transformers and liquid-filled transformers. In general, transformers and transformer vaults shall be installed in locations that are readily accessible to qualified personnel. Transformers require some maintenance and may even need to be replaced. For this reason, they cannot be installed where they are blocked or otherwise obstructed in a manner that quick access is denied. The NEC® makes distinctions in the installation requirements depending on the location and size of the transformer.

Ceiling Mounting. Dry-type transformers 600 V, nominal, or less are not required to be installed in a readily accessible space. NEC® Section 450.13 permits installation in spaces above hung ceilings provided the transformers are rated 50 kVA or less, adequate ventilation is provided, and proper clearances from combustible materials are maintained. **See Figure 12-14.**

Ceiling Mounting

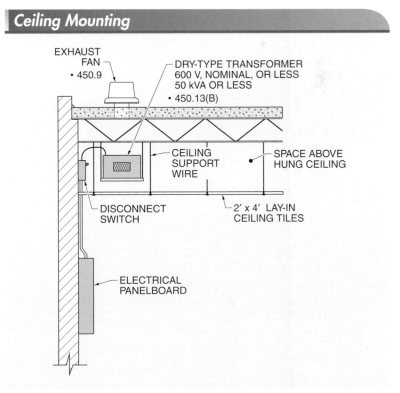

EXHAUST FAN
• 450.9

DRY-TYPE TRANSFORMER 600 V, NOMINAL, OR LESS 50 kVA OR LESS
• 450.13(B)

CEILING SUPPORT WIRE

SPACE ABOVE HUNG CEILING

DISCONNECT SWITCH

2′ x 4′ LAY-IN CEILING TILES

ELECTRICAL PANELBOARD

Figure 12-14. *Transformers are permitted to be installed in spaces above hung ceilings provided adequate ventilation is provided.*

Indoors 112½ kVA or Less. Indoor requirements for dry-type transformers depend on the size of the transformer. Dry-type transformers installed indoors, and not over 112½ kVA, shall have a clearance from combustible materials of at least 12″ per NEC® Section 450.21(A). **See Figure 12-15.** The exception to this section does permit the separation requirement to be omitted provided the transformer is rated 600 V or less and is completely enclosed, with or without ventilation openings.

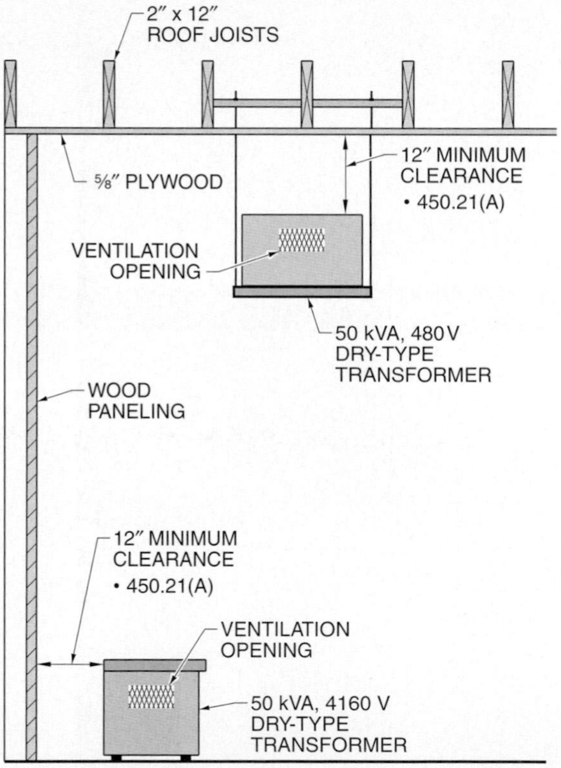

Figure 12-15. Dry-type transformers rated at 112½ kVA or less and installed indoors shall have a minimum clearance of 12″ from combustible materials.

A separation of less than 12″ from combustible materials is permitted when a fire-resistant, heat-insulating barrier is installed between the transformer and the combustible material. For the purposes of this section, "fire-resistant" means the equivalent of a 1-hour minimum fire rating.

NEC® Section 450.9 requires that transformers with ventilating openings shall be installed so that obstructions are not allowed to block the openings. In addition, the required clearances must be marked on the transformer.

Indoors over 112½ kVA. In general, transformers rated over 112½ kVA are required to be installed in a fire-resistant transformer room per NEC® Section 450.21(B). Any construction methods or materials that provide at least a 1-hour minimum fire rating are suitable.

Two exceptions permit transformers rated over 112½ kVA to be installed in non-fire-resistant transformer rooms. **See Figure 12-16.** Per NEC® Section 450.21(B), Ex. 1, transformers with Class 155 or higher insulation systems which are separated from combustible materials by a fire-resistant, heat-insulating barrier need not be installed in a transformer room constructed of fire-resistant materials. In addition, these transformers need not be in a transformer room if they are separated from combustible materials by a distance of not less than 6′ horizontally and 12′ vertically.

Per NEC® Section 450.21(B), Ex. 2, transformers with Class 155 or higher insulation systems which are totally enclosed, other than ventilation openings, need not be installed in transformer rooms constructed of fire-resistant materials. The insulation system temperature class is required by NEC® Section 450.11 to be marked on the nameplate of each dry-type transformer.

Outdoors. Dry-type transformers installed outdoors shall meet the installation requirements of NEC® Section 450.22. In general, the transformer enclosure for transformers installed outdoors shall be weatherproof. The enclosure shall be constructed so that exposure to weather does not interfere with transformer operation.

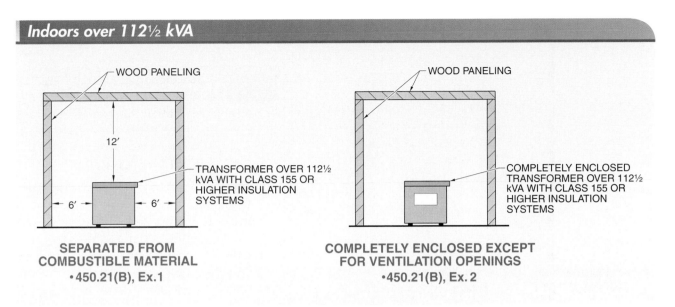

Indoors over 112½ kVA

SEPARATED FROM
COMBUSTIBLE MATERIAL
• 450.21(B), Ex.1

COMPLETELY ENCLOSED EXCEPT
FOR VENTILATION OPENINGS
• 450.21(B), Ex. 2

Figure 12-16. Transformers larger than 112½ kVA, with Class 155 or higher insulation systems, are permitted to be installed indoors if separated from combustible materials or completely enclosed.

Additionally, if the transformer rating exceeds 112½ kVA, the transformer is not permitted to be installed within 12″ of any combustible materials used in the construction of the building. However, if the transformer has Class 155 or higher insulation systems and is completely enclosed, other than the ventilation openings, the 12″ separation is not required. **See Figure 12-17.**

Less-Flammable Liquid-Insulated (LFLI) Transformers

LFLI transformers are filled with a liquid that has reduced flammability characteristics but is not completely nonflammable. LFLI transformers may be installed indoors or outdoors.

Indoors. Per NEC® Section 450.23, LFLI transformers installed indoors shall meet one of three general requirements. **See Figure 12-18.** Per NEC® Section 450.23(A)(1), LFLI transformers are permitted to be installed indoors in Type I or Type II buildings when the transformer is rated at 35,000 V or less, the liquid is listed and installed per the listing, a liquid

confinement area is provided for leaks or spills, and no combustible materials are stored in the area. Type I and II buildings generally are constructed of noncombustible materials or methods. See ANSI/NFPA 220-1999, *Types of Building Construction.*

Outdoors

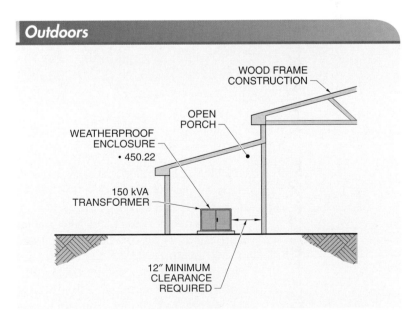

Figure 12-17. Transformers larger than 112½ kVA installed outdoors shall be located at least 12″ from combustible materials unless the transformer has a Class 155 or higher insulation system and is totally enclosed.

LFLI Transformers – Indoors

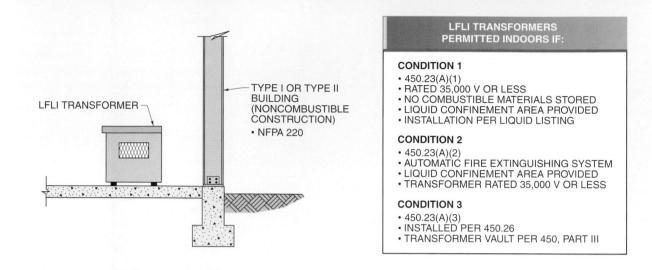

LFLI TRANSFORMER

TYPE I OR TYPE II
BUILDING
(NONCOMBUSTIBLE
CONSTRUCTION)
• NFPA 220

**LFLI TRANSFORMERS
PERMITTED INDOORS IF:**

CONDITION 1
• 450.23(A)(1)
• RATED 35,000 V OR LESS
• NO COMBUSTIBLE MATERIALS STORED
• LIQUID CONFINEMENT AREA PROVIDED
• INSTALLATION PER LIQUID LISTING

CONDITION 2
• 450.23(A)(2)
• AUTOMATIC FIRE EXTINGUISHING SYSTEM
• LIQUID CONFINEMENT AREA PROVIDED
• TRANSFORMER RATED 35,000 V OR LESS

CONDITION 3
• 450.23(A)(3)
• INSTALLED PER 450.26
• TRANSFORMER VAULT PER 450, PART III

Figure 12-18. LFLI transformers installed indoors shall meet one of three general requirements.

Dry-type transformers usually have air vents to allow air to circulate and cool the transformer.

Outdoors. LFLI transformers are installed outdoors per NEC® Section 450.23(B). In general, these transformers can be installed outdoors where attached to, adjacent to, or on the roof of a building if one of two conditions is met. **See Figure 12-19.** Per NEC® Section 450.23(B)(1), LFLI transformers are permitted to be installed outdoors if the building is classified as a Type I or Type II building and the installation complies with all of the instructions included in the listing of the liquid.

Additional safeguards may be required if the transformer is installed adjacent to window and door openings, fire escapes, and other combustible materials per NEC® Section 450.23(B)(1), FPN. While FPNs are not mandatory, in this case, additional consideration should be given because of the potential hazards involved.

Per NEC® Section 450.23(B)(2), the second condition under which LFLI transformers can be installed outdoors is when they are installed in accordance with the requirements for oil-insulated transformers installed outdoors. See NEC® Section 450.27. Basically, LFLI transformers can be installed outdoors with one of the following as appropriate for the hazard involved: space separations, fire-resistant barriers, automatic fire suppression systems, or enclosures.

Nonflammable Fluid-Insulated (NFFI) Transformers

NFFI transformers contain a liquid that is nonflammable. Such a liquid does not have a flash or fire point; therefore, NFFI transformers can be installed in either indoor or outdoor locations with very few limitations.

Limitations only apply to NFFI transformers rated over 35,000 V. These transformers, when installed indoors, shall be installed in a vault, a liquid confinement area shall be provided, and the transformer shall be provided with a pressure-relief vent. Gases can be a by-product of these types of transformers, so the transformer shall provide means for absorbing any of these gases, or the pressure-relief vent shall be connected to a chimney or flue to allow for the removal of the gases. **See Figure 12-20.**

Substations

When building a substation, the first piece of equipment to be placed is the control room that contains all the protective devices for the substation. These rooms are typically prefabricated and shipped to the job site. The transformer is placed on the slab and conduit is run. A very large transformer or control room may require experienced riggers to place it properly. A large framework is used to mount the high-voltage wires. The height of the frame keeps the high voltage away from anyone entering the substation. **See Figure 12-21.**

A unique ability of a substation is the voltage regulation it provides. Autotransformers are connected to the secondary of the main transformer to compensate for changes in loads. As the loads are increased, a regulator selects a tap that provides a boost to the output so the voltage remains fairly constant. As the loads are removed, the regulator changes taps and the output voltage is lowered, still maintaining a steady voltage level.

TECH FACT

Designers and installers of transformers should consider the requirements of NEC® Section 250.30, Grounding Separately Derived Alternating-Current Systems, when installing electrical distribution systems.

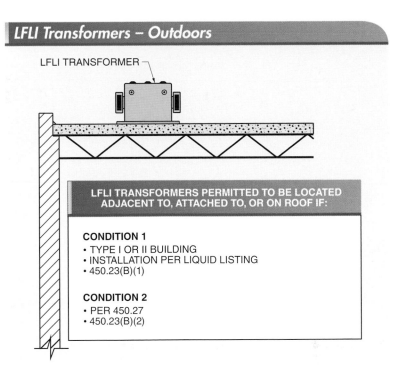

Figure 12-19. LFLI transformers installed outdoors shall meet one of two general requirements.

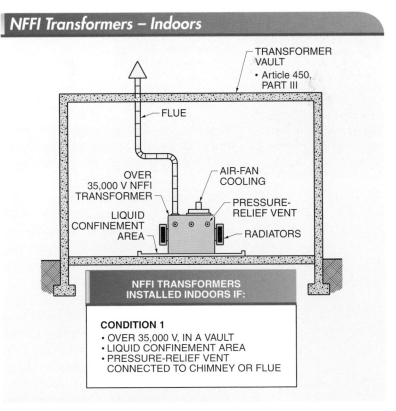

Figure 12-20. NFFI transformers over 35,000 V and installed indoors shall be installed in a transformer vault.

Substation Frame

Figure 12-21. *The large frame in a substation is used to protect people and property from the high voltages present in the substation.*

DEFINITION

*A **supervised instal-lation** is an electrical installation in which the conditions of main-tenance are such that only qualified persons monitor or service the electrical equipment.*

OVERCURRENT PROTECTION

Most electrical devices require overcurrent protection and transformers are no exception. Transformer overcurrent protection is required to protect the primary windings from short circuits and overloads and the secondary windings from overloads. NEC® Section 450.3 contains the requirements for overcurrent protection of transformers. Overcurrent protection requirements depend upon several factors.

The first factor is the voltage at which the transformer operates. Section 450.3(A) contains the rules for transformers rated over 600 V, nominal. Section 450.3(B) contains the rules for transformers operating at 600 V, nominal, or less. The second factor is the location of the overcurrent device. Requirements vary depending on whether only the primary winding is pro-

tected, or both the primary and secondary windings are protected.

The last factor applies to transformers rated over 600 V only. If conditions of maintenance ensure that only qualified personnel will work on the transformers, NEC® Table 450.3(A) permits primary protection only for these supervised installations.

The rules of these sections are intended to provide protection for the transformers only. Conductors on both the primary and secondary side will still need to be protected in accordance with the provisions of NEC® Sections 240.4 (F) and 240.21 (C).

Over 600 V, Nominal

The requirements for transformer overcurrent protection for transformers rated over 600 V depend upon whether the installation is nonsupervised or supervised. A *supervised installation* is an electrical installation in which the conditions of maintenance are such that only qualified persons monitor or service the electrical equipment. See Note 3 to NEC® Table 450.3(A).

Primary and Secondary. For nonsupervised transformer installations, each transformer shall have primary and secondary OCPDs in accordance with NEC® Table 450.3(A). The Table is organized around the permissible percentages for fuses and circuit breaker ratings on both the primary and secondary sides, the location, and the impedance of the transformer. For the purposes of this section of the NEC®, electronically actuated fuses are rated in accordance with the circuit breaker percentages. Per NEC® Table 450.3(A), Note 1, the next higher standard rating of a fuse or circuit breaker is permitted when the

calculated value does not correspond to a standard rating of a fuse or circuit breaker. **See Figure 12-22.**

Supervised Installations. For supervised transformer installations, there are two options for providing the required overcurrent protection. The first option is based on primary protection only. In these installations, each transformer shall be protected with a primary overcurrent device that does not exceed 250% for fuses or 300% for circuit breakers of the rated primary current of the transformer. If the calculated value does not correspond to a standard rating of a fuse or circuit breaker, the next higher standard size shall be permitted. **See Figure 12-23.** As with transformer installations of 600 V or less, individual overcurrent protection is not required at the transformer if the primary circuit OCPD provides the necessary protection.

Sizing Primary OCPDs – Nonsupervised, over 600 V Transformers

What size primary OCPD, using a fuse, is needed to protect the nonsupervised transformer?

Current: $I = \dfrac{kVA \times 1000}{V \times \sqrt{3}}$ $I = \dfrac{50 \times 1000}{4160 \times 1.73}$ $I = \dfrac{50,000}{7205} = 6.94$ A

Table 450.3(A): 6.94 × 300% = 20.82 A
Table 450.3(A), Note 1; 240.6(A): Next higher standard size: 20.82 = 25 A
Primary OCPD: **25 A fuse**

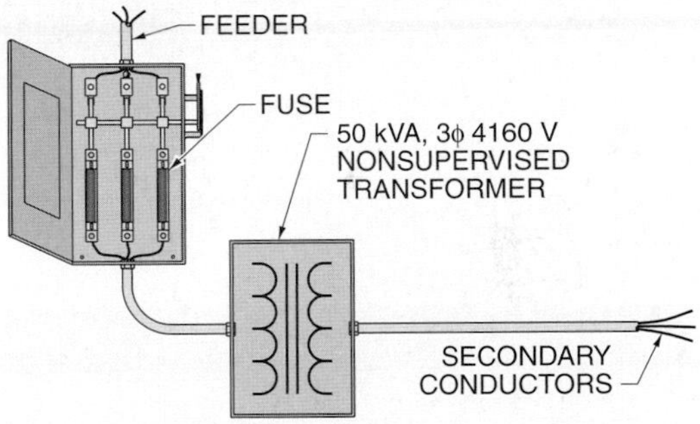

FEEDER
FUSE
50 kVA, 3φ 4160 V NONSUPERVISED TRANSFORMER
SECONDARY CONDUCTORS

Figure 12-22. Nonsupervised transformers over 600 V shall be protected with primary and secondary OCPDs.

Sizing Primary OCPDs – Supervised, over 600 V Transfers

What size primary OCPD, using an adjustable trip CB, is needed to protect the supervised transformer?

Current: $I = \dfrac{kVA \times 1000}{V \times \sqrt{3}}$ $I = \dfrac{25 \times 1000}{2400 \times 1.73}$ $I = \dfrac{25,000}{4157} = 6.01$ A

Table 450.3(A): 6.01 × 300% = 18.03 A
Table 450.3(A), Note 1: Next higher standard size: 18.03 = 20 A
Primary OCPD: **20 A ATCB**

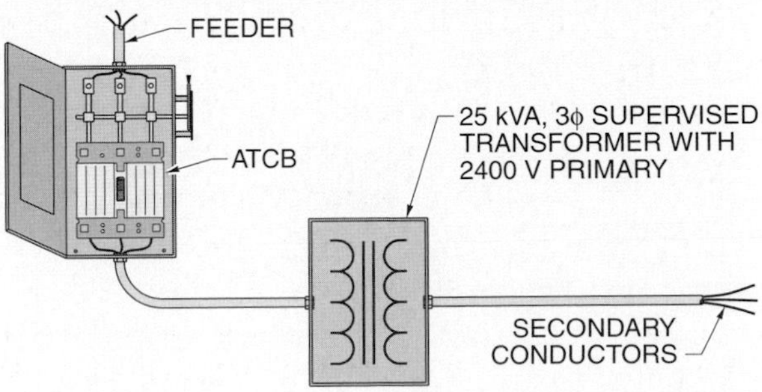

FEEDER
ATCB
25 kVA, 3φ SUPERVISED TRANSFORMER WITH 2400 V PRIMARY
SECONDARY CONDUCTORS

Figure 12-23. Supervised transformers over 600 V shall be provided with primary OCPDs not exceeding 250% for fuses and 300% for circuit breakers.

FLIR Systems

Overheating contacts inside the load tap changer caused an increase in the oil temperature, as shown in this infrared photograph.

The second option for transformers rated over 600 V is to provide both primary and secondary protection. NEC® Table 450.3(A) provides the permissible percentages for fuses and circuit breaker ratings on both the primary and secondary side based upon the impedance of the transformer. For transformers with an impedance of not more than 6%, the maximum rating or setting of the primary feeder OCPD can range from 300% to 600%, depending on the voltage of the primary and the secondary and the type of OCPD used, either fuse or circuit breaker **See Figure 12-24.**

600 V, Nominal, or Less

The requirements for overcurrent protection for transformers rated at 600 V, nominal, or less are given in NEC® Table 450.3(B) and accompanying notes. The overcurrent protection is necessary to protect the windings of the transformer and is independent of the overcurrent protection required for conductors. The secondary overcurrent device is permitted, as it is with services, to consist of not more than six fuses or circuit breakers grouped in one location. The total rating of the six OCPDs cannot exceed that of the required value for a single OCPD. In installations where the six OCPDs consist of both fuses and circuit breakers, the rating shall not exceed the value permitted for fuses.

Sizing Primary and Secondary OCPDs – Supervised, over 600 V Transformers

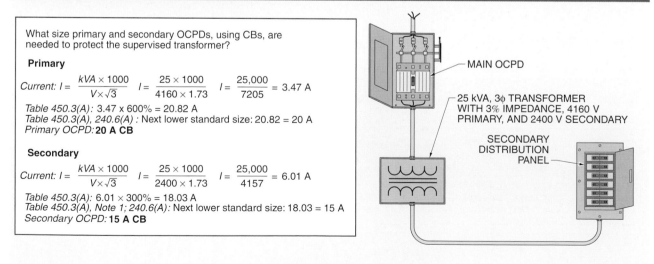

What size primary and secondary OCPDs, using CBs, are needed to protect the supervised transformer?

Primary

Current: $I = \dfrac{kVA \times 1000}{V \times \sqrt{3}}$ $I = \dfrac{25 \times 1000}{4160 \times 1.73}$ $I = \dfrac{25,000}{7205} = 3.47$ A

Table 450.3(A): 3.47 x 600% = 20.82 A
Table 450.3(A), 240.6(A) : Next lower standard size: 20.82 = 20 A
Primary OCPD: **20 A CB**

Secondary

Current: $I = \dfrac{kVA \times 1000}{V \times \sqrt{3}}$ $I = \dfrac{25 \times 1000}{2400 \times 1.73}$ $I = \dfrac{25,000}{4157} = 6.01$ A

Table 450.3(A): 6.01 × 300% = 18.03 A
Table 450.3(A), Note 1; 240.6(A): Next lower standard size: 18.03 = 15 A
Secondary OCPD: **15 A CB**

MAIN OCPD

25 kVA, 3ɸ TRANSFORMER WITH 3% IMPEDANCE, 4160 V PRIMARY, AND 2400 V SECONDARY

SECONDARY DISTRIBUTION PANEL

Figure 12-24. *Supervised transformers over 600 V shall be provided with primary and secondary OCPDs per NEC® Table 450.3(A).*

Primary. The general rule for transformers rated 600 V, nominal, or less is to protect the primary windings of the transformer at not more than 125% of the rated primary current of the transformer. Where this calculated value does not correspond to a standard rating for a fuse or a circuit breaker, per NEC® Section 240.6, and the rated primary current of the transformer is 9 A or more, NEC® Table 450.3(B), Note 1 permits the next higher standard rating to be used. **See Figure 12-25.**

If the rated primary current of the transformer is less than 9 A, the rating of the OCPD is permitted to be set at not more than 167% of the rated primary current. For transformer installations where the rated primary current is less than 2 A, the rating of the OCPD shall be permitted to be set at not more than 300% of the primary current. With the lower primary current values, the impedance of the transformer will act to limit potential fault current and the rating of the primary OCPD can be increased.

Sizing Primary OCPDs, 600 V or Less

What size primary OCPD, using a fuse, is needed to protect the transformer?

$$Amps: \ I = \frac{kVA \times 1000}{V \times \sqrt{3}} \qquad I = \frac{15 \times 1000}{208 \times 1.73} \qquad I = \frac{15,000}{360} = 41.67 \ A$$

Table 450.3(B): 41.67 × 125% = 52.09 A
Table 450.3(B), Note 1; 240.6(A): Next higher standard size: 52.09 = 60 A
Primary OCPD: **60 A fuse**

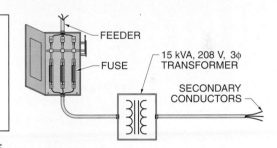

9 A OR MORE

What size primary OCPD, using a fuse, is needed to protect the transformer?

$$Amps: \ I = \frac{kVA \times 1000}{V \times \sqrt{3}} \qquad I = \frac{6 \times 1000}{480 \times 1.73} \qquad I = \frac{6000}{831} = 7.22 \ A$$

Table 450.3(B): 7.22 × 167% = 12.06 A
Table 450.3(B); 240.6(A): Next lower standard size: 12.06 = 10 A
Primary OCPD: **10 A fuse**

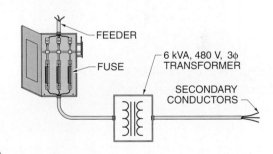

LESS THAN 9 A

What size primary OCPD, using a fuse, is needed to protect the transformer?

$$Amps: \ I = \frac{kVA \times 1000}{V} \qquad I = \frac{0.250 \times 1000}{240} \qquad I = \frac{250}{240} = 1.04 \ A$$

Table 450.3(B): 1.04 × 300% = 3.12 A
Table 450.3(B); 240.6(A): Next lower standard size: 3.125 = 3 A
Primary OCPD: **3 A fuse**

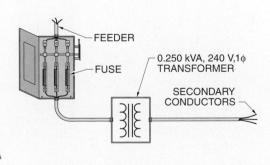
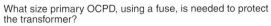

LESS THAN 2 A

Figure 12-25. Transformers at 600 V or less shall be provided with primary OCPDs at not more than 125% of the rated primary current, with exceptions.

NEC® Table 450.3(B) establishes maximum settings for transformer overcurrent protection as a percentage of transformer-rated current. A separate fused disconnecting means could be provided ahead of the transformer that complies with the maximum percentages of NEC® Table 450.3(B). If, however, the circuit overcurrent device is sized so that it also provides the necessary overcurrent protection for the transformer, a separate fused disconnecting means may not be required. **See Figure 12-26.**

Motor control circuit transformers installed per NEC® Section 430.72(C)(1–5) are permitted to be installed without primary overcurrent protection. These transformers are generally under 50 VA and have low or limited primary current rating.

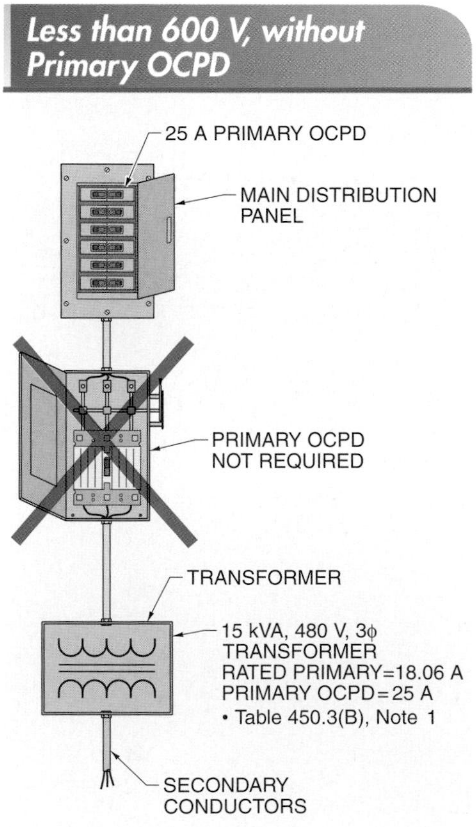

Primary and Secondary. Another option in providing overcurrent protection for transformers rated 600 V, nominal, or less is to protect both the primary and the secondary of the transformer. In these cases, NEC® Table 450.3(B) permits the primary feeder OCPD, set at not more than 250% of the rated primary current of the transformer, to protect the transformer primary, provided the secondary OCPD is set at a value which does not exceed 125% of the rated secondary current of the transformer. As with the primary-only protection requirements, if the rated secondary current of the transformer is 9 A or more and the calculated value does not correspond to a standard rating of a fuse or circuit breaker from NEC® Section 240.6, the next higher standard rated OCPD is permitted. **See Figure 12-27.**

Unlike the requirements for primary-only protection, there is no permission to increase the rating of the primary OCPD when it does not correspond to a standard rating of a fuse or circuit breaker. If the calculated value at 250% does not correspond to a standard rating, the next lower standard rating shall be used. If the rated secondary is less than 9 A, the rating of the secondary OCPD is permitted to be increased to a maximum value of 167% of the rated secondary current of the transformer. This rule is useful for applications in which several transformers are supplied from a single primary OCPD. Sizing at 250% prevents nuisance tripping due to the transformer current inrush characteristics.

Less than 600 V, without Primary OCPD

25 A PRIMARY OCPD

MAIN DISTRIBUTION PANEL

PRIMARY OCPD NOT REQUIRED

TRANSFORMER

15 kVA, 480 V, 3φ
TRANSFORMER
RATED PRIMARY=18.06 A
PRIMARY OCPD=25 A
• Table 450.3(B), Note 1

SECONDARY CONDUCTORS

Figure 12-26. *Transformers less than 600 V are permitted to be installed without individual primary OCPDs when the circuit OCPDs provide this protection.*

TECH FACT

Control transformers operating at 24 V and supplied from 120 V or 240 V sources are common in commercial buildings. These control transformers supply low voltage to the controllers for building automation systems that manage heating, ventilation, and air conditioning units and other electrical devices.

Sizing Primary and Secondary OCPDs

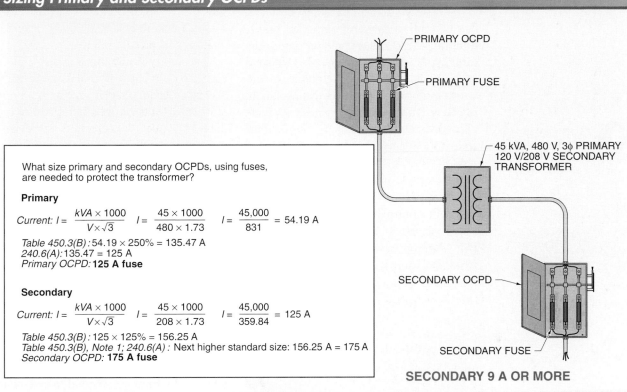

What size primary and secondary OCPDs, using fuses, are needed to protect the transformer?

Primary

Current: $I = \dfrac{kVA \times 1000}{V \times \sqrt{3}}$ $I = \dfrac{45 \times 1000}{480 \times 1.73}$ $I = \dfrac{45,000}{831} = 54.19$ A

Table 450.3(B): 54.19 × 250% = 135.47 A
240.6(A): 135.47 = 125 A
Primary OCPD: **125 A fuse**

Secondary

Current: $I = \dfrac{kVA \times 1000}{V \times \sqrt{3}}$ $I = \dfrac{45 \times 1000}{208 \times 1.73}$ $I = \dfrac{45,000}{359.84} = 125$ A

Table 450.3(B): 125 × 125% = 156.25 A
Table 450.3(B), Note 1; 240.6(A): Next higher standard size: 156.25 A = 175 A
Secondary OCPD: **175 A fuse**

PRIMARY OCPD

PRIMARY FUSE

45 kVA, 480 V, 3φ PRIMARY
120 V/208 V SECONDARY
TRANSFORMER

SECONDARY OCPD

SECONDARY FUSE

SECONDARY 9 A OR MORE

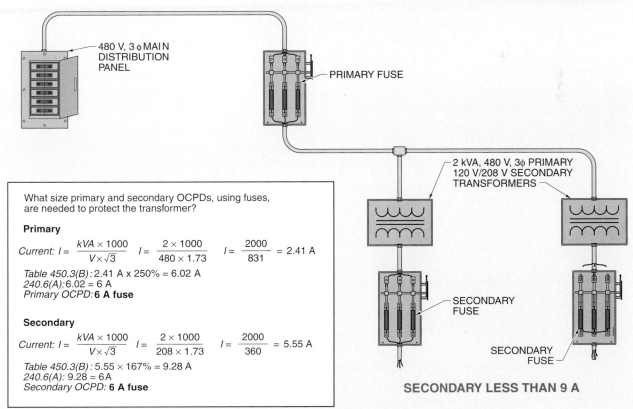

480 V, 3φ MAIN
DISTRIBUTION
PANEL

PRIMARY FUSE

What size primary and secondary OCPDs, using fuses, are needed to protect the transformer?

Primary

Current: $I = \dfrac{kVA \times 1000}{V \times \sqrt{3}}$ $I = \dfrac{2 \times 1000}{480 \times 1.73}$ $I = \dfrac{2000}{831} = 2.41$ A

Table 450.3(B): 2.41 A x 250% = 6.02 A
240.6(A): 6.02 = 6 A
Primary OCPD: **6 A fuse**

Secondary

Current: $I = \dfrac{kVA \times 1000}{V \times \sqrt{3}}$ $I = \dfrac{2 \times 1000}{208 \times 1.73}$ $I = \dfrac{2000}{360} = 5.55$ A

Table 450.3(B): 5.55 × 167% = 9.28 A
240.6(A): 9.28 = 6A
Secondary OCPD: **6 A fuse**

2 kVA, 480 V, 3φ PRIMARY
120 V/208 V SECONDARY
TRANSFORMERS

SECONDARY
FUSE

SECONDARY
FUSE

SECONDARY LESS THAN 9 A

Figure 12-27. Transformers under 600 V may be protected by primary and secondary OCPDs.

Arresters

A transformer, with all its metal hardware, can be damaged by lightning strikes. If lightning strikes a feeder and enters the primary, the transformer can be damaged or destroyed. A large voltage spike can destroy the dielectric quality of the insulation, resulting in a failure. An *arrester* is a device used to shunt a voltage spike to ground or from line to line. **See Figure 12-28.** Arresters are also used to control transient overvoltage situations that might affect the input to the transformer. A transient overvoltage or overcurrent is the result of sudden changes or surges on the line due to a variety of causes.

The arresters used today are constructed from a stack of metal oxide varistors (MOVs). A *metal oxide varistor (MOV)* is a nonlinear resistor that can change resistance with changes in voltage. At low voltage, an MOV has very high resistance, allowing very little current to flow. At high voltage, such as during a voltage spike, an MOV has very low resistance, allowing a transient current to flow through the MOV to ground.

DEFINITION

An *arrester* is a device used to shunt a voltage spike to ground or from line to line.

A *metal oxide varistor (MOV)* is a nonlinear resistor that can change resistance with changes in current.

Voltage Spike Arresters

Figure 12-28. Arresters are used to protect transformers from voltage surges.

SUMMARY...

- The primary factors used in selecting a transformer are the voltage, the phase requirements, the loads, the purpose and type, the noise level, and the cooling and ventilation requirements.

- The transformer primary and secondary usually have multiple possible wiring configurations.

- A 3-phase transformer can consist of all the coils wound on one core or it may be constructed from three single-phase transformers.

- The potential future loads that a transformer may need to supply should be taken into account when sizing a transformer.

- An open delta can be used as a less expensive way to design a system to allow for future expansion.

- A two-winding transformer has no electrical connection between the primary and secondary.

- An autotransformer has all or part of the windings shared by the primary and secondary.

...SUMMARY

- Dry-type transformers are cooled by circulating air.

- Liquid-filled transformers are cooled by an oil or other liquid that surrounds the coils.

- Typical liquids used for liquid-filled transformers are mineral oil, askarels, less-flammable liquids, and nonflammable liquids.

- A new transformer should be inspected for shipping damage to the bushings, insulators, and exterior finish.

- Common electrical tests that should be done on new transformers include an insulation resistance test, a winding resistance test, and a turns ratio test.

- Transformers are required to be protected from physical damage.

- The NEC® makes a distinction between the installation of dry-type transformers that are 112½ kVA or less and transformers that are over 112½ kVA.

- When installed indoors, dry-type transformers over 112½ kVA are generally required to be installed in a fire-resistant transformer room.

- Less-flammable liquid-filled transformers may be installed indoors or out, with appropriate fire protection and spill containment.

- Nonflammable fluid-insulated transformers have no restrictions on location unless they are rated for over 35,000 V.

- The NEC® makes a distinction between the overcurrent protection required for transformers over 600 V and transformers at 600 V or less.

- NEC® Table 450.3 gives the requirements for overcurrent protection devices.

- Supervised locations are those locations where conditions of maintenance and supervision ensure that only qualified persons service the equipment.

- For transformers rated at over 600 V, one option is to protect the primary winding at 250% of rated current for fuses and 300% for circuit breakers.

- For transformers rated at over 600 V, one option is to protect the primary and secondary windings. The amount of protection depends on the impedance.

- For transformers rated at 600 V, nominal, or less, one option is to protect the primary winding at 125% of the rated primary current.

- For transformers rated at 600 V, nominal, or less, one option is to protect the primary and secondary windings. This permits the primary feeder OCPD, set at not more than 250% of the rated primary current, to protect the transformer primary under most circumstances.

DEFINITIONS

- A *two-winding transformer* is a transformer with no electrical connection between the primary and the secondary.

- An *insulation resistance test* is a test performed to measure the leakage current through the coil insulation.

- A *winding resistance test* is a test performed to measure the electrical resistance of the transformer windings.

- A *supervised installation* is an electrical installation in which the conditions of maintenance are such that only qualified persons monitor or service the electrical equipment.

- An *arrester* is a device used to shunt a voltage spike to ground or from line to line.

- A *metal oxide varistor (MOV)* is a nonlinear resistor that can change resistance with changes in current.

REVIEW QUESTIONS

1. List at least four considerations when selecting a transformer.

2. Explain how an open delta can be used to allow for future expansion.

3. What is the difference in the windings between a two-winding transformer and an autotransformer?

4. What is the difference in the cooling method between a dry-type transformer and a liquid-insulated transformer?

5. What is the difference between the less-flammable liquids and nonflammable liquids that are used in transformers?

6. What is the purpose of inspecting the finish on a new transformer before installation?

7. What is the purpose of an insulation resistance test?

8. What is the purpose of a winding resistance test?

9. What are the clearance requirements around a dry-type transformer rated at 112½ kVA or less?

10. What are the restrictions on the placement of nonflammable fluid-insulated (NFFI) transformers?

11. What is a supervised installation?

12. What are the OCPD requirements for the primary winding of a transformer rated at 600 V, nominal, or less?

13. What are the OCPD requirements for the primary and secondary windings of transformers rated at 600 V, nominal, or less?

Maintenance and Troubleshooting

Without proper maintenance, any piece of equipment has an increased risk of failure. Transformer inspections are used to observe the operation of the transformer. The enclosures and internals of dry-type transformers need to be inspected and cleaned. For liquid-immersed transformers, the liquid level, temperature and pressure gauges, relief devices, and relays need to be inspected. Transformer maintenance includes many electrical tests as well as routine oil testing to analyze the operating condition of the transformer.

INSPECTIONS

A transformer maintenance program begins when the transformer is installed. After the installation, periodic inspection and maintenance are required to prevent costly failures and interruption of service. In general, clean the transformer as necessary and verify that it is in normal operating condition.

Accurate recordkeeping is the key to maintaining smooth operation of any equipment. The results of maintenance tests should be recorded and saved. Many of the test interpretations depend on the trend of a measured value. Maintenance and shutdowns are scheduled depending on the trends.

Dry-Type Transformers

A dry-type transformer that provides power for lighting and equipment rarely needs much in the way of maintenance. However, scheduled inspections and preventive maintenance can discover problems before they cause damage or downtime. The appearance, sound, and smell of a transformer are keys to a good inspection.

Enclosures. An enclosure protects the transformer from the elements. At the same time, the enclosure must allow the heat from the transformer to dissipate. The enclosure itself should be inspected. Dented enclosure walls can result in reduced air circulation. Restricted access around the enclosure can result in reduced airflow. Rodents and insects can build nests and block openings for cooling. These conditions can lead to overheating. **See Figure 13-1.**

Transformer Enclosures

INSPECT ENCLOSURE FOR DAMAGE

VERIFY OPENINGS ARE CLEAR

Figure 13-1. Transformer enclosures should be inspected for damage and blocked ventilation openings.

Some environments contain abrasive and conductive dust that can build up on the unit and cause catastrophic failures. A dust buildup can create a conductive path between taps and bushings. This is especially true when a transformer is exposed to water, such as spray or rain. Any buildup should be removed from the enclosure to prevent this type of problem.

While the enclosure is being visually inspected, the technician should also listen to the transformer. Any abnormal noises may indicate impending failure. Any unusual smells, such as a burnt odor, may indicate overheating. The transformer internals should be inspected as a follow-up.

Some of the dry-type transformers are open, with no enclosure, and are limited in

the locations they may occupy. This type of transformer may be installed in a large electrical panel or in a motor drive enclosure.

Internal Inspection. After the enclosure is inspected, the next step is to open up the enclosure and inspect the transformer itself. The technician needs to follow the NFPA 70E rules about personal protective equipment and flash boundaries, and OSHA rules about lockout/tagout.

The core should be examined. The laminations should be tight and secure. Loose laminations cause excessive noise and eventually cause damage to the transformer. All other hardware should be inspected. Nuts and bolts used to prevent transformer vibration can work loose. Loose nuts and bolts can also cause noise and should be repaired before the transformer is damaged. **See Figure 13-2.**

The coils should be inspected to verify that they are in the correct position on the core. Telescoping, or axial collapse, of the windings can be caused by high currents. The coil taps should be examined for damage. The other internals should be inspected for a buildup of dust or other contaminants. This buildup can interfere with cooling and allow overheating to occur.

Temperatures. Heat is often the first sign of problems in a dry-type transformer. A loose connection or damaged coil can be significantly hotter than other areas. A higher than normal temperature of the air leaving the enclosure when the transformer is operating is an indicator of impending failure.

Ground Connections. The noncurrent-carrying metal parts of a transformer installation are required by the NEC® to be effectively grounded. Conductive materials enclosing electric conductors or equipment, or forming part of such equipment, are grounded to prevent a voltage above the ground on these materials. Circuits and enclosures are grounded to help overcurrent devices operate in case of insulation failure or a ground fault.

Internal Inspection

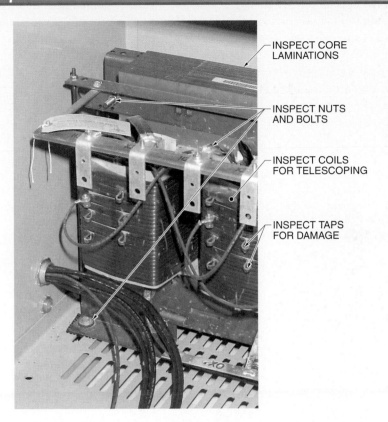

INSPECT CORE LAMINATIONS

INSPECT NUTS AND BOLTS

INSPECT COILS FOR TELESCOPING

INSPECT TAPS FOR DAMAGE

Figure 13-2. Transformer internals should be inspected for damage, loose laminations, loose nuts and bolts, and coil damage.

Ground connections should be inspected and the results recorded. The inspection should include examination for loose, corroded, or broken connections. These connections should be observed for any unusual heating at the connection. **See Figure 13-3.** Periodic resistance-to-ground measurements should be made and recorded.

Lightning Arresters. Lightning arresters should be regularly inspected and cleaned. This inspection should include a visual observation of dirt and foreign deposits on the arrester or any undue mechanical strain from insufficient slack in the line. Periodically, an inspection should be made that includes a complete overhaul and cleaning of the arresters. All connections should be clean and tight, and all broken or damaged parts should be replaced.

Grounding Connections

INSPECT GROUNDING STRAPS

Figure 13-3. Grounding connections need to be free from corrosion and damage and must be securely fastened.

Load Tap Changers. Most transformer installations include load tap changers that allow changes in the turns ratio of the transformer. The load tap changer should be rotated over its entire range and a careful inspection made to ensure that a complete contact surface and tension exist at each position. The exposed external mechanical parts should be carefully examined for wear, lubrication, corrosion, dirt, looseness, adjustment, and seal leakage. **See Figure 13-4.**

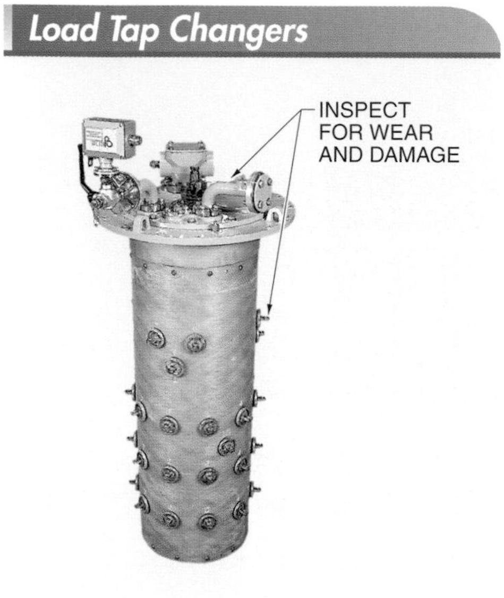

Load Tap Changers

INSPECT FOR WEAR AND DAMAGE

ABB, Inc.

Figure 13-4. Load tap changers should be inspected for wear and damage.

Liquid-Immersed Transformers

Liquid-immersed transformers should be inspected for the same things as a dry-type transformer, where appropriate. In addition, there are many things that are unique to liquid-immersed transformers that require attention during an inspection.

Liquid Level. The proper level of the insulating liquid must be maintained to prevent a transformer from overheating. The liquid level should be checked regularly. If there is any loss from evaporation or leakage, it should be replaced immediately and any repairs made to prevent further loss. **See Figure 13-5.**

Liquid Temperatures. The temperature of the liquid surrounding the coils should be checked. The temperature is normally about 15°F to 25°F (8°C to 14°C) less than the hottest-spot winding temperature. The hottest-spot winding temperature should be monitored continuously by an alarm system. If the liquid temperature increases from one maintenance inspection to another, there may be problems that should be investigated.

Some larger transformers have heat exchangers. The heat exchangers have hot oil on one side and air or water on the other side. The pumps, motors, and intake screens should be examined to ensure they are operating properly. The oil and water temperatures should be checked. At the same load on different days, the amount of cooling should be the same. The temperature drop of the oil and the temperature rise of the water should be constant as they go through the heat exchanger. If the amount of temperature rise of the water increases over time, the heat exchanger may be getting clogged or dirty with scale and sediment and should be inspected and cleaned.

Liquid Level

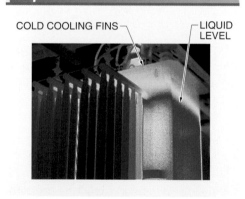

COLD COOLING FINS — LIQUID LEVEL

Figure 13-5. When the liquid level is low, the liquid does not flow properly to the cooling fins. This infrared photograph shows cold cooling fins and hot oil because of the low oil level.

Pressure/Vacuum Gauges. Pressurized nitrogen is often used as a blanket over the liquid in liquid-filled transformers. Most liquid-immersed transformers with a sealed tank are equipped with a pressure/vacuum gauge. **See Figure 13-6.** The gauge should be inspected and calibrated periodically. The gauge should also be monitored during normal operation. As the oil temperature changes, there should be a change in the pressure gauge reading. If the reading does not change, an air leak may be present above the oil level in the tank.

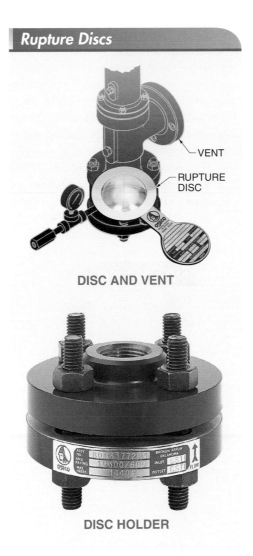

Tank Gauges

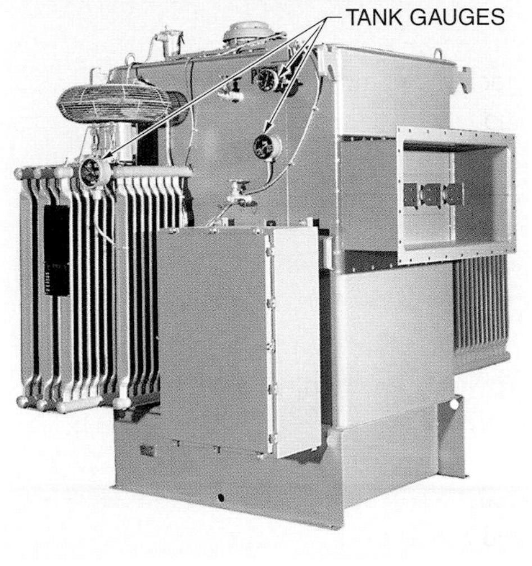

TANK GAUGES

ABB, Inc.

Figure 13-6. *Tank gauges should be inspected and calibrated.*

Rupture Discs

VENT

RUPTURE DISC

DISC AND VENT

DISC HOLDER

Oseco, Inc.

Figure 13-7. *Pressure relief devices are vented to the outside.*

Pressure Relief Devices. Gas is generated when an arc occurs under the liquid inside a transformer. If this arc is severe, such as that caused by a short circuit, the amount of gas generated could create sufficient pressure to rupture the transformer tank. Most liquid-immersed transformers are equipped with pressure relief devices, such as pressure relief diaphragms or mechanical relief devices to relieve excessive pressure before serious damage is done to the transformer tank.

Pressure relief devices are typically vented to the outside. **See Figure 13-7.** The operation of the pressure relief devices should be tested according to the manufacturer's specifications. The vent should be inspected to ensure that it is clear of obstructions.

Rupture discs are sometimes used as a pressure relief device. A rupture disc typically consists of a thin layer of metal mounted in a disc holder. Rupture discs need to be inspected on a regular basis. The presence of any rust, corrosion, cracks, or other wear or contamination indicates that the disc needs to be replaced.

Mechanical relief devices are normally furnished as standard equipment with transformers that use an inert gas cushion and are optional with liquid-immersed transformers. Mechanical relief devices should be inspected for evidence of damage or leaks.

Gas Detector and Pressure Relays. Gas detector relays are normally used with expansion-tank transformers to detect developing faults or localized overheating of insulating materials. **See Figure 13-8.** Gas detector relays are located in the pipe between the expansion tank and the transformer and are designed to trap any gas that rises through the oil. The ability of this relay to measure a small accumulation of gas enables it to detect faults in their beginning stage. The relays and their associated alarms should be inspected regularly.

Pressure relays are designed to operate on a rapid rise in pressure in the transformer tank, normally caused by an arcing fault in the transformer. These relays and their associated alarms should be inspected regularly.

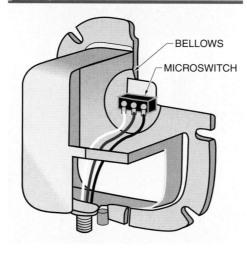

Relays

BELLOWS

MICROSWITCH

Figure 13-8. Gas detector and pressure relays are used to detect developing faults.

TECH FACT

Transformer windings are designed to be able to withstand a voltage higher than the nameplate rating to allow them to survive faults like short circuits or lightning strikes. Exposure to these faults causes insulation aging and premature failure.

MAINTENANCE AND TROUBLESHOOTING

All electrical tests performed on transformers must be performed carefully to ensure that there is no risk from a potentially powered circuit. Most tests are performed with the power removed from the transformer. The primary and secondary must be locked out. Anyone performing a test must stand clear of all parts of the circuits. All terminals must be treated as if there may be a potential shock hazard. Always disconnect test leads from the transformer before disconnecting the leads from the test set. The ground connection must be the first made and the last removed.

In some applications, parallel transformers can cause a backfeed that keeps not only the low-voltage circuits hot, but also the primary or high-voltage windings. A voltmeter and an audible voltage detector, such as a tic tracer, should be used to verify that the circuits are not energized.

Insulation Testing

Transformers may have cellulose insulation between the coils. **See Figure 13-9.** The wires are insulated to prevent shorts between wires. In spite of this, all insulation allows a small leakage current to flow through the insulation. Typically, the leakage current is so small the current does not cause any problems and is ignored until the leakage reaches a point where the leakage starts causing electrical shocks, unwanted operating temperatures, or equipment damage. Insulation can be tested to identify problems before failure.

Insulation Resistance Testing. An insulation resistance test is a test performed to measure the electrical resistance of insulation between two conductors or between a conductor and ground. The higher the insulation resistance, the less leakage current flows through the insulation. Insulation has the highest resistance when first placed in service. Ultimately, all insulation deteriorates over time and the resistance of the insulation decreases. An insulation resistance test is often used to check for degradation of the insulation over time.

Insulation

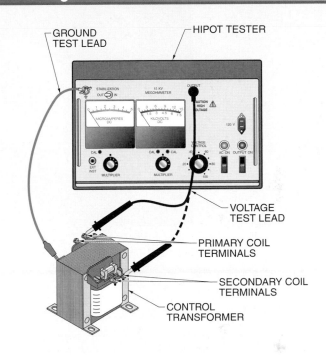

Figure 13-9. *Transformers have cellulose insulation between the coils.*

Insulation resistance testing must be performed with the transformer primary and secondary locked out. A *hipot tester* is a test instrument that can be used to evaluate insulation resistance by measuring leakage current. Hipot testers apply a high test voltage between two different conductors or between a conductor and ground and measure the leakage current. **See Figure 13-10.** A hipot tester indicates the condition of insulation by displaying the amount of leakage current through the insulation.

A hipot tester uses very high voltage for testing. The technician must follow the recommended procedures and safety procedures provided by the manufacturer of the hipot tester. After performing insulation tests with a hipot tester, the transformer must be discharged with the discharge function built into the tester. Alternatively, specialized discharge jumpers can be used to discharge the transformer.

A megohmmeter performs the same basic test as a hipot tester in that the meter also applies a test voltage to the circuit or component being tested. A megohmmeter indicates the condition of insulation by displaying the amount of resistance of the insulation in ohms (typically MΩ) when a test voltage is applied to the conductors. **See Figure 13-11.**

Hipot Testing

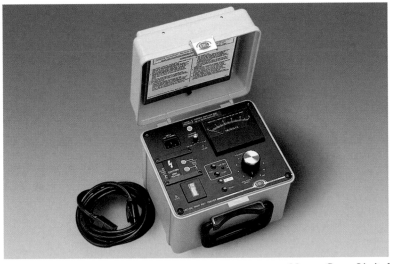

Figure 13-10. *Hipot testers directly measure leakage current through insulation.*

A hipot tester is used to directly measure the leakage current.

Megohmmeter Testing

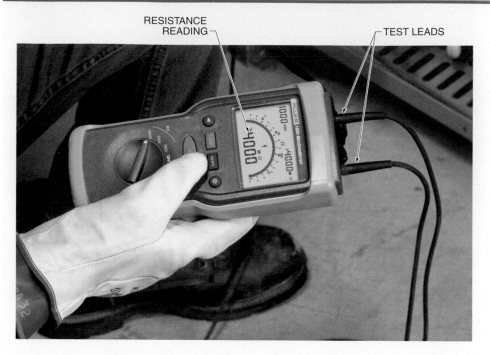

RESISTANCE READING

TEST LEADS

Fluke Corporation

Figure 13-11. Megohmmeters directly measure insulation resistance.

A *hipot tester* is a test instrument that can be used to evaluate insulation resistance by measuring leakage current.

Conductive leakage current is a very small current (μA) that flows through insulation between conductors or from conductor to ground.

Capacitive leakage current is the current that flows into insulation to charge the insulation like the plates of a capacitor.

Polarization absorption current is current caused by the polarization of molecules within the dielectric material.

An *insulation resistance spot test* is a test where one insulation resistance reading is used to evaluate the integrity of insulation.

Insulation Resistance. The insulation resistance values vary with the temperature and the amount of moisture in the insulation. The actual value of the insulation resistance is less important than the trend over time. **See Figure 13-12.** The temperature and humidity should be recorded at the time the test is made. These factors should be compared when making future tests and evaluating the trend of the insulating resistance.

Leakage Current. There are three independent causes of leakage current. *Conductive leakage current* is a very small current (μA) that flows through insulation between conductors or from conductor to ground. This current increases as insulation ages and deteriorates. This current is steady and is quickly established after a voltage is applied. **See Figure 13-13.**

When two or more conductors are near each other, charge builds up in the insulation between the wires and acts as a capacitor.

Due to this capacitance, capacitive leakage current flows through the insulation between the wires. *Capacitive leakage current* is the current that flows into insulation to charge the insulation like the plates of a capacitor. This current lasts only until the insulation has been charged to its full test voltage. When testing high-capacitance equipment, the capacitive leakage current can last for a very long time before settling out.

Polarization absorption current is current caused by the polarization of molecules within the dielectric material. The initial current is high and decreases over time. In high-capacitance equipment, it takes a long time for the polarization absorption current to decrease and settle out.

Insulation Resistance Spot Test. An *insulation resistance spot test* is a test where one insulation resistance reading is used to evaluate the integrity of insulation. A new transformer should have an insulation resistance spot test performed when it is installed. This test should be repeated on a regular basis and whenever there is reason to believe that there is a problem. This test is recommended even if the transformer has not been subjected to any of the factors that contribute to the deterioration of the insulating material within the transformer.

If the transformer has been exposed to an excessive voltage (such as lightning) or overload, has been overheated, or through normal inspection excessive moisture content is found in the insulating liquid, immediate tests should be made to determine the condition of the transformer insulation.

During an insulation resistance spot test, the test leads of the meter are used to apply a test voltage between the part of the circuit that will be carrying current and the parts of the circuit that are insulated from the current. The test voltage is applied for 60 sec to allow for the most accurate measurements. The insulation is weak or damaged if the resistance is lower than measurements taken previously or lower than the transformer manufacturer recommends.

Dielectric Absorption Test. A *dielectric absorption test* is a variation of the standard insulation resistance spot test where the voltage is applied for 10 min. The voltage is recorded every 15 sec for the first minute and every minute after that. **See Figure 13-14.** A curve that increases steadily over the test time indicates that the insulation is in good condition. A curve that flattens out over the test time indicates that the insulation is in poor condition.

The *polarization index (PI)* is the ratio of the 10 min reading to the 1 min reading taken on a dielectric absorption test. For example, for good insulation the 10 min reading may be 500 MΩ and the 1 min reading may be 110 MΩ. In this case, the PI is 4.5 (500 ÷ 110 = 4.5). For poor insulation the 10 min reading may be 85 MΩ and the 1 min reading may be 70 MΩ. In this case, the PI is 1.2 (85 ÷ 70 = 1.2).

Insulation Resistance Trend

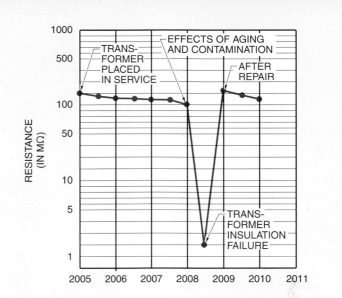

Figure 13-12. The trend of measured insulation resistance is more important than the actual measured value.

TECH FACT

Predictive maintenance is the monitoring of electrical and equipment characteristics against a predetermined tolerance to detect possible malfunctions and failures.

DEFINITION

A *dielectric absorption test* is a variation of the standard insulation resistance spot test where the voltage is applied for 10 min.

Leakage Current

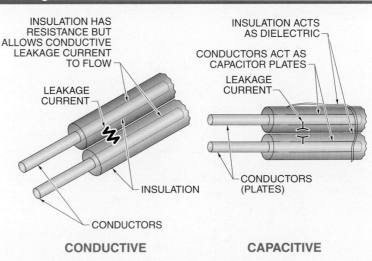

Figure 13-13. Leakage current is current that passes through insulation.

Dielectric Absorption Testing

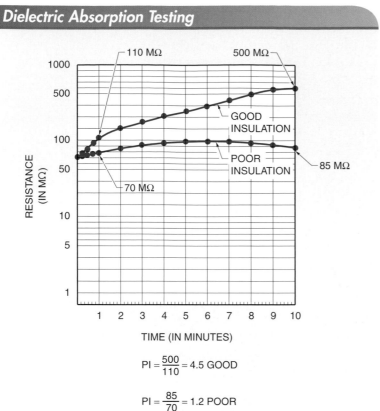

$$PI = \frac{500}{110} = 4.5 \ GOOD$$

$$PI = \frac{85}{70} = 1.2 \ POOR$$

Figure 13-14. *Dielectric absorption testing is used to indicate the condition of the insulation.*

 **DEFINITION**

*The **polarization index** (PI) is the ratio of the 10 min reading to the 1 min reading taken on a dielectric absorption test.*

__Insulation power factor__ is the ratio of the power dissipated in the insulation, in watts, to the apparent power, in volt-amps.

A low polarization index may indicate wet or damaged insulation. However, the value of the polarization index cannot be used as an absolute measure of the condition of the insulation. Historically, a measured PI of less than 2.0 was used to indicate a problem with the insulation. However, the release of IEEE Standard 43-2000 placed limitations on this measurement.

Modern insulation systems may have an initial resistance of gigaohms (GΩ) or even teraohms (TΩ). With a resistance this high, the leakage current becomes so small that accurate field measurement is no longer possible. The standard cautions against using any measurement where the 1 min insulation resistance reading is above 5000 MΩ.

A dielectric absorption test performed on insulation at different temperatures gives different results. The new standard also requires that the test must be corrected

to 40°C (104°F). In addition, the test must be performed above the dewpoint to ensure that there is no condensation on the insulation. Because of the complexity of interpreting these results, the standard says that the PI can be used as an indicator to estimate the suitability of a transformer. It cannot be used as an absolute measure of insulation quality.

Some manufacturers and transformer specialists recommend the use of the polarization ratio (PR), which is the ratio of the 3 min reading to the 15 sec reading, or the dielectric absorption ratio (DAR), which is the ratio of the 60 sec reading to the 30 sec reading.

Insulation Power Factor (Dissipation Factor) Test. An insulation power factor test measures the dielectric loss from winding to winding or winding to ground. *Insulation power factor* is the ratio of the power dissipated in the insulation, in W, to the apparent power, in VA. An insulation power factor test is sometimes called a dissipation factor test. The power factor value increases as the insulation deteriorates. An insulation power factor value less than 0.5% is generally considered to be a sign of adequate insulation. However, the history and trends are more important than the actual value of the insulation power factor.

There are two types of test equipment that can be used for this measurement. A low-voltage tester uses 100 V. A high-voltage tester uses up to 10 kV. **See Figure 13-15.** When the transformer can be isolated from all electrical noise, the low-voltage tester gives adequate results. When the transformer cannot be isolated, such as during field tests near other equipment, the high-voltage tester gives more reliable results. It is very important not to exceed the voltage rating of the winding being tested.

A problem with transformers is moisture in the insulation. Moisture can be indirectly measured with insulation power factor testing. Therefore, clear, dry weather conditions are recommended when performing a field power factor test. Humidity and contamination on the bushings can cause significant inaccuracies in the test.

To perform an insulation power factor test, each winding is tested separately with all the other windings grounded. The secondary windings should be grounded at one end only. The insulators and bushings must be cleaned. Any residue from the cleaner must be removed to prevent creating a conductive path. The test kit must be attached to the leads and the test performed according to the directions provided with the kit.

Power Factor Tester

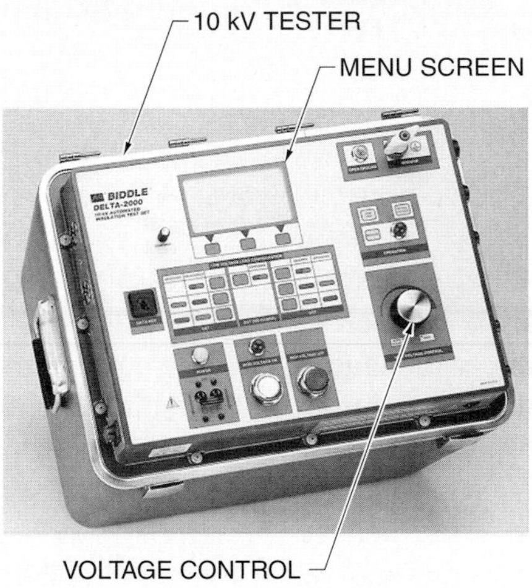

Megger Group Limited
Figure 13-15. *A power factor test is used to measure the dielectric loss.*

Winding Resistance Testing

A new transformer should have a winding resistance test performed when it is installed. This test should be repeated on a regular basis and whenever there is reason to believe that there is a problem.

A winding resistance test uses a low-resistance, high-precision ohmmeter. **See Figure 13-16.** This test can take a considerable amount of time to saturate the circuit. The measured value is compared to previous measurements to check for changes in the winding resistance that may indicate a problem, such as a loose connection or a damaged wire.

Low-resistance ohmmeters have their own power supply and some models are capable of testing windings at up to about 15 kV. These meters use battery power with an oscillator to transform the low voltage to the high levels needed to test for leakage currents to ground.

A new transformer that has a primary of 600 V or less should be tested at twice the rated voltage plus 50%. For a transformer with the primary rated at 480 V, twice the rated voltage is 960 V and 50% of 480 V is 240 V. The sum of 960 V and 240 V is 1200 V. Therefore, a new transformer with the primary rated at 480 V should be tested at a maximum of 1200 V.

A used transformer that has a primary of 600 V or less should be tested at the rated voltage plus 50%. Therefore, a used transformer with the primary rated at 480 V should be tested at a maximum of 720 V.

Winding Resistance Tester

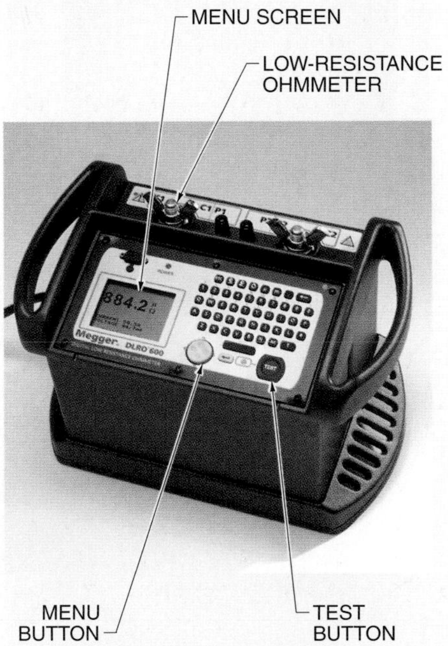

Megger Group Limited
Figure 13-16. *A winding resistance test is used to measure the resistance of the transformer winding to identify loose connections or damaged coils.*

Turns Ratio Testing

A new transformer should have a turns ratio test performed when it is installed. **See Figure 13-17.** This test may also be called a turns-to-turns ratio (TTR) test. A turns ratio test should be repeated on a regular basis and whenever there is reason to believe that there is a problem.

The turns ratio test is used to verify the turns ratio between the primary and the secondary. A load tap changer must also be cycled during the turns ratio test to check all the windings. A change from the original turns ratio test result indicates there is a problem, such as a short within the windings.

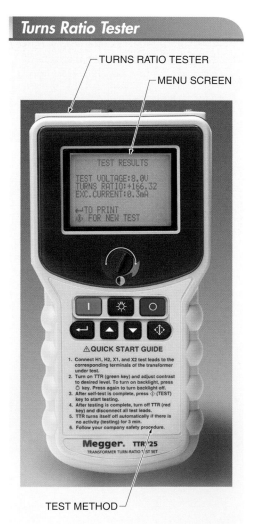

Megger Group Limited

Figure 13-17. *A turns ratio tester is used to verify the turns ratio of a transformer.*

Relays

Relays are the protective devices that take a transformer off-line in the event of a fault. Relays must be tested and calibrated periodically. Equipment manufacturers offer test equipment for making adjustments and calibrations. **See Figure 13-18.** A modern universal protective relay test set can test a wide variety of electromechanical and solid state protective relays. The test set contains a menu screen that lets the technician select the type of relay, and the display lists all the necessary functions needed to test that particular type of relay.

There are three common types of relays used to protect transformers. Differential relays are used to detect unexpected differences between the primary and secondary currents. Overcurrent relays are used to detect unexpectedly high phase currents. Overvoltage and undervoltage relays are used to detect unexpectedly high and low voltages.

Differential Relays. A differential relay is used to monitor the difference between the currents measured on a transformer primary and on a transformer secondary, taking into account the voltage ratio. For example, a transformer primary with 4800 V and 10 A may have a secondary at 480 V and 100 A, assuming no losses. A differential relay is used to monitor the current and respond if the difference becomes greater than allowed in the specifications. **See Figure 13-19.** For example, an internal short within the transformer could cause this difference in current. This causes the primary current in increase dramatically with little or no change in the secondary current.

Typical differential relays contain a differential unit to measure the difference between the currents, a harmonic restraint unit, an indicating instantaneous trip unit, and an indicating contactor switch. These are high-speed relays that operate on extremely high inrush currents to the transformer. A harmonic restraint unit is used to differentiate between high inrush current and fault current. Differential relays need to be tested and calibrated periodically.

Overcurrent Relays. Overcurrent protection can be provided with fuses and breakers for small transformers. For larger transformers, relays are commonly used. Overcurrent relays are a single-phase, nondirectional time-delay overcurrent device. When the overcurrent relay is installed, a maximum current is determined and the relay is set to this value. When the current exceeds the predetermined setting, the relay trips the transformer off-line by opening the breaker. There is a time curve that prevents nuisance tripping by short-duration loads. Overcurrent relays need to be tested and calibrated periodically.

Overvoltage and Undervoltage Relays. A sudden fall or rise in the voltage activates an undervoltage or an overvoltage relay. An undervoltage relay is switched when the voltage drops below a setpoint. An overvoltage relay is switched when the voltage rises above a setpoint. Overvoltage and undervoltage relays in the control room should be tested and calibrated periodically.

Other Electrical Tests

There are a few other electrical tests that should be conducted on a regular basis. Loose connections can cause high resistance and result in overheating. Infrared thermography or temperature measurement can be used to identify hot spots. **See Figure 13-20.**

Unbalanced loads are a common cause of early failure of transformers. A simple ammeter measurement on each phase can identify any significant unbalance. Harmonic currents can cause overheating and damage to a transformer. A power quality analyzer can be used to identify the presence and amount of harmonics. In addition, a reading from a true-rms ammeter can be compared to the reading from an analog ammeter. If there is a large difference between the two readings, there are probably significant harmonics on the line.

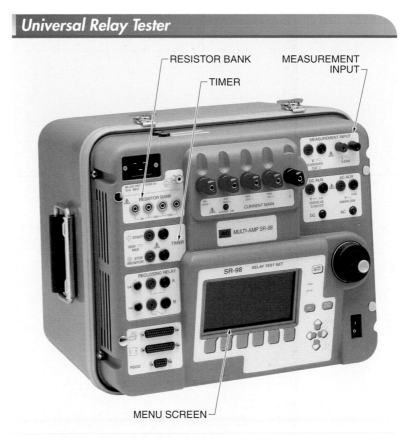

Universal Relay Tester

RESISTOR BANK
TIMER
MEASUREMENT INPUT

MENU SCREEN

Figure 13-18. *A relay tester is used to test protective relays.*

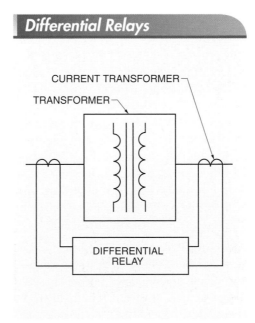

Differential Relays

CURRENT TRANSFORMER
TRANSFORMER
DIFFERENTIAL RELAY

Figure 13-19. *A differential relay measures the difference between the primary and secondary currents to identify faults.*

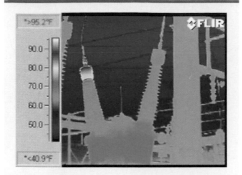

Infrared Thermography

FLIR Systems
Figure 13-20. *Infrared thermography is used to identify hot spots in electrical circuits.*

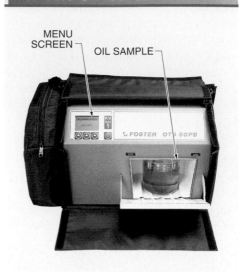

Portable Oil Test Set

MENU
SCREEN
OIL SAMPLE

Megger Group Limited
Figure 13-21. *A portable oil test set is used to perform field tests of transformer liquid.*

Liquid Testing

Proper inspection, testing, and maintenance of transformer insulating liquids is one of the best ways to keep a transformer in proper operating condition and to extend the life of the transformer. Although new systems of fluids are being developed, mineral oil is the most common fluid in use today.

The liquid serves two functions in a transformer. The liquid helps draw heat away from the core, keeping temperatures low and extending the life of the insulation. The liquid also acts as a dielectric material. To keep the transformer operating properly, both of these qualities must be maintained.

One of the jobs of maintenance, testing, and troubleshooting a transformer is to take samples of the insulating liquid. These samples are usually sent to an outside laboratory for analysis. Larger facilities that do frequent transformer liquid testing may invest in oil test sets to do some of the testing in-house. These test sets are also available as portable units that can be carried into the field. **See Figure 13-21.**

Sampling Procedures. The liquid circulates continuously while a transformer is hot and the liquid must be allowed to settle before samples are taken. It normally takes about 8 hours for the oil to settle in a small transformer containing a barrel of liquid. Several days may be necessary for the liquid to settle in a large transformer.

Insulating mineral oil is lighter than water. Therefore, an oil sample should be taken from the bottom of the tank to ensure that any water is included in the sample. Other liquids may be heavier than water. Consult the manufacturer before sampling.

Care must be taken when taking liquid samples, as cleanliness is extremely important. A small first sample should be drawn to purge the valve and associated fittings of anything other than fresh oil. The samples are then drawn off and prepared for shipment to a test laboratory. The test laboratory should provide the sample containers.

Special syringes are used to take samples for the moisture and dissolved gas analysis tests. The instructions from the test laboratory must be followed exactly to ensure that no atmospheric moisture or gas enters the sample before it is tested.

Dielectric Testing. The dielectric strength of an insulating liquid is a measure of the ability of the oil to withstand electrical stress. The dielectric strength is measured as the voltage required to break down the liquid in a 0.1″ gap between two electrodes of standard dimensions under specifically controlled testing conditions. The dielectric strength is lowered by the presence of contaminants in the oil. The minimum acceptable breakdown voltage for mineral oil is 22 kV. If the breakdown voltage is lower than this, maintenance is required. An oil sample is normally sent to an outside lab. However, portable test kits are available to perform this test in the field. **See Figure 13-22.**

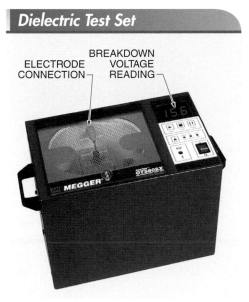

Dielectric Test Set

ELECTRODE CONNECTION
BREAKDOWN VOLTAGE READING

Megger Group Limited

Figure 13-22. *A dielectric test set is used to perform a test of the dielectric strength of a transformer liquid.*

Moisture Testing. Moisture present in the insulating oil forms conducting paths on nonconducting parts such as bushing shanks and insulated leads. This creates a definite hazard to the safe operation of the equipment. Moisture reduces the dielectric strength of all insulating liquids and may also be absorbed by the insulating materials used throughout the transformer. The dissolved moisture should be below about 15 parts per million (ppm). This test can be performed locally with a Karl Fischer moisture tester. **See Figure 13-23.**

Acidity. The ability to transfer heat depends on the ability of the oil to flow in and around the windings. Insulating oil can oxidize when oxygen from the air comes in contact with hot oil. This results in the formation of acids and sludge. The acid content of the oil is a measure of its deterioration and its tendency to form sludge.

Sludge is extremely harmful to a transformer because it settles on the internal parts of the transformer and greatly impedes cooling. A sludge deposit of ¼″ on the core and coils may increase the transformer temperature by about 30°F. The sludge can also raise the viscosity of the oil. Higher viscosity oil is thicker than lower viscosity oil. The thicker oil does not flow as well and the cooling ability of the oil is reduced.

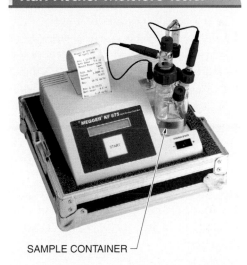

Karl Fischer Moisture Tester

SAMPLE CONTAINER

Megger Group Limited

Figure 13-23. *A Karl Fischer moisture tester is used to measure the dissolved moisture in a transformer liquid.*

Dissolved Gas Analysis. The primary mechanisms for the breakdown of insulating fluids are heat and contamination. Most tests of the transformer insulating liquid only tell whether or not degradation has occurred. Dissolved gas analysis (DGA) can often provide information about the cause of the degradation.

The types of gases present and the relative ratios of the gases can give indications about the cause of failure. Gases commonly present in the insulating liquid are nitrogen, oxygen, hydrogen, methane, acetylene, ethylene, ethane, carbon monoxide, and carbon dioxide. **See Figure 13-24.**

Nitrogen and oxygen are normally present. High levels of these gases can indicate the presence of water, rust, leaky bushings, or poor seals. For example, if the oxygen is below 50,000 ppm (5%), it is generally not a cause for concern. If the oxygen is above 50,000 ppm, there may be leaks in the oil tank seals.

Carbon monoxide and carbon dioxide reflect the demand on the transformer. Moderate levels of each can show that the transformer is experiencing minor overload conditions. High levels of each can show that the transformer is overheating.

Presence of the hydrocarbon gases, such as methane, acetylene, ethylene, and ethane, indicates problems with the integrity of the transformer internal functions. For example, if acetylene is present, this is proof that arcing is occurring in the transformer. The trans-

former should be shut down and repaired. The presence of the other hydrocarbon gases may indicate cellulose breakdown in the insulation or corona discharge. It is very important to record the results of these tests and watch for trends over time.

Dissolved Gas Analysis

Gas	Suggested Limits*
Hydrogen (H_2)	100
Oxygen (O_2)	50,000
Methane (CH_4)	120
Acetylene (C_2H_2)	35
Ethylene (C_2H_4)	50
Ethane (C_2H_6)	65
Carbon monoxide (CO)	350
Carbon dioxide (CO_2)	2500

* in ppm

Figure 13-24. *Dissolved gas analysis is used to troubleshoot transformers.*

SUMMARY

- Transformer enclosures should be inspected for damage and blocked air vents.

- An internal inspection includes an examination of the core and laminations, coils and taps, and nuts and bolts used to restrain vibration.

- Ground connections, lightning arresters, and load tap changers should be inspected periodically.

- Liquid-immersed transformer inspections should include inspecting the liquid level, pressure/vacuum gauges, pressure relief devices, and pressure and gas relays.

- Insulation testing includes resistance spot tests, dielectric absorption tests, and insulation power factor tests.

- Tests to be run on the liquid from liquid-immersed transformers include a dielectric test, moisture test, acidity test, and dissolved gas analysis.

DEFINITIONS

- A *hipot tester* is a test instrument that can be used to evaluate insulation resistance by measuring leakage current.

- *Conductive leakage current* is a very small current (μA) that flows through insulation between conductors or from conductor to ground.

- *Capacitive leakage current* is the current that flows into insulation to charge the insulation like the plates of a capacitor.

- *Polarization absorption current* is current caused by the polarization of molecules within the dielectric material.

- An *insulation resistance spot test* is a test where one insulation resistance reading is used to evaluate the integrity of insulation.

- A *dielectric absorption test* is a variation of the standard insulation resistance spot test where the voltage is applied for 10 min.

- The *polarization index (PI)* is the ratio of the 10 min reading to the 1 min reading taken on a dielectric absorption test.

- *Insulation power factor* is the ratio of the power dissipated in the insulation, in watts, to the apparent power, in volt-amps.

1. Why is it necessary to inspect transformer enclosures?

2. Why is it a problem to have dust buildup on transformer internals?

3. Why must the proper liquid level be maintained in a transformer?

4. What is the purpose of a pressure relief device on a liquid-immersed transformer?

5. What are some common safety procedures to be followed before working on a transformer?

6. What test instrument is used to directly measure leakage current? How does it work?

7. What test instrument is used to directly measure insulation resistance? How does it work?

8. What is the difference between an insulation resistance spot test and a dielectric absorption test?

9. What is the meaning of a low polarization index?

10. What is the purpose of a winding resistance test?

11. Why is it a problem to have moisture in the transformer insulating oil?

12. What does it mean when acetylene is found in transformer insulating oil?

Review Questions Answer Key

Chapter 1: Magnets and Magnetism

1. The molecular theory of magnetism states that ferromagnetic materials are made up of a very large number of molecular domains that can be arranged in either an organized or disorganized manner. A material is magnetic if it has organized molecular magnets and the individual fields add together. A material is nonmagnetic if it has disorganized molecular magnets and the individual fields cancel each other.

3. All magnets have a north (N) and a south (S) pole. The lines of magnetic flux leave the north pole and enter the south pole of a magnet or magnetic field.

5. The left-hand rule for coils can be used to determine the polarity of a coil. When a coil is wrapped by the left hand with the fingers in the direction of the current flow, the thumb points in the direction of the north pole.

7. Magnetomotive force is the force that produces magnetic lines of flux in a circuit as a result of current in a conductor. Magnetomotive force is analogous to voltage in an electric circuit. Field flux is the total quantity of a magnetic field. Field flux is analogous to current in an electric circuit. Reluctance is the opposition to magnetic flux in a magnetic circuit in a given volume of space or material. Reluctance is analogous to resistance in an electric circuit.

9. Saturation is where a magnetic core has substantially all the magnetic domains aligned with the field, and any increases in current no longer result in a stronger electromagnet.

11. When a transformer core approaches saturation, a larger and larger amount of magnetomotive force is required to deliver increases in magnetic field flux. The magnetomotive force is proportional to the current through the magnetizing coil. Therefore, larger and larger amounts of magnetomotive force require proportionally larger and larger increases in the coil current. The coil current increases dramatically at the peaks in order to maintain the shape of the magnetomotive force waveform.

Chapter 2: Operating Principles

1. The three requirements for induction are a conductor, a magnetic field, and relative motion between the conductor and the magnetic field. In a transformer, the conductor is the wire making up the coil. The AC power flowing through a conductor generates an expanding and collapsing magnetic field. The expanding and contracting magnetic field flows through the laminated core and provides the relative motion between the conductor in the secondary and the magnetic field.

3. 24.7 Ω

$$X_L = 2\pi fL$$
$$X_L = 2 \times 3.14 \times 60 \times 0.06$$
$$X_L = 22.6\ \Omega$$
$$Z = \sqrt{10^2 + 22.6^2}$$
$$Z = \mathbf{24.7\ \Omega}$$

5. No. The turns ratio is correct, but the inductance of the coils is too low to provide the current limits in the coil because of the lower reactance.

7. Four types of losses are resistive (I^2R) loss, eddy current loss, hysteresis, and flux loss.

9. When exposed to very high currents from short circuits, transformer coils have a tendency to move, or telescope, in opposite directions. To minimize the probability of telescoping, coils are designed so that the electrical centers of the two coils are in an identical position. Second, both windings are rigidly clamped in place to prevent movement from this axial force.

11. A transformer operating under no-load conditions has a low power factor because the circuit is almost purely reactive. As the load on a transformer increases, the reactance decreases and the power factor increases and approaches 1.

Chapter 3: Electrical Safety

1. The purpose of the NEC® is to protect people and property from hazards that arise from the use of electricity. Improper procedures when working with electricity can cause injury or death.

3. A ground fault circuit interrupter (GFCI) is a device that protects against electrical shock by detecting an imbalance of current in the normal conductor pathways and opening the circuit. When current in the two conductors of an electrical circuit varies by more than 5 mA, a GFCI opens the circuit. A GFCI is rated to trip quickly enough (1/40 of a second) to prevent electrocution.

5. The NFPA requires facility owners to perform a flash hazard analysis. A flash protection boundary must be established around electrical devices. This boundary is determined by calculations that estimate the maximum energy released and the distance that energy travels before dissipating to a safe level.

7. Electrical power and other potential sources of energy must be removed when electrical equipment is inspected, serviced, or repaired. To ensure the safety of personnel working with the equipment, power is removed and the equipment must be locked

out and tagged out. Lockout is the process of removing the source of electrical power and installing a lock that prevents the power from being turned on. To ensure the safety of personnel working with equipment, all electrical, pneumatic, and hydraulic power is removed and the equipment must be locked out and tagged out. Tagout is the process of placing a danger tag on the source of electrical power, which indicates that the equipment may not be operated until the danger tag is removed.

9. Article 500 of the National Electrical Code® (NEC®) and Article 440 of NFPA 70E cover the requirements of working in hazardous locations.

Chapter 4: Transformer Connections

1. Three transformers may be connected together to provide power for both motor and lighting loads. However, it is more common to connect the two coils of the secondary winding of a transformer in series for 3-wire service. The third wire (neutral) of the 3-wire circuit is connected to the secondary winding at the point where the series connection of the coils is made. This permits the use of the 240 V circuit for a motor load and two 120 V circuits for lighting loads.

3. The line current of a 3-phase system must be equal to the current in each of the individual coils. However, the line current, as it reaches the junction point of the two windings, has two paths through which it flows. Therefore, the current in the windings of a delta-connected system is less than the line current. The winding current is actually 57.7% of the line current (equal to the line current divided by 1.73). In a delta-connected system, the line current is equal to the winding current multiplied by 1.73. Therefore, the line current is the vector sum of the two coil currents.

5. In a 3-phase wye-connected system, the current in the line is the same as the current in the coil (phase) windings. This is because the current in a series circuit is the same throughout all parts of the circuit. In a balanced circuit, there is no current flow in the neutral wire because the sum of all the currents is zero.

Chapter 5: Harmonics

1. A linear load draws current in a waveform that matches the shape of the voltage waveform. A nonlinear load draws current in a waveform that does not match the voltage waveform.

3. Common nonlinear devices include copiers, electronic equipment, lighting ballasts, dimming equipment, computers, printers, and variable frequency motor drives.

5. The standard places a limit of 32% current THD with specific limits on other harmonics.

7. Displacement power factor is the ratio of real power to apparent power due to the phase displacement between the current and voltage. The phase displacement is caused by inductive and capacitive elements in a circuit. Distortion power factor is the ratio of real power to apparent power due to THD. The total power factor is the product of the displacement power factor and the distortion power factor.

9. A circuit K-factor is a measure of the increase in heating effects due to harmonics. A circuit with a K-factor of 2 has double the heating effect of a circuit with a K-factor of 1.

Chapter 6: Power Generation and Transmission

1. When the incoming frequency is higher than the operating frequency, the needle of the synchroscope turns clockwise. The faster the rate of rotation, the greater the difference in frequency.

3. Common sources of energy used to turn steam turbines are natural gas, nuclear energy, coal, and oil. The energy of water behind a dam is used to turn turbines. Engines fueled by natural gas, propane, diesel fuel, or gasoline often power standby power units.

5. In the three-lights-out method, a lamp is connected as a load across each phase of the two sources, A to A, B to B, and C to C. When all three lamps are dark, the potentials of both sources are the same and the second source can be brought on line. The phase connection is incorrect if the lights do not all turn on and off at the same time.

Chapter 7: Isolation Transformers, Reactors, and Chokes

1. The three basic functions performed by drive isolation transformers are voltage change, reduction of drive-induced ground currents, and common-mode noise reduction.

3. With 3-phase motor drives using SCR six-pulse (six-diode) rectifiers, there are short intervals of time when more than one SCR in ON. This causes a transient short circuit to flow six times per line power cycle. This short circuit current flow causes nonlinear voltage to drop across the system impedance and results in distortion in the voltage waveform called a notch.

5. All two-winding transformers are isolation transformers. However, general-purpose transformers are not fully rated for motor drive applications because they cannot supply the required distorted current at full load without exceeding their design temperature rise. In addition, many standard transformers have shortened life expectancy due to mechanical stress when these transformers supply DC motor drive current transients. Another contributor to shortened life expectancy is the severe cyclic nature of motor drive process control applications.

A drive isolation transformer is designed with additional reactance to reduce power quality problems associated with the effects of nonlinear loads. A drive isolation transformer must be able to withstand the heating effects of the nonlinear drive loads and operate well above the ambient temperature. Drive isolation transformers must also withstand the thermal and mechanical stresses caused by the highly cyclic load demands of both AC and DC motor drive process applications.

Chapter 8: Autotransformers

1. An autotransformer can be made from a two-winding transformer by connecting the two windings in series to form one continuous winding. When tap H2 is connected to tap X1, the windings are in series and the transformer operates as an autotransformer. The two windings are on the same core and connected in series.

3. Autotransformers are designed to be overloaded during motor starting. This allows a less expensive design to be used. However, the autotransformer must be allowed time to cool down before it is used again. A typical duty cycle is 10 seconds ON and 10 minutes OFF.

5. An autotransformer can be installed in combination with the tap changer to split the percentage of change as the changer is operated. The position of the tap changer determines how much of the high-voltage winding is in the circuit. As the switches in the tap changer close, varying amounts of the high-voltage winding provide flux for the secondary.

7. Autotransformers are used in starting large synchronous and induction motors, changing the taps on large transformers, and as variable transformers.

Chapter 9: Buck-Boost Transformers

1. Buck-boost transformers are often used to adjust the voltage at the end of long transmission lines. However, buck-boost transformers should not be used to correct the voltage drop when the load fluctuates. The voltage drop on the line varies with load. When the buck-boost transformers are designed and installed for high load conditions, unexpected high voltages may result during lightly loaded conditions.

3. 0.15 kVA. The current is 12.5 A (1500 / 120 = 12.5). The voltage boost required is 12 V. The buck-boost transformer is sized as follows:

$$P = \frac{I \times E}{1000}$$

$$P = \frac{12.5 \times 12}{1000}$$

$$P = \frac{150}{1000}$$

$$P = \textbf{0.15 kVA}$$

Chapter 10: Special Transformers

1. It is common industry practice to protect a control transformer on both the primary and secondary sides. The purpose for fusing both sides of a control transformer is to protect the primary if there is a short in the secondary.

3. Processes often change over time, with the result that control system circuits are modified to match the changes. Additional relays and contactors may be installed without considering the size of the control transformer. If the control transformer is not replaced, it may be undersized for the new circuits. When the total load connected to a control transformer exceeds the rating of the transformer, the voltage output of the transformer decreases.

5. Potential transformers can be made with almost any turns ratio so that the voltage can always be reduced to 120 V.

7. The three types of current transformers in general use are the window, bar, and wound.

9. A variable transformer usually consists of a wiper or brush that can be rotated across the windings to create a variable turns ratio. Variable transformers are often built with less than

1 V/turn, permitting fine voltage adjustment. The winding turns are evenly spaced so the output voltage is proportional to position of the knob. A transformer with multiple taps has a limited number of fixed turns ratios, and only a limited number of discrete outputs are available.

Chapter 11: Special Connections

1. The requirements for satisfactory operation of two or more transformers connected in parallel include having the same polarity, angular phase displacement and phase sequence, voltage ratio, and percent impedance.

3. A main transformer is one of the transformers in a T-connected transformer bank or used for a Scott connection, and is provided with a 50% voltage tap.

5. A Scott connection is a method of wiring transformers to transform from 3-phase to 2-phase or 2-phase to 3-phase.

7. A 60 Hz transformer that has a voltage rating of more than twice the rating of a 25 Hz transformer can be used as a substitute. Alternatively, two transformers with a voltage rating matching the 25 Hz transformer can be connected in series.

Chapter 12: Selection and Installation

1. The primary factors used in selecting a transformer are the voltage, the phase requirements, the loads, the purpose and type, the noise level, and the cooling and ventilation requirements.

3. A two-winding transformer is a transformer with no electrical connection between the primary and the secondary. A two-winding transformer is the most common type of transformer. All two-winding transformers have high-voltage leads and low-voltage leads connected to the windings. An autotransformer has all or part of the windings shared by the primary and secondary. When selecting a transformer, a two-winding transformer can be interconnected for use as an autotransformer by connecting the primary in series with the secondary.

5. Liquids for use in transformers have been developed that are flammable but do have a relatively high fire point. These are less-flammable liquids. Nonflammable-liquid transformers are filled with a liquid that has no flash point or fire point and does not burn in air.

7. A relatively common problem with transformers is insulation failure between the coils. An insulation resistance test is a test performed to measure the leakage current through the coil insulation. When the leakage current is too high, the insulation can fail and short out the coil.

9. Dry-type transformers installed indoors, and not over 112½ kVA, shall have a clearance from combustible materials of at least 12″ per NEC® Section 450.21(A). A separation of less than 12″ from combustible materials is permitted when a fire-resistant, heat-insulating barrier is installed between the transformer and the combustible material.

11. A supervised installation is an electrical installation in which the conditions of maintenance are such that only qualified persons monitor or service the electrical equipment.

13. NEC® Table 450.3(B) permits the primary feeder OCPD, set at not more than 250% of the rated primary current of the transformer, to protect the transformer primary, provided the secondary OCPD is set at a value which does not exceed 125% of the rated secondary current of the transformer. If the rated secondary is less than 9 A, the rating of the secondary OCPD is permitted to be increased to a maximum value of 167% of the rated secondary current of the transformer.

Unlike the requirements for primary-only protection, there is no permission to increase the rating of the primary OCPD when it does not correspond to a standard rating of a fuse or circuit breaker. If the calculated value at 250% does not correspond to a standard rating, the next lower standard rating shall be used.

Chapter 13: Maintenance and Troubleshooting

1. An enclosure protects the transformer from the elements. At the same time, the enclosure must allow the heat from the transformer to dissipate. Dented enclosure walls can result in reduced air circulation. Restricted access around the enclosure can result in reduced air flow. Rodents and insects can build nests and block openings for cooling. All these conditions can lead to overheating.

Some environments contain abrasive and conductive dust that can build up on the unit and cause catastrophic failures. A dust buildup can create a conductive path between taps and bushings. This is especially true when a transformer is exposed to water, such as spray or rain. Any buildup should be removed from the enclosure.

3. The proper level of the insulating liquid must be maintained to prevent a transformer from overheating. The liquid level should be checked regularly. If there is any loss from evaporation or leakage, it should be replaced immediately and any repairs made to prevent further loss.

5. Follow all NFPA 70E and OSHA requirements. The primary and secondary must be locked out. Anyone performing a test must stand clear of all parts of the circuits. All terminals must be treated as if there may be a potential shock hazard. Always disconnect test leads from the transformer before disconnecting the leads from the test set. The ground connection must be the first made and the last removed.

In some applications, parallel transformers can cause a backfeed that keeps not only the low-voltage circuits hot, but also the primary or high-voltage windings. A voltmeter and an audible voltage detector, such as a tic tracer, should be used to verify that the circuits are not energized.

7. A megohmmeter performs the same basic test as a hipot tester in that the meter also applies a test voltage to the circuit or component being tested. A megohmmeter indicates the condition of insulation by displaying the amount of resistance in ohms (typically $M\Omega$) of the insulation when a test voltage is applied to the conductors.

9. A low polarization index may indicate wet or damaged insulation. However, the value of the polarization index cannot be used as an absolute measure of the condition of the insulation. Historically, a measured PI of less than 2.0 was used to indicate a problem with the insulation. However, the release of IEEE Standard 43-2000 placed limitations on this measurement.

11. Moisture present in the insulating oil forms conducting paths on nonconducting parts, such as bushing shanks and insulated leads. This creates a definite hazard to the safe operation of the equipment. Moisture reduces the dielectric strength of all insulating liquids and may also be absorbed by the insulating materials used throughout the transformer. The dissolved moisture should be below about 15 parts per million (ppm).

Appendix

Maintenance Inspection Sheet

Manufacturer_____ kVA_____ Serial #_____

Equipment # _____Date_____Pri Volts_____ Sec Volts_____

    Condition of enclosure: Dents_____ Rust_____ Wet_____

    Enclosure mounting: Loose_____ Missing_____

    Conditions of location: Clean_____ Cluttered_____ Dirty_____

    Openings for cooling: Clean_____ Blocked_____

    Connections: A ph_____ B ph_____ C ph_____ Grd_____

    Volts phase-to-phase: A-B_____ B-C_____ C-A_____

    Volts-to-ground: A-Grd_____ B-Grd_____ C-Grd_____

    Currents: A ph_____ B ph_____ C ph_____ Neu_____

    Coil Temps: A ph_____ B ph_____ C ph_____

    Core condition and internal hardware:_____

    _____

    Overcurrent protection. Open and check connections for heat:_____

    _____

    Ambient air temp:_____ Ending air temp:_____

    Need repairs: Yes_____No_____

    Repair date if needed_____

Comments or Repairs_____

_____

_____

Technician_____

Full-Load Currents — 1φ Transformers

Voltage kVA Rating	120	208	240	480	600	2400
1	8.34	4.8	4.16	2.08	1.67	0.42
3	25	14.4	12.5	6.25	5	1.25
5	41.7	24	20.8	10.4	8.35	2.08
7.5	62.5	36.1	31.2	15.6	12.5	3.12
10	83.4	48	41.6	20.8	16.7	4.16
15	125	72	62.5	31.2	25	6.25
25	208	120	104	52	41.7	10.4
37.5	312	180	156	78	62.5	15.6
50	417	240	208	104	83.5	20.8
75	625	361	312	156	125	31.2
100	834	480	416	208	167	41.6
125	1042	600	520	260	208	52
167.5	1396	805	698	349	279	70
200	1666	960	833	416	333	83.3
250	2080	1200	1040	520	417	104
333	2776	1600	1388	694	555	139
500	4170	2400	2080	1040	835	208

For other kVA ratings or voltages: $I = \dfrac{kVA \times 1000}{V}$

Full-Load Currents — 3φ Transformers

Voltage kVA Rating	208	240	480	600	2400	4160
3	8.3	7.2	3.6	2.9	0.72	0.415
6	16.6	14.4	7.2	5.8	1.44	0.83
9	25	21.6	10.8	8.7	2.16	1.25
15	41.6	36	18	14.4	3.6	2.1
30	83	72	36	29	7.2	4.15
45	125	108	54	43	10.8	6.25
75	208	180	90	72	18	10.4
100	278	241	120	96	24	13.9
150	416	350	180	144	36	20.8
225	625	542	271	217	54	31.2
300	830	720	360	290	72	41.5
500	1390	1200	600	480	120	69.4
750	2080	1800	900	720	180	104
1000	2775	2400	1200	960	240	139
1500	4150	3600	1800	1440	360	208
2000	5550	4800	2400	1930	480	277
2500	6950	6000	3000	2400	600	346
5000	13,900	12,000	6000	4800	1200	694
7500	20,800	18,000	9000	7200	1800	1040
10,000	27,750	24,000	12,000	9600	2400	1386

For other kVA ratings or voltages: $I = \dfrac{kVA \times 1000}{V \times \sqrt{3}}$

Transformer Connections . . .

1φ

3φ, Group 1, 0° Angular Displacement

. . . Transformer Connections . . .

3φ, Group 2, 180° Angular Displacement

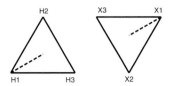

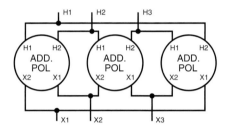

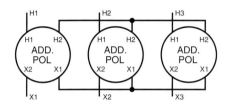

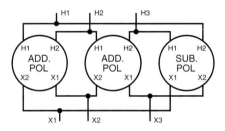

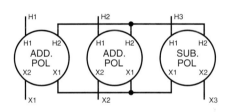

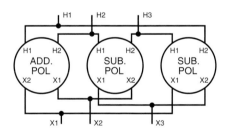

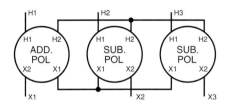

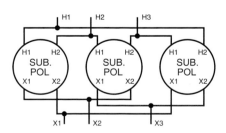

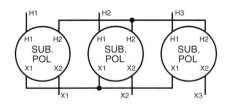

. . . *Transformer Connections* . . .

3φ, Group 3, 30° Angular Displacement

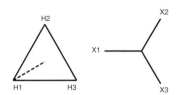

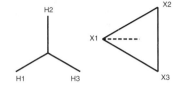

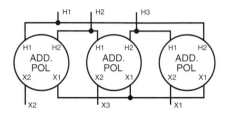

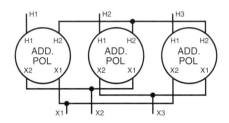

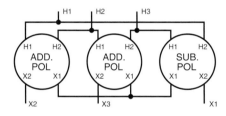

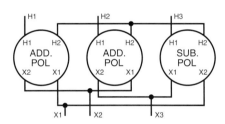

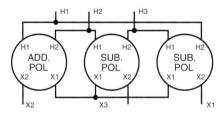

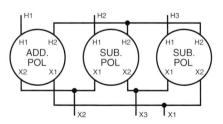

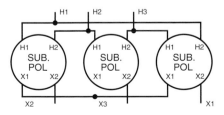

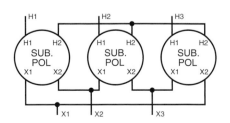

... *Transformer Connections* ...

3φ to 6φ, Group 4, 0° Angular Displacement

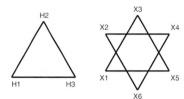

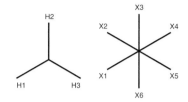

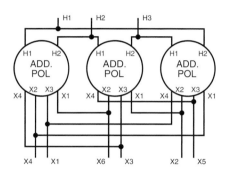

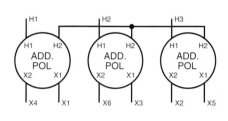

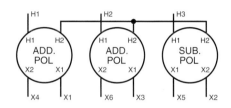

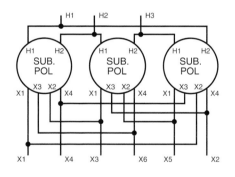

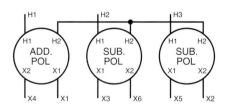

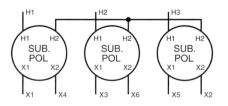

. . . *Transformer Connections*

3φ to 6φ, Group 5, 30° Angular Displacement

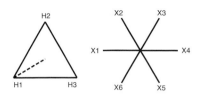

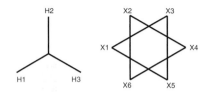

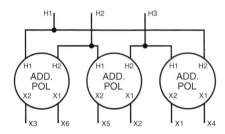

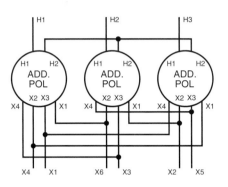

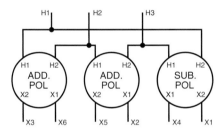

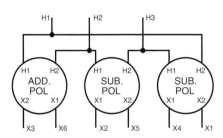

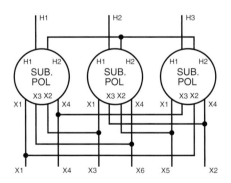

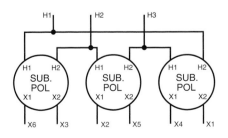

Enclosure Selection

Type	Use	Service Conditions	Tests	Comments	Type
1	Indoor	No unusual	Rod entry, rust resistance		
3	Outdoor	Windblown dust, rain, sleet, and ice on enclosure	Rain, external icing, dust, and rust resistance	Do not provide protection against internal condensation or internal icing	
3R	Outdoor	Falling rain and ice on enclosure	Rod entry, rain, external icing, and rust resistance	Do not provide protection against dust, internal condensation, or internal icing	
4	Indoor/outdoor	Windblown dust and rain, splashing water, hose-directed water, and ice on enclosure	Hosedown, external icing, and rust resistance	Do not provide protection against internal condensation or internal icing	
4X	Indoor/outdoor	Corrosion, windblown dust and rain, splashing water, hose-directed water, and ice on enclosure	Hosedown, external icing, and corrosion resistance	Do not provide protection against internal condensation or internal icing	
6	Indoor/outdoor	Occasional temporary submersion at a limited depth			
6P	Indoor/outdoor	Prolonged submersion at a limited depth			
7	Indoor locations classified as Class I, Groups A, B, C, or D, as defined in the NEC®	Withstand and contain an internal explosion of specified gases, sufficiently contain an explosion so an explosive gas-air mixture in the atmosphere is not ignited	Explosion, hydrostatic, and temperature	Enclosed heat-generating devices shall not cause external surfaces to reach temperatures capable of igniting explosive gas-air mixtures in the atmosphere	
9	Indoor locations classified as Class II, Groups E or G, as defined in the NEC®	Dust	Dust penetration, temperature, and gasket aging	Enclosed heat-generating devices shall not cause external surfaces to reach temperatures capable of igniting explosive gas-air mixtures in the atmosphere	
12	Indoor	Dust, falling dirt, and dripping noncorrosive liquids	Drip, dust, and rust resistance	Do not provide protection against internal condensation	
13	Indoor	Dust, spraying water, oil, and noncorrosive coolant	Oil explosion and rust resistance	Do not provide protection against internal condensation	

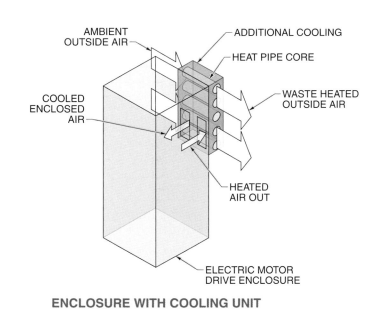

ENCLOSURE WITH COOLING UNIT

Baldor Motors and Drives

NEMA 4X ENCLOSURE

Test Instrument Terminology . . .

Term	Symbol	Definition
AC		Continually changing current that reverses direction at regular intervals. Standard U.S. frequency is 60 Hz
AC COUPLING		Device that passes an AC signal and blocks a DC signal. Used to measure AC signals that are riding on a DC signal
AC/DC		Indicates ability to read or operate on alternating and direct current
ACCURACY ANALOG METER		Largest allowable error (in percent of full scale) made under normal operating conditions. The reading of a meter set on the 250 V range with an accuracy rating of ± 2% could vary ± 5 V. Analog meters have greater accuracy when readings are taken on the upper half of the scale
ACCURACY DIGITAL METER		Largest allowable error (in percent of reading) made under normal operating conditions. A reading of 100.0 V on a meter with an accuracy of ±2% is between 98.0 V and 102.0 V. Accuracy may also include a specified number of digits (counts) that are added to the basic accuracy rating. For example, an accuracy of ± 2% (± 2 digits) means that a display reading of 100.0 V on the meter is between 97.8 V and 102.2 V
ALLIGATOR CLIP		Long-jawed, spring-loaded clamp connected to the end of a test lead. Used to make temporary electrical connections
AMBIENT TEMPERATURE		Temperature of air surrounding a meter or equipment to which the meter is connected
AMMETER		Meter that measures electric current
AMMETER SHUNT		Low-resistance conductor that is connected in parallel with the terminals of an ammeter to extend the range of current values measured by the ammeter
AMPLITUDE		Highest value reached by a quantity under test
ATTENUATION		Decrease in amplitude of a signal
AUDIBLE		Sound that can be heard
AUTORANGING		Function that automatically selects a meter's range based on signals received
AVERAGE VALUE		Value equal to .637 times the amplitude of a measured value

. . . Test Instrument Terminology . . .

Term	Symbol	Definition
BACKLIGHT		Light that brightens the meter display
BANANA JACK		Meter jack that accepts a banana plug
BANANA PLUG		Long, thick terminal connection on one end of a test lead used to make a connection to a meter
BATTERY SAVE		Feature that enables a meter to shut down when battery level is too low or no key is pressed within a set time
BNC		Coaxial-type input connector used on some meters
CAPTURE		Function that records and displays measured values
CELSIUS	°C	Temperature measured on a scale for which the freezing point of water is 0° and the boiling point is 100°
CLOSED CIRCUIT		Circuit in which two or more points allow a predesigned current to flow
COUNTS		Unit of measure of meter resolution. A 1999 count meter cannot display a measurement of $1/10$ of a volt when measuring 200 V or more. A 3200 count meter can display a measurement of $1/10$ of a volt up to 320 V
DC		Current that constantly flows in one direction
DECIBEL (dB)		Measurement that indicates voltage or power comparison in a logarithmic scale
DIGITS		Indication of the resolution of a meter. A $3\frac{1}{2}$ digit meter can display three full digits and one half digit. The three full digits display a number from 0 to 9. The half digit displays a 1 or is left blank. A $3\frac{1}{2}$ digit meter displays readings up to 1999 counts of resolution. A $4\frac{1}{2}$ digit meter displays readings up to 19,999 counts of resolution
DIODE		Semiconductor that allows current to flow in only one direction
DISCHARGE		Removal of an electric charge
DUAL TRACE		Feature that allows two separate waveforms to be displayed simultaneously
EARTH GROUND		Reference point that is directly connected to ground

. . . *Test Instrument Terminology* . . .

Term	Symbol	Definition
EFFECTIVE VALUE		Value equal to 0.707 of the amplitude of a measured quantity
FAHRENHEIT	°F	Temperature measured on a scale for which the freezing point of water is 32° and the boiling point is 212°
FREEZE		Function that holds a waveform (or measurement) for closer examination
FREQUENCY		Number of complete cycles occurring per second
FUNCTION SWITCH		Switch that selects the function (AC voltage, DC voltage, etc.) that a meter is to measure
GLITCH		Momentary spike in a waveform
GLITCH DETECT		Function that increases the meter sampling rate to maximize the detection of the glitch(es)
GROUND		Common connection to a point in a circuit whose potential is taken as zero
HARD COPY		Function that allows a printed copy of the displayed measurement
HOLD BUTTON	HOLD H	Button that allows a meter to capture and hold a stable measurement
LIQUID CRYSTAL DISPLAY (LCD)		Display that uses liquid crystals to display waveforms, measurements, and text on its screen
MEASURING RANGE		Minimum and maximum quantity that a meter can safely and accurately measure
NOISE		Unwanted extraneous electrical signals
OPEN CIRCUIT		Circuit in which two (or more) points do not provide a path for current flow
OVERFLOW		Condition of a meter that occurs when a quantity to be measured is greater than the quantity the meter can display
OVERLOAD	OL	Condition of a meter that occurs when a quantity to be measured is greater than the quantity the meter can safely handle for the meter range setting
PEAK		Highest value reached when measuring
PEAK-TO-PEAK		Highest and lowest voltage value of a waveform

... Test Instrument Terminology

Term	Symbol	Definition
POLARITY		Orientation of the positive (+) and negative (−) side of direct current or voltage
PROBE		Pointed metal tip of a test lead used to make contact with the circuit under test
PULSE		Waveform that increases from a constant value, then decreases to its original value
PULSE TRAIN		Repetitive series of pulses
RANGE		Quantities between two points or levels
RECALL		Function that allows stored information (or measurements) to be displayed
RESOLUTION		Sensitivity of a meter. A meter may have a resolution of 1 V or 1 mV
RISING SLOPE		Part of a waveform displaying a rise in voltage
ROOT-MEAN-SQUARE		Value equal to 0.707 of the amplitude of a measured value
SAMPLE		Momentary reading taken from an input signal
SAMPLING RATE		Number of readings taken from a signal every second
SHORT CIRCUIT		Two or more points in a circuit that allow an unplanned current flow
TERMINAL		Point to which meter test leads are connected
TERMINAL VOLTAGE		Voltage level that meter terminals can safely handle
TRACE		Displayed waveform that shows the voltage variations of the input signal as a function of time
TRIGGER		Device which determines the starting point of a measurement
WAVEFORM		Pattern defined by an electrical signal
ZOOM		Function that allows a waveform (or part of waveform) to be magnified

Industrial Electrical Symbols . . .

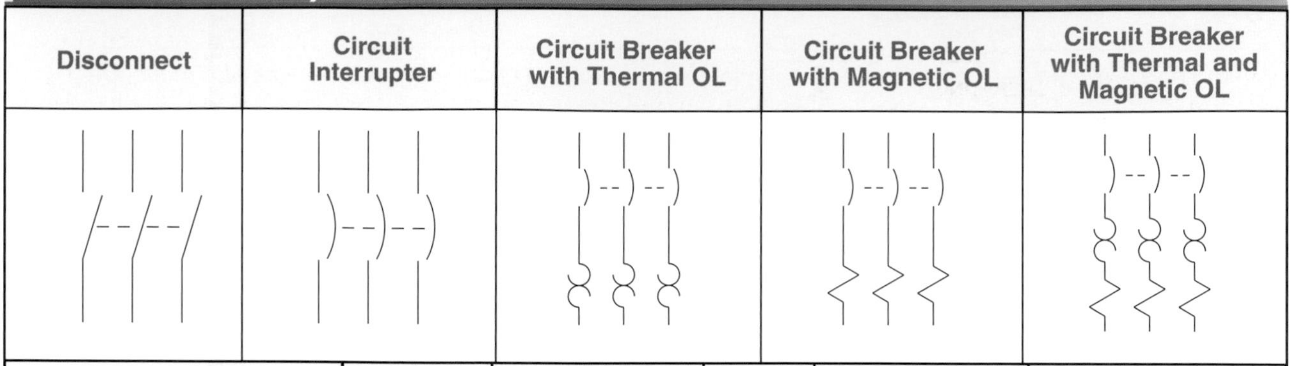

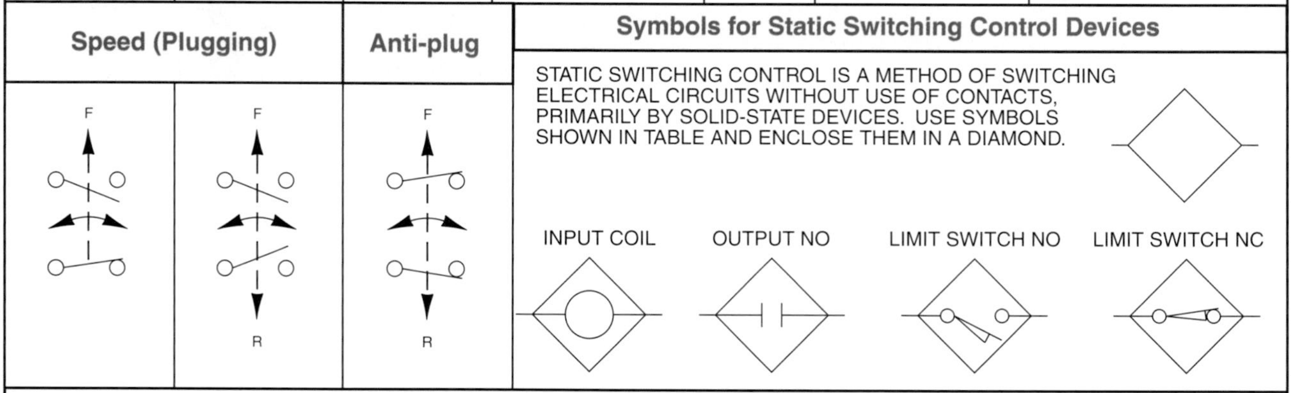

Disconnect	Circuit Interrupter	Circuit Breaker with Thermal OL	Circuit Breaker with Magnetic OL	Circuit Breaker with Thermal and Magnetic OL

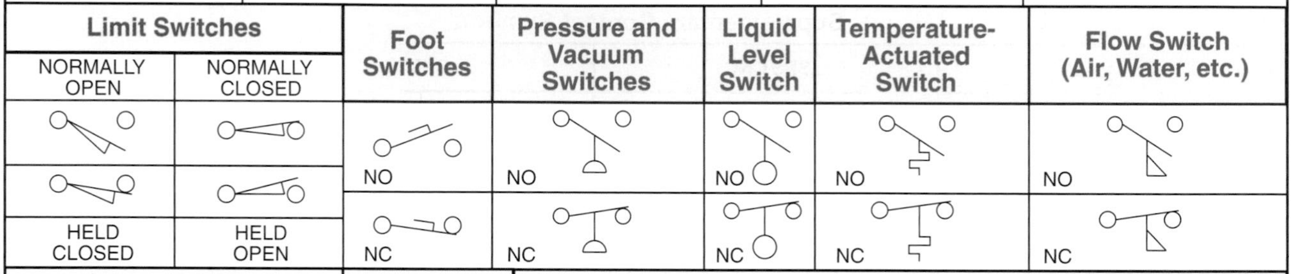

Limit Switches		Foot Switches	Pressure and Vacuum Switches	Liquid Level Switch	Temperature-Actuated Switch	Flow Switch (Air, Water, etc.)
NORMALLY OPEN	NORMALLY CLOSED					
		NO	NO	NO	NO	NO
		NC	NC	NC	NC	NC
HELD CLOSED	HELD OPEN					

Speed (Plugging)	Anti-plug	Symbols for Static Switching Control Devices

STATIC SWITCHING CONTROL IS A METHOD OF SWITCHING ELECTRICAL CIRCUITS WITHOUT USE OF CONTACTS, PRIMARILY BY SOLID-STATE DEVICES. USE SYMBOLS SHOWN IN TABLE AND ENCLOSE THEM IN A DIAMOND.

INPUT COIL   OUTPUT NO   LIMIT SWITCH NO   LIMIT SWITCH NC

Selector

TWO-POSITION	THREE-POSITION	TWO-POSITION SELECTOR PUSHBUTTON

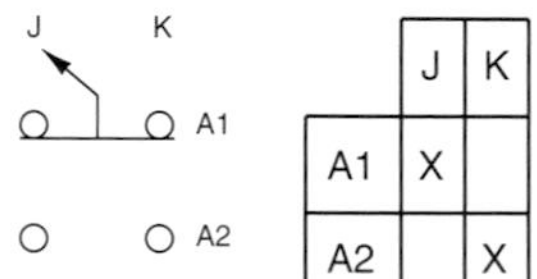

TWO-POSITION

	J	K
A1	X	
A2		X

X-CONTACT CLOSED

THREE-POSITION

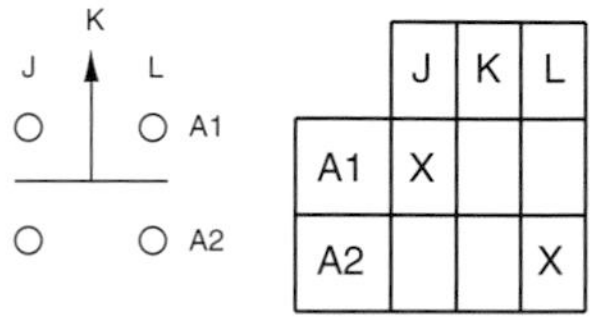

	J	K	L
A1	X		
A2			X

X-CONTACT CLOSED

TWO-POSITION SELECTOR PUSHBUTTON

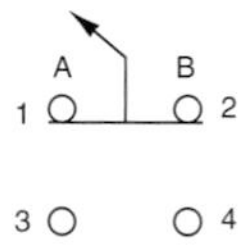

CONTACTS	SELECTOR POSITION			
	A		B	
	BUTTON		BUTTON	
	FREE	DEPRESSED	FREE	DEPRESSED
1-2	X			
3-4		X	X	X

X - CONTACT CLOSED

Pushbuttons

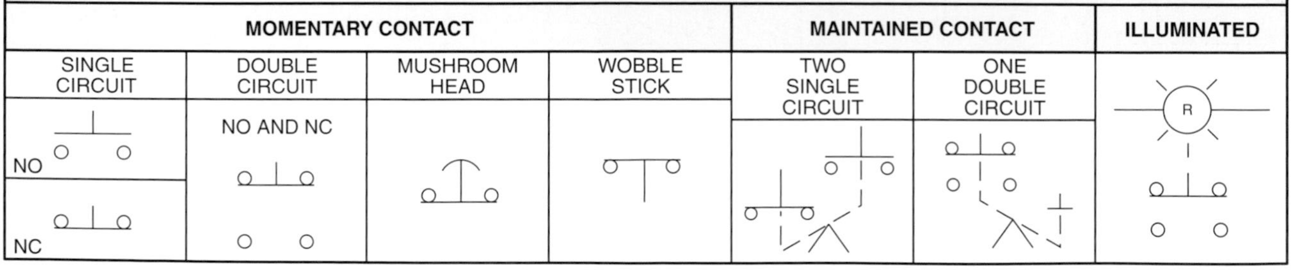

MOMENTARY CONTACT				MAINTAINED CONTACT		ILLUMINATED
SINGLE CIRCUIT	DOUBLE CIRCUIT	MUSHROOM HEAD	WOBBLE STICK	TWO SINGLE CIRCUIT	ONE DOUBLE CIRCUIT	
NO	NO AND NC					
NC						

. . . Industrial Electrical Symbols . . .

Contacts

INSTANT OPERATING				TIMED CONTACTS - CONTACT ACTION RETARDED AFTER COIL IS:				Overload Relays	
WITH BLOWOUT		WITHOUT BLOWOUT		ENERGIZED		DE-ENERGIZED		THERMAL	MAGNETIC
NO	NC	NO	NC	NOTC	NCTO	NOTO	NCTC		

Supplementary Contact Symbols

SPST NO		SPST NC		SPDT		TERMS
SINGLE BREAK	DOUBLE BREAK	SINGLE BREAK	DOUBLE BREAK	SINGLE BREAK	DOUBLE BREAK	SPST SINGLE-POLE, SINGLE-THROW
DPST, 2NO		DPST, 2NC		DPDT		
SINGLE BREAK	DOUBLE BREAK	SINGLE BREAK	DOUBLE BREAK	SINGLE BREAK	DOUBLE BREAK	

SPDT SINGLE-POLE, DOUBLE-THROW

DPST DOUBLE-POLE, SINGLE-THROW

DPDT DOUBLE-POLE, DOUBLE-THROW

NO NORMALLY OPEN

NC NORMALLY CLOSED

Meter (Instrument)

INDICATE TYPE BY LETTER

TO INDICATE FUNCTION OF METER OR INSTRUMENT, PLACE SPECIFIED LETTER OR LETTERS WITHIN SYMBOL.

AM or A	AMMETER	VA	VOLTMETER
AH	AMPERE HOUR	VAR	VARMETER
µA	MICROAMMETER	VARH	VARHOUR METER
mA	MILLAMMETER	W	WATTMETER
PF	POWER FACTOR	WH	WATTHOUR METER
V	VOLTMETER		

Pilot Lights

INDICATE COLOR BY LETTER

NON PUSH-TO-TEST	PUSH-TO-TEST

Inductors

IRON CORE

AIR CORE

Coils

DUAL-VOLTAGE MAGNET COILS

HIGH-VOLTAGE	LOW-VOLTAGE
LINK	LINKS

BLOWOUT COIL

. . . Industrial Electrical Symbols . . .

Transformers

AUTO	AIR CORE	CURRENT	CONTROL TRANSFORMER		AUTOTRANSFORMER FOR REDUCED-VOLTAGE STARTING
			SINGLE-VOLTAGE	DUAL-VOLTAGE	

Single-voltage: H1, H2 (top); X2, X1 (bottom)
Dual-voltage: H3, H2 (top), H1, H4; X2, X1 (bottom)
Autotransformer: % 50, 65, 80, 100, 0

AC Motors

SINGLE-PHASE	SEPARATE PHASE, TWO-SPEED	THREE-PHASE	SEPARATE WINDING, TWO-SPEED	CONSTANT-TORQUE, TWO-SPEED
T1, T2	HIGH T1, COM T2, LOW T3	T1, T2, T3	T1, T11 / T3, T2, T13, T12	T4, T3, T1, T5, T2, T6

VARIABLE-TORQUE, TWO-SPEED	CONSTANT-HORSEPOWER, TWO-SPEED	WYE/DELTA, REDUCED-VOLTAGE	WYE-CONNECTED, PART WINDING, REDUCED-VOLTAGE
T4, T1, T3, T5, T2, T6	T4, T3, T1, T5, T2, T6	T6, T1, T3, T4, T5, T2	T1 T2 T3, T5, T7 T8 T9, T4 T6

DC Motors / Wiring / Connections

				DC Motors				Wiring			Connections

ARMATURE	SHUNT FIELD	SERIES FIELD	COMM OR COMPENS FIELD	NOT CONNECTED	POWER	WIRING TERMINAL	MECHANICAL
ARM	SHOW 4 LOOPS	SHOW 3 LOOPS	SHOW 2 LOOPS	CONNECTED	CONTROL	GROUND	MECHANICAL INTERLOCK

Control and Power Connections — 600 V or Less Across-the-Line Starters

		1φ	2φ, 4-WIRE	3φ
LINE MARKINGS		L1, L2	L1, L3 PHASE 1 / L2, L4 PHASE 2	L1, L2, L3
GROUND WHEN USED		L1 IS ALWAYS UNGROUNDED	—	L2
MOTOR RUNNING OVERCURRENT UNITS IN	1 ELEMENT	L1	—	—
	2 ELEMENT	—	L1, L4	—
	3 ELEMENT	—	—	L1, L2, L3
CONTROL CIRCUIT CONNECTED TO		L1, L2	L1, L3	L1, L2
FOR REVERSING INTERCHANGE LINES		—	L1, L3	L1, L3

. . . Industrial Electrical Symbols

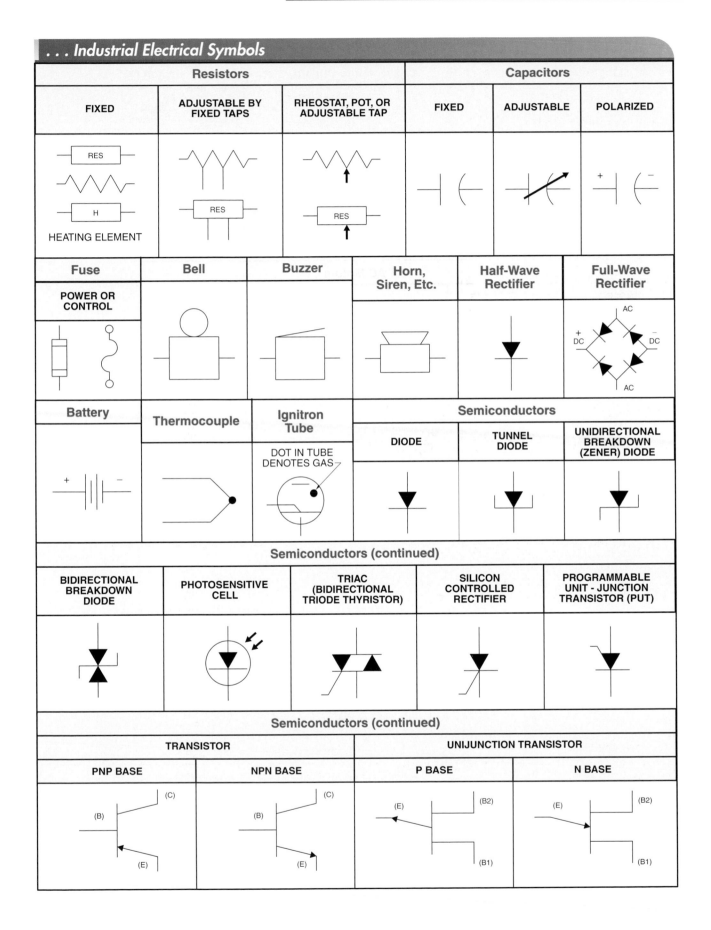

Residential Electrical Symbols . . .

Lighting Outlets

OUTLET BOX AND INCANDESCENT LIGHTING FIXTURE
CEILING   WALL

INCANDESCENT TRACK LIGHTING

BLANKED OUTLET   (B)   (B)

DROP CORD   (D)

EXIT LIGHT AND OUTLET BOX. SHADED AREAS DENOTE FACES.

OUTDOOR POLE-MOUNTED FIXTURES

JUNCTION BOX   (J)   (J)

LAMPHOLDER WITH PULL SWITCH   (L)PS   (L)PS

MULTIPLE FLOODLIGHT ASSEMBLY

EMERGENCY BATTERY PACK WITH CHARGER

INDIVIDUAL FLUORESCENT FIXTURE

OUTLET BOX AND FLUORESCENT LIGHTING TRACK FIXTURE

CONTINUOUS FLUORESCENT FIXTURE

SURFACE-MOUNTED FLUORESCENT FIXTURE

Panelboards

FLUSH-MOUNTED PANELBOARD AND CABINET

SURFACE-MOUNTED PANELBOARD AND CABINET

Convenience Outlets

SINGLE RECEPTACLE OUTLET

DUPLEX RECEPTACLE OUTLET

TRIPLEX RECEPTACLE OUTLET

SPLIT-WIRED DUPLEX RECEPTACLE OUTLET

SPLIT-WIRED TRIPLEX RECEPTACLE OUTLET

SINGLE SPECIAL-PURPOSE RECEPTACLE OUTLET

DUPLEX SPECIAL-PURPOSE RECEPTACLE OUTLET

RANGE OUTLET   R

SPECIAL-PURPOSE CONNECTION   DW

CLOSED-CIRCUIT TELEVISION CAMERA

CLOCK HANGER RECEPTACLE   (C)

FAN HANGER RECEPTACLE   (F)

FLOOR SINGLE RECEPTACLE OUTLET

FLOOR DUPLEX RECEPTACLE OUTLET

FLOOR SPECIAL-PURPOSE OUTLET

UNDERFLOOR DUCT AND JUNCTION BOX FOR TRIPLE, DOUBLE, OR SINGLE DUCT SYSTEM AS INDICATED BY NUMBER OF PARALLEL LINES

Busducts and Wireways

SERVICE, FEEDER, OR PLUG-IN BUSWAY   B   B   B

CABLE THROUGH LADDER OR CHANNEL   C   C   C

WIREWAY   W   W   W

Switch Outlets

SINGLE-POLE SWITCH S

DOUBLE-POLE SWITCH S_2

THREE-WAY SWITCH S_3

FOUR-WAY SWITCH S_4

AUTOMATIC DOOR SWITCH S_D

KEY-OPERATED SWITCH S_K

CIRCUIT BREAKER S_{CB}

WEATHERPROOF CIRCUIT BREAKER S_{WCB}

DIMMER S_{DM}

REMOTE CONTROL SWITCH S_{RC}

WEATHERPROOF SWITCH S_{WP}

FUSED SWITCH S_F

WEATHERPROOF FUSED SWITCH S_{WF}

TIME SWITCH S_T

CEILING PULL SWITCH S

SWITCH AND SINGLE RECEPTACLE S

SWITCH AND DOUBLE RECEPTACLE S

A STANDARD SYMBOL WITH AN ADDED LOWERCASE SUBSCRIPT LETTER IS USED TO DESIGNATE A VARIATION IN STANDARD EQUIPMENT   a,b   a,b $S_{a,b}$

. . . Residential Electrical Symbols

Commercial and Industrial Systems

PAGING SYSTEM DEVICE

FIRE ALARM SYSTEM DEVICE

COMPUTER DATA SYSTEM DEVICE

PRIVATE TELEPHONE SYSTEM DEVICE

SOUND SYSTEM

FIRE ALARM CONTROL PANEL — FACP

Signaling System Outlets for Residential Systems

PUSHBUTTON

BUZZER

BELL

BELL AND BUZZER COMBINATION

COMPUTER DATA OUTLET

BELL RINGING TRANSFORMER — BT

ELECTRIC DOOR OPENER — D

CHIME — CH

TELEVISION OUTLET — TV

THERMOSTAT — T

Underground Electrical Distribution or Electrical Lighting Systems

MANHOLE — M

HANDHOLE — H

TRANSFORMER-MANHOLE OR VAULT — TM

TRANSFORMER PAD — TP

UNDERGROUND DIRECT BURIAL CABLE

UNDERGROUND DUCT LINE

STREET LIGHT STANDARD FED FROM UNDERGROUND CIRCUIT

Aboveground Electrical Distribution or Lighting Systems

POLE

STREET LIGHT AND BRACKET

PRIMARY CIRCUIT

SECONDARY CIRCUIT

DOWN GUY

HEAD GUY

SIDEWALK GUY

SERVICE WEATHERHEAD

Panel Circuits and Miscellaneous

LIGHTING PANEL

POWER PANEL

WIRING – CONCEALED IN CEILING OR WALL

WIRING – CONCEALED IN FLOOR

WIRING EXPOSED

HOME RUN TO PANEL BOARD
Indicate number of circuits by number of arrows. Any circuit without such designation indicates a two-wire circuit. For a greater number of wires indicate as follows: (3 wires) (4 wires), etc.

FEEDERS
Use heavy lines and designate by number corresponding to listing in feeder schedule

WIRING TURNED UP

WIRING TURNED DOWN

GENERATOR — G

MOTOR — M

INSTRUMENT (SPECIFY) — I

TRANSFORMER — T

CONTROLLER

EXTERNALLY-OPERATED DISCONNECT SWITCH

PULL BOX

AC/DC Formulas

To Find	DC	AC		
		1ϕ, 115 V or 220 V	1ϕ, 208 V, 230 V, or 240 V	3ϕ – All Voltages
I, HP known	$\dfrac{HP \times 746}{E \times Eff}$	$\dfrac{HP \times 746}{E \times Eff \times PF}$	$\dfrac{HP \times 746}{E \times Eff \times PF}$	$\dfrac{HP \times 746}{1.73 \times E \times Eff \times PF}$
I, kW known	$\dfrac{kW \times 1000}{E}$	$\dfrac{kW \times 1000}{E \times PF}$	$\dfrac{kW \times 1000}{E \times PF}$	$\dfrac{kW \times 1000}{1.73 \times E \times PF}$
I, kVA known		$\dfrac{kVA \times 1000}{E}$	$\dfrac{kVA \times 1000}{E}$	$\dfrac{kVA \times 1000}{1.763 \times E}$
KW	$\dfrac{I \times E}{1000}$	$\dfrac{I \times E \times PF}{1000}$	$\dfrac{I \times E \times PF}{1000}$	$\dfrac{I \times E \times 1.73 \times PF}{1000}$
kVA		$\dfrac{I \times E}{1000}$	$\dfrac{I \times E}{1000}$	$\dfrac{I \times E \times 1.73}{1000}$
HP (output)	$\dfrac{I \times E \times Eff}{746}$	$\dfrac{I \times E \times Eff \times PF}{746}$	$\dfrac{I \times E \times Eff \times PF}{746}$	$\dfrac{I \times E \times 1.73 \times Eff \times PF}{746}$

Eff = efficiency

Horsepower Formulas

To Find	Use Formula	Example		
		Given	Find	Solution
HP	$HP = \dfrac{I \times E \times Eff}{746}$	240 V, 20 A, 85% Eff	HP	$HP = \dfrac{I \times E \times Eff}{746}$ $HP = \dfrac{20\,A \times 240\,V \times 85\%}{746}$ $HP = \textbf{5.5}$
I	$I = \dfrac{HP \times 746}{E \times Eff \times PF}$	10 HP, 240 V, 90% Eff, 88% PF	I	$I = \dfrac{HP \times 746}{E \times Eff \times PF}$ $I = \dfrac{10\,HP \times 746}{240\,V \times 90\% \times 88\%}$ $I = \textbf{39 A}$

Voltage Drop Formulas — 1ϕ, 3ϕ

Phase	To Find	Use Formula	Example		
			Given	Find	Solution
1ϕ	VD	$VD = \dfrac{2 \times R \times L \times I}{1000}$	240 V, 40 A, 60′ L, .764 R	VD	$VD = \dfrac{2 \times R \times L \times I}{1000}$ $VD = \dfrac{2 \times .764 \times 60 \times 40}{1000}$ $VD = \textbf{3.67 V}$
3ϕ	VD	$VD = \dfrac{2 \times R \times L \times I}{1000} \times .866$	208 V, 110 A, 75′ L, .194 R, .866 multiplier	VD	$VD = \dfrac{2 \times R \times L \times I}{1000} \times .866$ $VD = \dfrac{2 \times .194 \times 75 \times 110}{1000} \times .866$ $VD = \textbf{2.77 V}$

Electrical/Electronic Abbreviations/Acronyms

Abbr/Acronym	Meaning	Abbr/Acronym	Meaning	Abbr/Acronym	Meaning
A	Ammeter; Ampere; Anode; Armature	FU	Fuse	PNP	Positive-Negative-Positive
AC	Alternating Current	FWD	Forward	POS	Positive
AC/DC	Alternating Current; Direct Current	G	Gate; Giga; Green; Conductance	POT.	Potentiometer
A/D	Analog to Digital	GEN	Generator	P-P	Peak-to-Peak
AF	Audio Frequency	GRD	Ground	PRI	Primary Switch
AFC	Automatic Frequency Control	GY	Gray	PS	Pressure Switch
Ag	Silver	H	Henry; High Side of Transformer; Magnetic Flux	PSI	Pounds Per Square Inch
ALM	Alarm			PUT	Pull-Up Torque
AM	Ammeter; Amplitude Modulation	HF	High Frequency	Q	Transistor
AM/FM	Amplitude Modulation; Frequency Modulation	HP	Horsepower	R	Radius; Red; Resistance; Reverse
		Hz	Hertz	RAM	Random-Access Memory
ARM.	Armature	I	Current	RC	Resistance-Capacitiance
Au	Gold	IC	Integrated Circuit	RCL	Resistance-Inductance-Capacitance
AU	Automatic	INT	Intermediate; Interrupt	REC	Rectifier
AVC	Automatic Volume Control	INTLK	Interlock	RES	Resistor
AWG	American Wire Gauge	IOL	Instantaneous Overload	REV	Reverse
BAT.	Battery (electric)	IR	Infrared	RF	Radio Frequency
BCD	Binary Coded Decimal	ITB	Inverse Time Breaker	RH	Rheostat
BJT	Bipolar Junction Transistor	ITCB	Instantaneous Trip Circuit Breaker	rms	Root Mean Square
BK	Black	JB	Junction Box	ROM	Read-Only Memory
BL	Blue	JFET	Junction Field-Effect Transistor	rpm	Revolutions Per Minute
BR	Brake Relay; Brown	K	Kilo; Cathode	RPS	Revolutions Per Second
C	Celsius; Capacitiance; Capacitor	L	Line; Load; Coil; Inductance	S	Series; Slow; South; Switch
CAP.	Capacitor	LB-FT	Pounds Per Foot	SCR	Silicon Controlled Rectifier
CB	Circuit Breaker; Citizen's Band	LB-IN.	Pounds Per Inch	SEC	Secondary
CC	Common-Collector Configuration	LC	Inductance-Capacitance	SF	Service Factor
CCW	Counterclockwise	LCD	Liquid Crystal Display	1 PH; 1φ	Single-Phase
CE	Common-Emitter Configuration	LCR	Inductance-Capacitance-Resistance	SOC	Socket
CEMF	Counter Electromotive Force	LED	Light Emitting Diode	SOL	Solenoid
CKT	Circuit	LRC	Locked Rotor Current	SP	Single-Pole
CONT	Continuous; Control	LS	Limit Switch	SPDT	Single-Pole, Double-Throw
CPS	Cycles Per Second	LT	Lamp	SPST	Single-Pole, Single-Throw
CPU	Central Processing Unit	M	Motor; Motor Starter; Motor Starter Contacts	SS	Selector Switch
CR	Control Relay			SSW	Safety Switch
CRM	Control Relay Master	MAX.	Maximum	SW	Switch
CT	Current Transformer	MB	Magnetic Brake	T	Tera; Terminal; Torque; Transformer
CW	Clockwise	MCS	Motor Circuit Switch	TB	Terminal Board
D	Diameter; Diode; Down	MEM	Memory	3 PH; 3φ?	Three-Phase
D/A	Digital to Analog	MED	Medium	TD	Time Delay
DB	Dynamic Braking Contactor; Relay	MIN	Minimum	TDF	Time Delay Fuse
DC	Direct Current	MN	Manual	TEMP	Temperature
DIO	Diode	MOS	Metal-Oxide Semiconductor	THS	Thermostat Switch
DISC.	Disconnect Switch	MOSFET	Metal-Oxide Semiconductor Field-Effect Transistor	TR	Time Delay Relay
DMM	Digital Multimeter			TTL	Transistor-Transistor Logic
DP	Double-Pole	MTR	Motor	U	Up
DPDT	Double-Pole, Double-Throw	N; NEG	North; Negative	UCL	Unclamp
DPST	Double-Pole, Single-Throw	NC	Normally Closed	UHF	Ultrahigh Frequency
DS	Drum Switch	NEUT	Neutral	UJT	Unijunction Transistor
DT	Double-Throw	NO	Normally Open	UV	Ultraviolet; Undervoltage
DVM	Digital Voltmeter	NPN	Negative-Positive-Negative	V	Violet; Volt
EMF	Electromotive Force	NTDF	Nontime-Delay Fuse	VA	Volt Amp
F	Fahrenheit; Fast; Field; Forward; Fuse	O	Orange	VAC	Volts Alternating Current
FET	Field-Effect Transistor	OCPD	Overcurrent Protection Device	VDC	Volts Direct Current
FF	Flip-Flop	OHM	Ohmmeter	VHF	Very High Frequency
FLC	Full-Load Current	OL	Overload Relay	VLF	Very Low Frequency
FLS	Flow Switch	OZ/IN.	Ounces Per Inch	VOM	Volt-Ohm-Milliammeter
FLT	Full-Load Torque	P	Peak; Positive; Power; Power Consumed	W	Watt; White
FM	Fequency Modulation	PB	Pushbutton	w/	With
FREQ	Frequency	PCB	Printed Circuit Board	X	Low Side of Transformer
FS	Float Switch	PH; ?	Phase	Y	Yellow
FTS	Foot Switch	PLS	Plugging Switch	Z	Impedance

Overcurrent Protection Devices

Motor Type	Code Letter	FLC (%)				
		Motor Size	TDF	NTDF	ITB	ITCB
AC*	—	—	175	300	150	700
AC*	A	—	150	150	150	700
AC*	B–E	—	175	250	200	700
AC*	F–V	—	175	300	250	700
DC	—	$\frac{1}{8}$ to 50 HP	150	150	150	150
DC	—	Over 50 HP	150	150	150	175

* full-voltage and resistor starting

Glossary

American National Standards Institute (ANSI): A U.S. national organization that helps identify industrial and public needs for standards.

apparent power: The power, in VA or kVA, that is the sum of true power, reactive power, and harmonic power.

arc blast: An explosion that occurs when the surrounding air becomes ionized and conductive.

arc flash: A short circuit through the air.

arrester: A device used to shunt a voltage spike to ground or from line to line.

autotransformer: Any transformer in which the primary and secondary circuits have a portion of their two windings in common.

ballast: A controller responsible for providing the initial startup voltage and maintaining a constant current through a lamp.

bar current transformer: A special type of window current transformer with a solid bar placed permanently through the window.

bell transformer: A transformer used to supply low-voltage, low-power circuits, such as for doorbells, annunciators, and similar systems.

buck-boost transformer: A small transformer designed to buck (lower) or boost (raise) line voltage.

Canadian Standards Association (CSA): A Canadian nonprofit membership association for standards development, sale of publications, training, and membership services.

capacitive leakage current: The current that flows into insulation to charge the insulation like the plates of a capacitor.

caution signal word: A word used to indicate a potentially hazardous situation which, if not avoided, may result in minor or moderate injury.

choke: A reactor used to restrict the current to AC or DC drives in the case of a short circuit in the drive itself.

coefficient of coupling: *See mutual inductance.*

coil tap: An extra electrical connection on a transformer coil that allows a varying number of turns of a coil to be part of a circuit.

common-mode choke: A reactor used to reduce common-mode noise current generated by the rapid switching of a motor drive or a signaling device.

common-mode noise: A type of electromagnetic interference induced on power or communications lines.

conductive leakage current: A very small current (µA) that flows through insulation between conductors or from conductor to ground.

confined space: A space large enough and so configured that an employee can physically enter and perform assigned work, that has limited or restricted means for entry and exit, and is not designed for continuous employee occupancy.

control transformer: A transformer that is used to step down the voltage to the control circuit of a circuit or machine.

copper loss: *See resistive loss.*

current crest factor: The peak value of a waveform divided by the rms value of the waveform.

current transformer: A transformer used to step down line current to make it easier to measure.

danger signal word: A word used to indicate a imminently hazardous situation which, if not avoided, results in death or serious injury.

decibel: The unit used to measure the intensity level of sound.

delta connection: A 3-phase connection method that has the wires from the ends of each coil connected end-to-end to form a closed loop.

dielectric absorption test: A variation of the standard insulation resistance spot test where the voltage is applied for 10 min.

displacement power factor: The ratio of true power to apparent power due to the phase displacement between the current and voltage.

distortion power factor: The ratio of true power to apparent power due to THD.

double-delta connection: A method of wiring transformers where both sets of secondary windings are connected in a delta connection, but one is reversed with respect to the other.

eddy current loss: Power loss in a transformer or motor due to currents induced in the metal field structure from the changing magnetic field.

electrical shock: A shock that results anytime a body becomes part of an electrical circuit.

electrical warning signal word: A word used to indicate a high-voltage location and conditions that could result in death or serious personal injury if proper precautions and procedures are not followed.

electromagnet: A magnet whose energy is produced by the flow of electric current.

electromagnetism: Magnetism produced when electricity passes through a conductor.

equipment grounding conductor (EGC): An electrical conductor that provides a low-impedance ground path between electrical equipment and enclosures within the distribution system and takes current back to the source.

exciting current: The no-load current through a primary core.

explosion vent: A pipe, 4″ in diameter or greater, that extends a few feet above the cover of a transformer and is curved toward the ground at the outlet end of the pipe.

explosion warning signal word: A word used to indicate locations and conditions where exploding parts may cause death or serious personal injury if proper precautions and procedures are not followed.

ferromagnetic material: A material that is easily magnetized.

field flux: *See magnetic flux.*

floating system: A control system without a grounded secondary in the control transformer.

flux loss: A power loss that occurs in a transformer when some of the lines of flux from the primary do not travel through the core to the secondary.

Fourier transform: The mathematical method of converting a time-based waveform, like a sine waveform, into frequency-based information.

generator: A machine that converts mechanical energy into electrical energy by means of electromagnetic induction.

grounded conductor: A conductor that has been intentionally grounded.

grounded system: A control system with a grounded secondary in the control transformer.

ground fault circuit interrupter (GFCI): A device that protects against electrical shock by detecting an imbalance of current in the normal conductor pathways and opening the circuit.

grounding: The connection of portions of the distribution system to earth in order to establish a common electrical reference and a low impedance fault path to facilitate operation of overcurrent protective devices.

grounding electrode conductor (GEC): A conductor that connects grounded parts of a power distribution system (equipment grounding conductors, grounded conductors, and all metal parts) to the grounding system.

ground loop: A circuit that has more than one point connected to earth ground, with a voltage potential difference between the two ground points.

harmonic: Voltage or current at a frequency that is an integer (whole number) multiple (2nd, 3rd, 4th, etc.) of the fundamental frequency.

harmonic filter: A device used to reduce harmonic frequencies and THD.

harmonic power: The power, in VA or kVA, lost to harmonic distortion.

harmonics mitigating transformer (HMT): A transformer designed to reduce the harmonics in a power distribution system.

harmonic sequence: The phasor rotation with respect to the fundamental (60 Hz) frequency.

hipot tester: A test instrument that can be used to evaluate insulation resistance by measuring leakage current.

hysteresis: The property of ferromagnetic materials where the magnetic induction of a coil lags the magnetic field that is charging the coil.

inductance: The property of a device or circuit that causes it to store charge in an electromagnetic field.

induction: The ability of a device or circuit to generate reactance to oppose a changing current (self-induction) or the ability to generate a current in a nearby circuit (mutual induction).

insulation power factor: The ratio of the power dissipated in the insulation, in watts, to the apparent power, in volt-amps.

insulation resistance spot test: A test where one insulation resistance reading is used to evaluate the integrity of insulation.

insulation resistance test: A test performed to measure the leakage current through the coil insulation.

interconnected wye connection: A method of wiring transformers where two separate windings are interconnected on each phase in a zigzag fashion.

International Electrotechnical Commission (IEC): An international organization that develops international safety standards for electrical equipment.

I²R loss: *See resistive loss.*

leather protectors: Gloves worn over rubber insulating gloves to prevent penetration of the rubber insulating gloves and to provide added protection against electrical shock.

line choke: *See choke.*

line current: The current flow through the lines to a load.

line reactor: *See reactor.*

line voltage: The voltage measured line-to-line.

lockout: The process of removing the source of electrical power and installing a lock that prevents the power from being turned on.

magnet: A substance that produces a magnetic field that attracts ferromagnetic materials.

magnetic field: A force produced by a magnet and interacts with other magnets or other magnetic fields.

magnetic field intensity: The amount of magnetomotive force distributed over the length of a magnet.

magnetic flux: Imaginary lines of force that make up the total quantity of an electromagnetic field.

magnetic flux density: The concentration of the magnetic flux in a given area.

magnetism: A force that acts at a distance and is caused by a magnetic field.

magnetizing current: *See exciting current.*

magnetomotive force: Force that produces magnetic lines of flux in a circuit as a result of current in a conductor.

main bonding jumper (MBJ): A connection at the service equipment that connects the equipment grounding conductor, the grounding electrode conductor, and the grounded conductor (neutral conductor).

main transformer: One of the transformers in a T-connected transformer bank and is provided with a 50% voltage tap. A main transformer is also used in a Scott connection.

metal oxide varistor (MOV): A nonlinear resistor that can change resistance with changes in current.

mutual inductance: A measure of the efficiency by which power is transferred from the primary to the secondary coils.

mutual induction: The ability of an inductor in one circuit to induce a voltage in another circuit.

National Electrical Code® (NEC®): A standard on practices for the design and installation of electrical products published by the National Fire Protection Association (NFPA).

National Electrical Manufacturers Association (NEMA): A U.S. national organization that assists with information and standards concerning proper selection, ratings, construction, and safety standards for electrical equipment.

National Fire Protection Association (NFPA): A national organization that provides guidance in assessing the hazards of the products of combustion.

notch: A distortion in a voltage waveform where the voltage quickly drops toward zero and then returns to the correct value.

Occupational Safety and Health Administration (OSHA): A federal agency that requires all employers to provide a safe environment for their employees.

parallel transformer connection: Two or more transformers wired in parallel that carry a common load.

permanent magnet: A magnet that can hold its magnetism for a long period of time.

permeability: A measure of the ability of a material to conduct magnetic flux.

permit-required confined space: A confined space that has specific health and safety hazards associated with it.

personal protective equipment (PPE): Clothing and/or equipment worn by a technician to reduce the possibility of injury in the work area.

phase current: The current flow through the individual windings.

phase voltage: The voltage measured across the windings.

phasor rotation: The order in which waveforms from each phase (A, B, and C) cross zero.

polarization absorption current: The current caused by the polarization of molecules within the dielectric material.

polarization index (PI): The ratio of the 10 min reading to the 1 min reading taken on a dielectric absorption test.

potential transformer: A precision two-winding transformer used to step down high voltage to allow safe voltage measurement.

pothead: A transition device between underground cable and overhead lines.

power factor: The ratio of true power to apparent power in a circuit or distribution system.

qualified person: A person who is trained in, and has specific knowledge of, the construction and operation, testing, and performance of electrical equipment or a specific task, and is trained to recognize and avoid electrical hazards that might be present with respect to the equipment or specific task.

reactance: The opposition to current flow in an AC current.

reactive power: The power, in VAR or kVAR, stored and released by inductors and capacitors.

reactor: A coil added in series with a load to reduce inrush current, voltage notching effects, and voltage spikes.

reluctance: The opposition to magnetic flux in a magnetic circuit in a given volume of space or material.

resistive loss: The power loss in a transformer caused by the resistance of the copper wire used to make the windings.

retentivity: A measure of the ability of a magnet to retain magnetism after the magnetizing force has been removed.

rubber insulating gloves: Gloves made of latex rubber and used to provide maximum isolation from electrical shock.

rubber insulating matting: A floor covering that provides technicians protection from electrical shock when working on live electrical currents.

safety label: A label that indicates areas or tasks that can pose a hazard to personnel and/or equipment.

saturable-core reactor: An inductor whose inductance is controlled through the use of a magnetic field created by a second winding wound around the same iron core as the primary winding.

saturation: The condition where a magnetic core has substantially all the magnetic domains aligned with the field and any increases in current no longer result in a stronger electromagnet.

Scott connection: A method of wiring transformers to transform from 3-phase to 2-phase or 2-phase to 3-phase.

self-induction: The ability of an inductor in a circuit to generate inductive reactance, which opposes change in the circuit.

separately derived system (SDS): A system that supplies electrical power derived (taken from) transformers, storage batteries, solar photovoltaic systems, or generators.

step-down transformer: A transformer with the source connected to the winding with the most turns and the loads connected to the winding with the fewest turns.

step-up transformer: A transformer with the source connected to the winding with the fewest turns and the load connected to the winding with the most turns.

supervised installation: An electrical installation in which the conditions of maintenance are such that only qualified persons monitor or service the electrical equipment.

switched-mode power supply (SMPS): A power supply for electronic devices that include an internal control circuit that quickly switches the load current ON and OFF in order to deliver a stable output voltage.

synchroscope: A device that indicates whether two AC sources to be connected in parallel are in the correct phase relationship.

2-phase system: A power supply consisting of two phases that are 90° out of phase.

Tagout: The process of placing a danger tag on the source of electrical power and installing a lock that prevents the power from being turned on.

T-connection: A low-cost method of wiring two transformers to provide 3-phase power.

teaser transformer: One of the transformers in a T-connected transformer bank and is connected to the 50% tap of the main transformer. A teaser transformer is also used in a Scott connection.

temperature rise: The difference between the hot-spot maximum core temperature at full load and the temperature when not operating.

temporary magnet: A magnet that retains only trace amounts of magnetism after a magnetizing force has been removed.

tertiary winding: A third winding that is often used in power transformers to provide station power requirements or a tie with synchronous condensers.

total demand distortion (TDD): The ratio of the current harmonics to the maximum load current.

total harmonic distortion (THD): The amount of harmonics on a line compared to the fundamental frequency of 60 Hz.

transformer: An electric device that uses electromagnetism to change voltage from one level to another or to isolate one voltage from another through the process of mutual induction.

transient voltage (voltage spike): An unwanted voltage of very short duration in an electrical circuit.

true power: The power, in W or kW, used by motors, lights, and other devices to produce useful work.

turns ratio: The number of the turns in the primary to the number of turns in the secondary.

turns-to-turns ratio: *See turns ratio.*

two-winding transformer: A transformer with no electrical connection between the primary and the secondary.

Underwriters Laboratories Inc. (UL): An independent organization that tests equipment and products to see if they confirm to national codes and standards.

variable transformer: A transformer used to make fine adjustments to the output voltage.

voltage surge: A higher-than-normal voltage that temporarily exists on one or more power lines.

volts per turn (V/turn): The voltage dropped across each turn of a coil or the voltage induced into each turn of the secondary coil.

warning signal word: A word used to indicate a potentially hazardous situation which, if not avoided, could result in death or serious injury.

winding resistance test: A test performed to measure the electrical resistance of the transformer windings.

winding current: The current flow through the individual windings.

winding voltage: The voltage measured across the phase windings.

window current transformer: A transformer that consists of a secondary winding wrapped around a core and the primary sent through the opening in the core.

wound current transformer: A transformer with separate primary and secondary windings wrapped around a laminated core.

wye connection: A 3-phase connection that has one end of each coil connected together and the other end of each coil left open for external connections.

zigzag connection: A method of wiring transformers where the windings are divided over several legs of the transformer core.

Index

USING THE *TRANSFORMER PRINCIPLES AND APPLICATIONS* CD-ROM

Before removing the CD-ROM from the protective sleeve, please note that the book cannot be returned for refund or credit if the CD-ROM sleeve seal is broken.

System Requirements

The *Transformer Principles and Applications* CD-ROM is designed to work best on a computer meeting the following hardware/software requirements:

- Intel® Pentium® (or equivalent) processor
- Microsoft® Windows® 95, 98, 98 SE, Me, NT®, 2000, or XP operating system
- 64 MB of free available system RAM (128 MB recommended)
- 90 MB of available disk space
- 800 × 600 16-bit (thousands of colors) color display or better
- Sound output capability and speakers
- CD-ROM drive
- Internet Explorer™ 3.0 or Netscape® 3.0 or later browser software

OPENING FILES

Insert the CD-ROM into the computer CD-ROM drive. Within a few seconds, the home screen will be displayed allowing access to all features of the CD-ROM. Information about the usage of the CD-ROM can be accessed by clicking on USING THIS CD-ROM. The Chapter Quick Quizzes™, Illustrated Glossary, Transformer Resources, Media Clips, and Reference Material can be accessed by clicking on the appropriate button on the home screen. Clicking on the American Tech web site button (www.go2atp.com) accesses information on related educational products. Unauthorized reproduction of the material on this CD-ROM is strictly prohibited.